W9-AOD-855

OPEN INNOVATION

OPEN INNOVATION

Researching a New Paradigm

Edited by
HENRY CHESBROUGH,
WIM VANHAVERBEKE,
and
JOEL WEST

OXFORD
UNIVERSITY PRESS

Great Clarendon Street, Oxford OX2 6DP

Oxford University Press is a department of the University of Oxford.
It furthers the University's objective of excellence in research, scholarship,
and education by publishing worldwide in

Oxford New York

Auckland Cape Town Dar es Salaam Hong Kong Karachi
Kuala Lumpur Madrid Melbourne Mexico City Nairobi
New Delhi Shanghai Taipei Toronto

With offices in

Argentina Austria Brazil Chile Czech Republic France Greece
Guatemala Hungary Italy Japan South Korea Poland Portugal
Singapore Switzerland Thailand Turkey Ukraine Vietnam

Oxford is a registered trade mark of Oxford University Press
in the UK and in certain other countries

Published in the United States
by Oxford University Press Inc., New York

First published 2006

A catalogue record for this title is available from the British Library

Library of Congress Cataloging in Publication Data
Data available

Typeset by SPI Publisher Services, Pondicherry, India
Printed in Great Britain
on acid-free paper by
Biddles Ltd., King's Lynn, Norfolk

ISBN 0–19–929072–5 978–0–19–929072–7

10 9 8 7 6 5 4 3 2 1

Preface

Open Innovation: The New Imperative for Creating and Profiting from Technology was published by Harvard Business School Press in the spring of 2003. It coined the term 'open innovation' and explained its application to managerial problems. But despite implicit (and often explicit) links to the large body of prior innovation research, the ideas only slowly gained acceptance among academic researchers.

This book is intended to offer a more theoretical view of open innovation in the context of innovation research, using both conceptual papers and original empirical studies. It also offers an inherently broader view of the phenomenon as interpreted by fifteen researchers representing fourteen universities across five countries.

It is different from most academic volumes in two ways. First, the chapters were commissioned and written specifically for the book rather than compiled from existing work. Second, each of the three main themes—firms, intellectual property, and networks—were developed by one of the editors who edited three chapters and wrote an integration chapter. In this regard, the book reflects an integration of components and modules into an overall system.

The genesis of this project began in the fall of 2003, only a few months after publication of Henry's earlier book, when Wim proposed an Open Innovation workshop to be held prior to the 2004 Academy of Management meeting. He enlisted help from Henry and later, Joel. After drawing a standing room crowd early on a Saturday morning, we discussed the idea of an academic book which was enthusiastically embraced by a number of publishers, including Oxford. Most of the Professional Development Workshop (PDW) presenters agreed to join the book, forming the nucleus of what you see before you today.

We want to thank all the authors for working so long and hard on their respective chapters, with most submitting four separate revisions. They also provided essential feedback for the integration chapters (1, 6, and 10). We appreciate the assistance provided by Nicoline Verbeek in compiling the references.

We also want to thank all those who believed in the project and helped us along the way. Our first thanks naturally go to Anita McGahan and Chris Tucci for giving us the workshop time in August 2004 Academy of Management meeting in New Orleans as well as the All-Academy Symposium a year later in

Honolulu. Our thanks go to our fellow panelists in New Orleans (Ashish Arora, Jens Frøslev Christensen, Kwanghui Lim, Markku Maula, and Gina O'Connor) as well as those in Honolulu (Jens Frøslev Christensen, Kira Fabrizio, Keld Laursen, and Ammon Salter) for advancing the dialogue we have attempted to capture in this book. We also appreciate the efforts at the Danish Research Unit for Industrial Dynamics (DRUID) Summer 2005 conference by Richard Langlois, Constance Helfat, Ammon Salter, and especially Jens Frøslev Christensen.

At Oxford, we particularly want to thank our acquisitions editor, David Musson. From the very first day the project was conceived and presented, his enthusiasm and support stood apart from his counterparts at other publishers. We also appreciate the editorial and production assistance of the very professional staff of OUP.

We continue our own efforts to advance Open Innovation through additional research and by helping firms apply the concept. At the same time, we are practising Open Innovation by building a network of researchers. We thank all of you who advance the understanding of Open Innovation through your own study and use of the concept, and invite you to continue the process by visiting our website and updating us with your own insights.

Henry Chesbrough
Wim Vanhaverbeke
Joel West

http:/www.OpenInnovation.net
http://www.OpenInnovation.eu

April 2006

List of Tables

Contributors

Henry Chesbrough is Executive Director of the Center for Open Innovation at the Haas School of Business, University of California at Berkeley. His research focuses on managing technology and innovation. His book *Open Innovation* was named the best book on innovation in 2003 on NPR's All Things Considered. *Scientific American* magazine named him one of the top fifty technology and business leaders for 2003. His academic work has been published in *Research Policy, Industrial and Corporate Change, Research-Technology Management, Business History Review,* and the *Journal of Evolutionary Economics*. More managerial work has been published in the *Harvard Business Review, Sloan Management Review,* and *California Management Review.*

Jens Frøslev Christensen is Professor in Management of Innovation at the Department of Industrial Dynamics and Strategy at Copenhagen Business School. His research interests focus on issues in management of innovation, especially as related to the relationship between large companies and small technology-based firms, and on new tendencies in the industrial dynamics of sectors such as consumer electronics, Internet services and IT security. He has published numerous books, the latest being *The Industrial Dynamics of the New Digital Economy* with Peter Maskell (Edward Elgar, 2003) and articles in journals such as *Industrial and Corporate Change, Managerial and Decision Economics*, and *Research Policy.*

Myriam Cloodt is a postdoctoral researcher in the field of International Business and Strategy, within the Organisation Science and Marketing group of the Department of Technology Management, Eindhoven University of Technology. Her research interests include international business and strategic management with a focus on open innovation and technological change, mergers and acquisitions, strategic alliances, and network analytical methods. She received her Ph.D. from the Maastricht Research School of Economics of Technology and Organizations (METEOR) at the University of Maastricht.

Kira R. Fabrizio is an Assistant Professor in the Organization and Management group at Goizueta Business School at Emory University. She holds a Ph.D. in Business Administration from the Haas School of Business, University of California at Berkeley. Professor Fabrizio's research examines technology transfer and knowledge management and the university-industry boundary with implications for industrial innovation.

Scott Gallagher is an Assistant Professor at James Madison University in Harrisonburg, Virginia. His work on strategy and innovation has appeared in the *Academy of Management Journal, IEEE Transactions on Engineering Management,* and the *Journal of Business Strategy.* He earned his Ph.D. in organization management from Rutgers University in 2000.

Stuart J. H. Graham is Assistant Professor of Strategic Management at the College of Management, Georgia Institute of Technology. Dr. Graham conducts research on the management of innovation and technology, intellectual property strategies, and the legal environment of business. He received his Ph.D. from the University of California at Berkeley, and holds advanced degrees in Law (JD), Business (MBA, MS), and Geographical Information Systems (MA). An attorney licensed to practice in New York State, he is a member of the American Bar Association, and has conducted his research under grants provided by the National Academies of Science and the OECD, among others.

Thomas Keil is an Assistant Professor of Entrepreneurship and Policy at the Schulich School of Business, York University in Toronto, Canada. His main research interests focus on corporate venturing, mergers and acquisitions, and technological standards in technology intensive industries. His work has been published in journals such as *Entrepreneurship Theory and Practice, Journal of Management Studies, Telecommunications Policy, Technovation, Technology Analysis and Strategic Management,* and *Computers in Industry* and has won several awards. He consults in the areas of strategic management, technology strategy and innovation, and is active in executive education.

Markku V. J. Maula is Professor of Venture Capital at the Institute of Strategy and International Business at Helsinki University of Technology. He graduated with a Doctor of Science in Technology (with distinction) from Helsinki University of Technology in 2001. He has received several international awards for outstanding scholarship including the Heizer Award from Academy of Management for the best doctoral dissertation in the field of new enterprise development. His research has been published in leading entrepreneurship journals. In addition to his research and teaching roles, he has acted as an adviser to firms and government agencies in issues related to venture capital, corporate venturing, and innovation policy.

David C. Mowery is the William A. & Betty H. Hasler Professor of New Enterprise Development at the Haas School of Business, University of California at Berkeley. He is the author, editor, or (co)editor of numerous research volumes on technology and innovation, including *The Oxford Handbook of Innovation* (Oxford, 2004), *Ivory Tower and Industrial Innovation* (Stanford, 2004), *The Sources* *of Industrial Leadership* (Cambridge, 1999), *Paths of Innovation* (Cambridge, 1998), *The International Computer Software Industry* (Oxford, 1996) and *Technology and the Wealth of Nations* (Stanford, 1992). He holds a B.A., M.A. and Ph.D. in economics from Stanford University.

Gina Colarelli O'Connor is Associate Professor in the Lally School of Management and Technology at Rensselaer Polytechnic Institute, and the Academic Director of the Radical Innovation Research Program. Dr. O'Connor earned her Ph.D. in Marketing and Corporate Strategy at New York University in 1990. Prior to that, she worked for McDonnell Douglas Corporation. Professor O'Connor's teaching and research efforts focus on how large established firms link advanced technology development to market opportunities, and how they create new markets. She has published more than twenty-five articles in refereed journals and is coauthor of the book *Radical Innovation: How Mature Firms Can Outsmart Upstarts* (Harvard Business School Press, 2000).

Jukka-Pekka Salmenkaita is Senior Business Development Manager at Nokia Multimedia. He received a Doctor of Science in Technology from Helsinki University of Technology in 2004. His research interests include analysis of innovation policy instruments and management of innovation processes in open systems. His professional interest is to apply these in the context of systemic technologies and complex organizations.

Caroline Simard is a researcher at the Stanford Graduate School of Business with the Stanford Project on the Evolution of Nonprofits. Her research interests include social networks and the circulation of ideas and knowledge, corporate and nonprofit management, regional clusters, and new media. Caroline holds a Ph.D. from Stanford's Department of Communication. She is currently working on extending her dissertation, which investigated the role of interorganizational networks in the creation and evolution of a regional cluster in wireless communication.

Timothy S. Simcoe is Assistant Professor of Strategic Management at the Joseph L. Rotman School of Management, University of Toronto. His current research focuses on the political economy of compatibility standards as well as their impact on economic and technological change. Tim earned his M.A. in Economics and his Ph.D. in Business Administration from the University of California at Berkeley in 2005. Prior to that, he worked as a consultant in the economic and information technology practices of Ernst & Young LLP, and as a research assistant for the US Council of Economic Advisers.

Wim Vanhaverbeke is Professor of Strategy and Organisation at the Hasselt University in Belgium, as well as a Research Fellow at the Eindhoven Center for Innovation Studies at the Technical University of Eindhoven in the Netherlands. Professor Vanhaverbeke is currently working on research about external sourcing of technological capabilities, alliance networks and alliance management. He has published in international journals such as *Organization Science* and *Organization Studies*. He serves on the editorial board of the *Journal of Engineering and Technology Management* and the *International Journal of Technology Marketing*.

Joel West is Associate Professor of Technology Management at the College of Business, San José State University. His research focuses on ICT industries, where he has studied how firms achieve proprietary gains from open intellectual property strategies in open source software, open standards and open innovation. Such work is part of a broader research agenda on the recognition and realization of new business in technology-driven industries. He holds a Ph.D. from the University of California at Irvine and an S.B. from the Massachusetts Institute of Technology; in between, he worked as a software engineer, manager, and entrepreneur.

1

Open Innovation: A New Paradigm for Understanding Industrial Innovation

Henry Chesbrough

1.1 Defining Open Innovation

The Open Innovation paradigm can be understood as the antithesis of the traditional vertical integration model where internal research and development (R&D) activities lead to internally developed products that are then distributed by the firm. If pressed to express its definition in a single sentence, Open Innovation is the use of purposive inflows and outflows of knowledge to accelerate internal innovation, and expand the markets for external use of innovation, respectively. Open Innovation is a paradigm that assumes that firms can and should use external ideas as well as internal ideas, and internal and external paths to market, as they look to advance their technology. Open Innovation processes combine internal and external ideas into architectures and systems. They utilize business models to define the requirements for these architectures and systems. The business model utilizes both external and internal ideas to create value, while defining internal mechanisms to claim some portion of that value. Open Innovation assumes that internal ideas can also be taken to market through external channels, outside the current businesses of the firm, to generate additional value.

The Open Innovation paradigm treats R&D as an open system. Open Innovation suggests that valuable ideas can come from inside or outside the company and can go to market from inside or outside the company as well. This approach places external ideas and external paths to market on the same level of importance as that reserved for internal ideas and paths to market in the earlier era.

Open Innovation is sometimes conflated with open source methodologies for software development. There are some concepts that are shared between the two, such as the idea of greater external sources of information to create value. However, open innovation explicitly incorporates the business model as the source of both value creation and value capture. This latter role of the business model enables the organization to sustain its position in the industry value chain over time. While open source shares the focus on value creation throughout an industry value chain, its proponents usually deny or downplay the importance of value capture. Chapter 5 will consider these points at greater length.

At its root, Open Innovation assumes that useful knowledge is widely distributed, and that even the most capable R&D organizations must identify, connect to, and leverage external knowledge sources as a core process in innovation. Ideas that once germinated only in large companies now may be growing in a variety of settings—from the individual inventor or high-tech start-up in Silicon Valley, to the research facilities of academic institutions, to spin-offs from large, established firms. These conditions may not be present in every business environment, and scholars must be alert to the institutional underpinnings that might promote or inhibit the adoption of open innovation.

1.2 The Open Innovation Paradigm

The book *Open Innovation* (Chesbrough 2003*a*) describes an innovation paradigm shift from a closed to an open model. Based on close observation of a small number of companies, the book documents a number of practices associated with this new paradigm. That book was written for managers of industrial innovation processes, and the work has received significant attention among managers. To the extent that such managers are able to assess the utility of new approaches, Open Innovation has achieved a certain degree of face validity within at least a small portion of high-tech industries. Open Innovation has taken on greater saliency in light of the debate about globalization and the potential for the R&D function itself to become outsourced, as the manufacturing function was twenty years earlier.[1]

Figure 1.1 shows a representation of the innovation process under the previous Closed Innovation model. Here, research projects are launched from the science and technology base of the firm. They progress through the process, and some of the projects are stopped, while others are selected for further work. A subset of these are chosen to go through to the market. This process is termed a 'closed' process because projects can only enter in one way, at the beginning, and can only exit in one way, by going into the market. AT&T's Bell Laboratories stands as an exemplar of this model, with many notable research achievements, but a notoriously inwardly focused culture.

Figure 1.2 shows a representation of an Open Innovation model. Here, projects can be launched from either internal or external technology sources, and new technology can enter into the process at various stages. In addition,

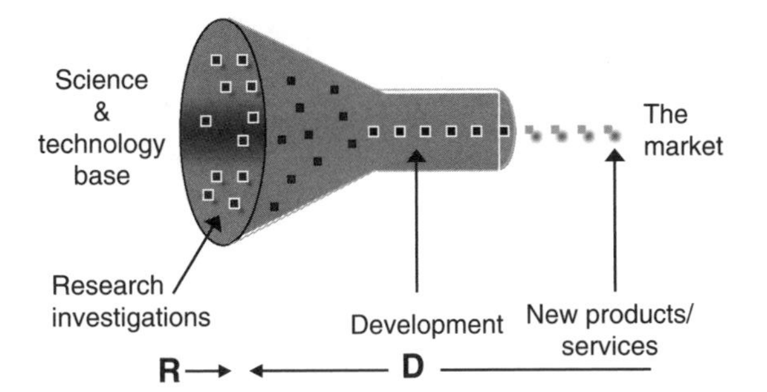

Figure 1.1. The current paradigm: a Closed Innovation model

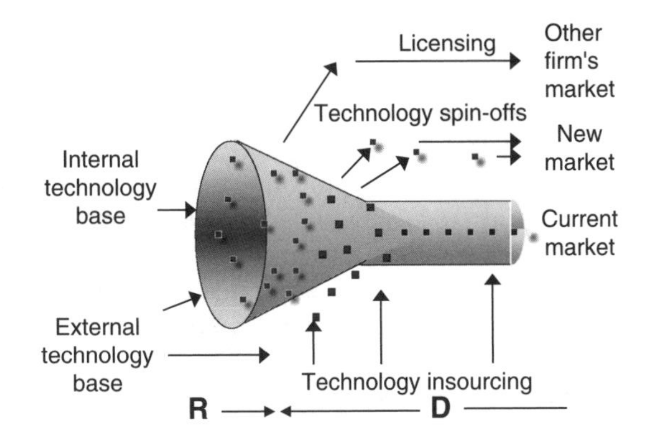

Figure 1.2. An Open Innovation paradigm

projects can go to market in many ways as well, such as through outlicensing or a spin-off venture company, in addition to going to market through the company's own marketing and sales channels. I labeled this model 'open' because there are many ways for ideas to flow into the process, and many ways for it to flow out into the market. IBM, Intel, and Procter & Gamble (P&G) all exemplify aspects of this open innovation model.

Academic scholars of innovation are trained to be rightly skeptical of new frameworks and concepts. Such concepts often consist of little more than fads and fashions (Abrahamson 1996). At best, such fads distract managers from more important activity, and at worst, fads can actually damage organizations and people. Scholars withhold their support of these novelties, unless and until they can demonstrate a more enduring contribution to the advancement of knowledge.

It is far too soon to claim that the paradigm of Open Innovation will make an enduring contribution to our understanding of innovation. However, it is not too soon to claim that it has already made an impact on our understanding of innovation. There is growing academic interest in the concept, as well as some nascent research activity that, when taken together, suggests that this may be a fruitful avenue for scholarly inquiry. It is the purpose of this book to document this early scholarly interest, and to point the way forward for further research that can develop the concept more fully.

1.3 Anomalies in Innovation

Any model that claims to be a new paradigm for industrial innovation must account for anomalies that are not well-explained in an earlier paradigm (Kuhn 1962; Feyerabend 1981). The evidence in *Open Innovation* offers numerous such explanations.[2,3] To take one example here, that will be discussed below, the field of innovation studies has long been aware of the difficulty of capturing spillovers from industrial R&D. These spillovers were regarded as a cost of doing business in the prior paradigm. Open Innovation treats spillovers as a consequence of the company's business model. These spillovers need not be a cost of doing business, they are an opportunity to expand a company's business model, or to spin off a technology outside the firm to locate a different business model.

A second example lies in the treatment of intellectual property. In the Closed model, companies historically accumulated intellectual property to provide design freedom to their internal staff. The primary objective was to avoid costly litigation. However, most patents are actually worth very little, and the vast majority are never used by the business that holds them.[4] In Open Innovation, intellectual property (IP) represents a new class of assets that can deliver additional revenues to the current business model, and also point the way towards entry into new businesses and new business models. A recent managerial book, *Rembrandts in the Attic* (Rivette and Klein 2000), proclaimed that companies needed to dust off their IP and offer it for sale to others. However, it did not provide an explanation for why those others would buy the IP. Open Innovation supplies a coherent rationale for why companies should be both active sellers and active buyers of IP.

1.3.1 *External Validity*

A new paradigm must also explain evidence beyond its initial area of inquiry if it is to have external validity (Yin 1988). In *Open Innovation*, the evidence adduced to support this model is taken almost exclusively from qualitative evidence in so-called 'high technology' industries, such as computers, information technology, and pharmaceuticals (Chesbrough 2003*a*, 2003*b*, 2003*c*, 2003*d*). Yet these industries represent only a few of the many sectors in an advanced industrial economy. It remains an open question whether the concepts of Open Innovation apply to lower-tech or more mature industries. Similarly, the evidence to date is taken from US-based companies.

The relevance of Open Innovation to companies operating outside the US remains to be demonstrated.

As will be seen in this book, progress is already being made on these questions of external validity. While the work is only the first wave of research in this area, there appears to be evidence that suggests that Open Innovation's explanatory power is not limited to a small number of companies operating in a small number of US high-tech industries.

1.4 Antecedents to Open Innovation

Open Innovation follows a long tradition of studying the processes of innovation, and stands on the shoulders of many previous scholars. Business historians have documented the extensive markets for innovation that predated the rise of the corporate R&D laboratory, and often predated the enforcement of intellectual property law (Lamoreaux et al. 1999; Lerner 2000). Innovation was at that time a rather open system. Joseph Schumpeter (1934) gave a powerful impetus to the study of innovation with his comparison of the entrepreneur and the entrenched incumbent firm, and in later work (1942) acknowledged the growing influence of corporations and their R&D activities in the innovation process.

Historical accounts suggest that early R&D activities grew out of the need in many industries to maintain and improve production activities (Chandler 1990). Because these activities were frequently unique for each firm, investments in R&D were firm-specific. David C. Mowery documented the rise of the corporate R&D laboratory in American manufacturing, and attributed this rise to the costs of organizing innovation inside the firm, relative to the costs of organizing innovation through the market (Mowery 1983). From the technology base created by internal R&D, firms naturally moved to exploit their accumulated knowledge to develop new products, thereby enhancing their economies of scope; in many industries large scale dedicated R&D functions emerged, providing a barrier to entry through economies of scale (Teece 1986; Chandler 1990).

The benefits of scale and scope for internal R&D (relative to the external market) gave rise to a vertically integrated innovation model where large enterprises internalized their firm-specific R&D activities, and commercialized them through internal development, manufacturing, and distribution processes. The managerial approach used for this proprietary model was summed up by Harvard president James Bryant Conant as 'picking a man of genius, giving him money, and leaving him alone' (Conant 2002).[5] Edison's Menlo Park, AT&T's Bell Labs, and Xerox's PARC were exemplars of this type of innovation model and brought about many inventions and innovations during the twentieth century.

To be sure, there were downsides noted to this model in the earlier literature. Richard Nelson observed back in 1959 that basic research generated many spillovers, and that firms who funded this research had only limited ability to appropriate value from these spillovers (Nelson 1959). Katz and Allen (1985) documented the Not Invented Here (NIH) syndrome that often accompanied the Chandlerian model of deep vertical integration of R&D for economies of scale and scope. Rosenbloom and Spencer (1996) argued that the leading industrial labs were in deep trouble, concluding that this model of innovation was 'at the end of an era'.

As noted above, these exemplary R&D organizations encountered difficulties when internal research generated spillovers that could not be internally commercialized. In some cases, such technology would be licensed to others, but in the majority of cases it 'sat on a shelf' waiting either for internal development or its research proponents to leave the firm and develop it on their own. This led to the Kuhnian anomaly of having the benefits of the innovation accrue *not* to the firm that financed its development, but instead to other firms who were able to capture the benefits of the innovation. Perhaps the best known contemporary example of such spillovers is Xerox PARC (Smith and Alexander 1988; Chesbrough and Rosenbloom 2002; Chesbrough 2002*b*). While these anomalies were documented, they were not adequately explained under the old model. They amounted to a regrettable but necessary cost of doing business. Indeed, some research scholarship on 'radical innovation' (Leifer et al. 2000) suggests that firms need to return to the long term, more patient approach to industrial research, even though there will be inevitable spillovers not captured.

Another rich source of antecedents has been substantial prior work on the importance of external technology, at least when it was 'inbound' to the organization. Nelson and Winter (1982) modeled the firm's decision to search for new technology outside of its own organization. Cohen and Levinthal (1990) wrote about the 'two faces' of R&D (which were inside and outside the firm) and also about the importance of investing in internal research in order to be able to utilize external technology, an ability they termed, 'absorptive capacity'. Nathan Rosenberg asked the question, why do firms conduct basic research with their own money (Rosenberg 1994), and answered that this research enhanced the firm's ability to use external knowledge. Firms that fail to exploit such external R&D may be at a severe competitive disadvantage (Rosenberg and Steinmueller 1988). Eric von Hippel (1988) identified four external sources of useful knowledge: (*a*) suppliers and customers; (*b*) university, government, and private laboratories; (*c*) competitors; and (*d*) other nations. Ove Granstrand and his colleagues (1997:13) note that:

The creation of corporate competencies in new fields was a dynamic process of learning, often requiring a combination of external technology acquisition and in-house technological activities and usually resulting in an increase in R&D expenditures. While technology sourcing was rarely a substitute for in-house R&D, it was an important complement to it.

Richard Langlois (2003*a*) has documented the 'post-Chandlerian firm', in which innovations develop in a less hierarchical fashion.

If firms cannot (or don't wish to) develop sufficient absorptive capacity themselves, they may utilize strategic alliances in order to gain such knowledge or utilize complementary resources to exploit that knowledge (see Gulati 1998 for a review of alliances, and also Nooteboom 1999). This alliance or 'network' approach is particularly common in technology intensive industries such as biotechnology (Powell et al. 1996; Mowery et al. 1996; Bekkers et al. 2002). Finally, geographic location has also been shown to result in knowledge spillovers between firms and from university research in many industries, especially high-tech (Porter 1990; Baptista and Swann 1998; Kenney 2000*a*).

Many models have been developed to explain how firms can exploit external knowledge. Perhaps the simplest method is to imitate a competitor: such free riding on the product and market investments of rivals is a common way for firms to overcome a first mover strategy (Lieberman and Montgomery 1998). Consulting with customers who are lead users can provide firms ideas about discovering, developing, and refining innovations (von Hippel 1988). Public sources are also an important source of knowledge, for example government R&D spending was identified almost fifty years ago as an important stimulus for private R&D (David et al. 2000). Similarly, university research is often explicitly funded by companies to generate external spillovers (Colyvas et al. 2002). Recently, open source software has emerged as an important phenomenon that utilizes external knowledge in a network structure (Lerner and Tirole 2002; O'Mahoney 2003; Dedrick and West 2004; von Hippel 2005).

Other scholars have studied the use of alliances (Gerlach 1992) and the construction of networks by firms (Gomes-Casseres 1996) as another means of actively seeking out and incorporating external knowledge into the innovation processes of firms. The work of Woody Powell and his colleagues (Powell et al. 1996) examines the costs of benefits of networks for innovating firms, while the work of Jeffrey Dyer applies the concept of networks and keiretsu formation to the automotive industry (Dyer 1996). Nooteboom (1999) examines the use of alliances in technology-based industries and companies.

Other recent research has called attention to the rise of intermediate markets in particular industries (Arora et al. 2001*a*). These intermediate markets alter the incentives for innovation, and also condition the mode of entry of new technologies and new firms into an industry (Gans et al. 2001). The presence of intermediate markets may interact with more networked structures to change the way in which innovation is organized.

1.5 What's New? Contributions of the Open Innovation Paradigm

Given this wealth of antecedents in the academic literature, is there anything new or different about the Open Innovation paradigm? Yes. The first difference is that external knowledge played a useful, but supplemental role in prior theorizing about innovation. The firm was the locus of innovation, and the internal activities of the firm were the central object of study (see, e.g. the works of Alfred Chandler, and his focus on 'the first mover' firm in different industries). The exemplars of the internal model were Bell Laboratories, and the many industrial R&D laboratories that emulated Bell Laboratories' organization. Even later theories of 'absorptive capacity' never specified what the balance between internal and external innovation sources ought to be. In Open Innovation, external knowledge plays an equal role to that afforded to internal knowledge in the earlier conception.

A second area of differentiation is the centrality of the business model in the Open Innovation paradigm. In the 'man of genius' mode of the Closed Innovation paradigm, one paid little or no attention to the business model in organizing for innovation. Instead, the focus was on securing 'the best and the brightest', and then trusting that world class research talent, when sufficiently funded, will come up with valuable new innovations that will somehow find a path to market. In Open Innovation, companies actively seek people of genius from both inside and outside the firm to provide fuel for the business model. In turn, Open Innovation suggests that inventive output from within the firm not be restricted to the current business model, but instead have the opportunity to go to market through a variety of channels (with the current business model perhaps having a right of first refusal).

A third distinction is that earlier innovation theories effectively assumed the absence of any measurement error (either of a Type I or Type II kind) in the evaluation of R&D projects (Chesbrough 2004). This evaluation is done in the context of the company's business model, and whether the project 'fits' their business model. If an R&D project was cancelled, there was nothing more to be done about it, and there was no reason to suspect that there was any systematic error in the assessment that led to the project's termination. Innovation processes, in fact, were managed so as to reduce the chance of a Type I or 'false positive' evaluation error, which would result when an R&D project went entirely through the process, went to market through the company's business model, and failed. The possibility of a Type II or 'false negative' error, where the project does not fit the company's business model, and is therefore not perceived as valuable to the firm, was not deemed important, even though statistical theory suggests that efforts to reduce a Type I error will inadvertently increase the chance of a Type II error (Judge et al. 1985). And firms typically lacked any process for managing false negative R&D projects.

In Open Innovation, the business model is the cognitive device that focuses the evaluation of R&D projects within the firm (Chesbrough and Rosenbloom 2002). As a cognitive aide, the business model filters in projects that 'fit' with the model, and selects against those that do not. This evaluation is not objective; biases can and do exist. While firms rightly seek to minimize the incidence of false positives, the alert firm also must incorporate additional processes to manage false negatives, in order to appropriate value from them and identify potential new markets and business models from them.

A fourth and related distinction is that prior concepts accorded little or no recognition to purposive outbound flows of knowledge and technology (in contrast to the unwitting outbound flows that are termed 'spillovers', which were discussed above). Even when firms went outside to absorb external knowledge, it was for the purpose of internal development, manufacture, and sales. In the Open Innovation paradigm, enabling outward flows of technologies allows firms to let technologies that lack a clear path to market internally seek such a path externally. In so doing, the internal businesses of the firm now compete with these external channels to market (such as licensing, ventures, and spin-offs that can create additional value) for new technologies. These external channels, in turn, can provide important evidence of emerging or neglected technical or market opportunities (in a way, a second opinion), constituting one means to manage 'false negative' R&D projects. These channels have to be managed as real options, as opposed to the more traditional net present value approach for allocating budgets to projects (Bower 1970).

A fifth point of departure lies in the assumptions of the underlying knowledge landscape. While the abundance of knowledge has been known since at least the time of Hayek (1945), this insight did not penetrate the door of the industrial R&D model. In the proprietary model of innovation, useful knowledge is scarce, hard to find, and hazardous to rely upon (a root cause of the NIH syndrome). In Open Innovation, useful knowledge is generally believed to be widely distributed, and of generally high quality. Even the most capable and sophisticated R&D organizations need to be well connected to these external sources of knowledge.

This is well illustrated in Merck's annual report for the year 2000. Although the company is widely respected for its excellent internal research, the report stated on page 8:

> Merck accounts for about 1 percent of the biomedical research in the world. To tap into the remaining 99 percent, we must actively reach out to universities, research institutions and companies worldwide to bring the best of technology and potential products into Merck. The cascade of knowledge flowing from biotechnology and the unraveling of the human genome—to name only two recent developments—is far too complex for any one company to handle alone.

Importantly, these external sources extend well beyond universities and national laboratories, to startup companies, specialized small companies, individual inventors, even retired technical staff or graduate students.

A sixth differentiation is the new and proactive role for IP management in the Open Innovation model. While the practice of proactive IP management is hardly new to certain industrial firms (think of Dolby Laboratories or Gore, Inc., the licensors of Goretex), prior theories of innovation treated IP as a byproduct of innovation, and its use was primarily defensive. This would enable firms to practice their (internal) technologies without being blocked or held up by external IP. Should such blockage arise, IP could be cross-licensed or bartered to restore the status quo ante. In Open Innovation, this is but one of many possible uses of IP. IP becomes a critical element of innovation, since IP flows in and out of the firm on a regular basis, and can facilitate the use of markets to exchange valuable knowledge. IP can sometimes even be given away through publication, or donation.

A seventh area of difference is the rise of intermediaries in innovation markets. While intermediaries have been observed in related areas such as technology alliances (Nooteboom 1999), they now play a direct role in innovation itself. As innovation becomes a more open process, intermediate markets have now arisen in which parties can transact at stages which previously were conducted entirely within the firm. At these junctures, specialist firms now provide information, access, and even financing to enable transactions to occur. The growing importance of intermediaries is perhaps most elaborated in the Pharmaceuticals industry (such as Innocentive, or Yet2.com), but it is emerging in many industries (such as NineSigma, or YourEncore). This is difficult to explain in the Closed Innovation model (or perhaps it is regarded as a curiosity of little research interest). It is a significant trend to understand in the new paradigm.

The eighth and last distinguishing point out of this new approach is the development of new and different metrics for assessing the performance of a firm's innovation process. Classical metrics include the percentage of sales spent on (internal) R&D, the number of new products developed in the past year, the percentage of sales from new products, and the number of patents produced per dollar of R&D. New metrics will expand or perhaps substitute for some of these measures. Questions of how much R&D is being conducted in within the firm's supply chain (rather than R&D occurring simply within the firm itself) become more important. What percentage of innovation activities originated outside of the firm—and how this compares to the industry in which the firm operates—may be another. The time it takes for ideas to get from the lab to the market, and how that varies by channel to market (internal, outlicense, spin-off, etc.) will be still another. The rate of utilization of patents owned by the firm will be still another, as unutilized patents may have alternate paths to monetizing value. Investments in outside firms may also become important (Chesbrough 2002*a*) (see Table 1.1).

Table 1.1. Points of differentiation for Open Innovation, relative to prior theories of innovation

1. Equal importance given to external knowledge, in comparison to internal knowledge
2. The centrality of the business model in converting R&D into commercial value
3. Type I and Type II measurement errors (in relation to the business model) in evaluating R&D projects
4. The purposive outbound flows of knowledge and technology
5. The abundant underlying knowledge landscape
6. The proactive and nuanced role of IP management
7. The rise of innovation intermediaries
8. New metrics for assessing innovation capability and performance

In sum, while Open Innovation draws extensively from an earlier body of academic scholarship, it offers a number of distinctive perspectives and interpretations of that prior scholarship. In our judgment, these are sufficient to warrant consideration as a new paradigm for understanding innovation. The final assessment will of course remain with the reader. We modestly hope that younger scholars will find inspiration in these pages, and take up the opportunity to work in this area. More experienced scholars may find important connections with their own work, which might enrich the Open Innovation approach, and shed additional light on their work in turn.

1.6 Conclusion

The field of innovation studies arguably operates in Pasteur's Quadrant (Stokes 1997), in that the processes and practices of industry actors often extend beyond the bounds predicted by academic theory. Close observation of the experiments that some of these firms have enacted reveals that the inwardly focused, vertically integrated model of industrial innovation so celebrated by Chandler (1990) and others has given way to a new, and not yet well-understood model (Langlois 2003*a*).

While the contours of the new model of innovation remain obscure, it is clear that any adequate understanding will require a more externally focused perspective, involving the actions of multiple actors in a far more distributed innovation environment. Such a new model will require close study of the innovation activities of the organization from multiple levels of analysis (including individual, group, organization, community, and institutional). It will likely provide a more satisfying account for current Kuhnian anomalies in our theories of innovation (such as spillovers, unutilized IP, NIH, spin-offs), even as it raises new research issues (the provision of long term research, the limits of business models, the links between innovation and IP management). It may even point the way to new innovation actions not yet witnessed much in industry. This could include the greater use of purposive outbound knowledge flows by companies, the emergence of a secondary market for IP, and the creation of new organizational roles and practices for identifying, incorporating, and adding value to external knowledge sources. Chapter 14 will consider these points at greater length.

Notes

1. For a alarmist assessment of the trend, see Forrester Research, which estimates that 3.3 million R&D jobs will move offshore over the next twelve years (http://www.nytimes.com/2003/10/05/business/05ECON.html?tntemail0). For a more hopeful assessment, see the McKinsey Global Research Institute, http://www.mckinsey.com/knowledge/mgi/offshore/, which estimates that the US will capture 78 percent of the value created from offshore R&D employment.
2. The entirety of Chapter 1 of that book examines the experience of Xerox's Palo Alto Research Center, and offers a different interpretation of the root cause of Xerox's problems with PARC. Xerox was judged to be effective in utilizing PARC technologies that fit with Xerox's copier and printer business model. The failure was that Xerox could not conceive of an alternate business model through which to commercialize technologies that did not comport with that model. By contrast, the profile of IBM in Chapter 5 showed a company that did reconceive its business model.
3. One paradox posed in Open Innovation was the surprising ability of Cisco to keep up with Lucent and its Bell Labs. As the book noted, 'Though they were direct competitors in a very technologically complex industry, Lucent and Cisco were not innovating in the same manner. Lucent devoted enormous resources to exploring the world of new materials and state of the art components and systems, to come up with fundamental discoveries that could fuel future generations of products and services. Cisco, meanwhile, did practically no internal research of this type.

   Instead, Cisco deployed a rather different weapon in the battle for innovation leadership. It scanned the world of startup companies that were springing up all around it, which were commercializing new products and services. Some of these startups, in turn, were founded by veterans of Lucent, or AT&T, or Nortel, who took the ideas they worked on at these companies, and attempted to build companies around them. Sometimes, Cisco would invest in these startups. Other times, it simply partnered with them. And more than occasionally, it would later acquire them. In this way, Cisco kept up with the R&D output of perhaps the finest industrial research organization in the world, without doing much internal research of its own.' (p. xviii)
4. While comprehensive evidence of these points is not yet available, some elements are already in the literature. Lemley (2001: 11–12) cites studies that report a large fraction of patents are neither used, nor licensed by firms. Davis and Harrison (2001) report that more than half of Dow's patents were unutilized. Sakkab (2002) states that less than 10 percent of Procter & Gamble's patents were utilized by one of its businesses.
5. I am grateful to Scott Gallagher and Joel West for identifying this example.

Part I

Firms Implementing Open Innovation

2

New Puzzles and New Findings

Henry Chesbrough

2.1 Introduction

This short chapter is intended to frame the subsequent three chapters within this volume. This part of the volume is focused on research that examines the implications of Open Innovation on innovation activities *within* the firm. Subsequent parts will address the implications of Open Innovation outside the firm, and in the surrounding environment, respectively.

Adopting a more open innovation model within a large organization invites the consideration of many puzzles. First, if external innovation is so helpful, why is there so much variation in whether and how much companies utilize it? Second, if many technologies go unused within a firm, why aren't more technologies offered for sale to outside organizations instead?

A more subtle puzzle lies in the domain of open source software. The Open Innovation model treats a company's business model as both necessary and sufficient for innovation success. How then are we to regard the open source software movement? By construction, many of the key elements of open source eschew the exclusionary aspects of intellectual property protection and traditional business models. Yet this lack of a business model does not seem to be impairing the advance of open source, which is growing in its impact. Does this growth contradict, or at least sharply qualify, the claims of Open Innovation with regard to the importance of business models?

I will develop each of these three issues to some degree, and provide some remarks on each of the three subsequent chapters that follow in this section. I will conclude by synthesizing the findings of the three chapters in light of these issues.

2.2 Puzzles in the Limited Use of External Innovation

Open Innovation (Chesbrough 2003*a*) argued that 'not all the smart people work for you', and maintained that there was an increasingly dispersed

Table 2.1. US industrial R&D spending by size of enterprise, selected years, 1981–2001

	1981	1989	1999	2001
Less than 1,000 employees	4.4	9.2	22.5	24.7%
1,000–4,999 employees	6.1	7.6	13.6	13.5%
5,000–9,999 employees	5.8	5.5	9	8.8%
10,000–24,999 employees	13.1	10	13.6	13.6%
25,000+ employees	70.7	67.7	41.3	39.4%

Sources: For 2001: National Science Foundation, Division of Science Resources Statistics, Research and Development in Industry: 2001 (Arlington, VA, forthcoming); Science & Engineering Indicators – 2004, Table 4–5; for prior years: *Open Innovation*, p. 48, citing earlier NSF reports from its Science Resource Studies unit.

distribution of useful knowledge in companies of all sizes, and outside the US as well as within the US. More recent data continue to strongly suggest a more level playing field for industrial innovation activity. Data from the National Science Foundation, in Table 2.1, show that small firms (defined here to be those firms with less than 1,000 employees) continue to increase their share of the total amount of industrial R&D spending, amounting to almost 25 percent of total industry spending in 2001. Large firms (defined here at firms with more than 25,000 employees) has seen their collective share of industrial R&D fall to under 40 percent of total industry spending in that year.

Data on patent awards shows a similar pattern, along with an increasingly global component, with foreign companies claiming a rising share of US patents, as shown in Table 2.2. Corporations of all sizes comprise about 88 percent of all US patents issued in 2003, but the percentage of patents held by organizations that received forty or more patents amounted to less than half of all issued patents.

These data, and other data such as the growth of employment in small enterprises, relative to employment in large firms, all combine to suggest that the playing field for innovation is becoming more level. Put differently,

Table 2.2. US PTO patent awards by type of recipient, calendar year 1995, 2003

	1995	2003
US corporation	43.4	44.6
US government	1.0	0.5
US individual	12.7	8.0
Foreign corporation	38.1	43.2
Foreign government	0.2	0.1
Foreign individual	4.5	3.6
Companies receiving 40 or more Patents as % of total patents	41.6*	48.0

Sources: for 1995, ftp://ftp.uspto.gov/pub/taf/topo_95.pdf; for 2003: ftp://ftp.uspto.gov/pub/taf/topo_03.pdf; for 1995, report shows organizations receiving 30 or more patents; Sterm (2004).

there appear to be fewer economies of scale in R&D in a growing number of industries than there were a generation ago.[1]

If this is indeed the case, this more level playing field has powerful implications for the organization of innovation. In a more distributed environment, where organizations of every size have potentially valuable technologies, firms would do well to make extensive use of external technologies. The limited large sample data currently available though (e.g. Gassmann and von Zedtwitz 2002*a*; Laursen and Salter 2006) suggest that there is substantial variation in the use of external technologies in a firm's innovation process. What might explain this variation?

Part of the explanation may lie in the behavioral response of internal employees to the introduction of external technologies, which has long gone by the name of 'the Not Invented Here' syndrome (Katz and Allen 1985). This 'NIH' syndrome is partly based upon an attitude of xenophobia: we cannot trust it, because it is not from us, and is therefore different from us. But there are more rational components that might induce internal employees to reject external technologies as well.

2.2.1 *Rational Reasons for Resisting the Incorporation of External Technologies*

One such component is the need to manage risk in executing R&D projects, especially when the cycle time to complete a project is accelerating (Fine 1998). When cycle times accelerate in a project, there is less time to evaluate and incorporate external technologies into a fast-moving project. More subtly, when projects are moving fast, project leaders seek to minimize the risk of unexpected outcomes in the project. Internally sourced technologies pose enough risk to the project meeting its scheduled ship date already. Externally sourced technologies, coming from a much wider variety of sources about whom much less is known (when compared to internally generated technologies), may greatly increase the perceived risk to the project. The expected value of an external technology may be as high—or even higher—than an internal technology. But the variance around that expected value likely may be much higher as well.

This suggests a researchable question: Do projects incorporating external technologies experience higher variance in project outcomes (e.g. cost, time, quality) than those that rely upon internal technologies? A more behavioral variant of that question would be: Are projects that incorporate external technologies *perceived* by project participants to increase the risk to project outcomes? If so, are those perceptions subsequently validated by the data, or do these perceptions shift as new outcomes appear?

The above line of inquiry presumes that internal employees are simply unaware of the real characteristics of externally sourced technologies, and that there are costs incurred to find out these characteristics. A more subtle challenge is the impact on the internal staff's subsequent actions if *and when externally sourced technologies prove to be highly effective*. In this instance, the overall project's success may be enhanced by the inclusion of externally sourced technology. But the top managers in the firm might infer from this experience that the firm doesn't need quite so many internal R&D staff to accomplish the *next* project, that the next project ought to rely more on external technology as well. In this case, the short term success of the project might be to the long term detriment of internal R&D staffing levels and internal research funding. This is also a researchable question: When companies employ external technologies successfully in their innovation process, do internal R&D staffing levels rise or fall in subsequent periods?

This suggests that there may be an asymmetry in the risks and rewards (from the perspective of the project leader and the project team) from greater utilization of external technologies in R&D projects. The project team must bear responsibility for the success or failure of the project, and therefore must have the final decision over whether and when to incorporate external technologies into the project as part of the project's development. If the external technology fails (and remember that it may have a higher variance in expected outcome), the project team bears the responsibility. But 'success' in the use of external technologies in this process may jeopardize internal staffing levels in future. So, the project team confronts a risky situation in which they bear full responsibility if the use of external technology 'fails', yet the team may bear other long-term costs if the use of external technology 'succeeds'.

This prompts a reexamination of two important cases cited by Open Innovation (Chesbrough 2003*a*): IBM and P&G. In IBM's case, the company operated with a highly vertically integrated and inwardly focused innovation model since the inception of its System 360 (Pugh 1995). IBM's shift towards a far more open, less vertically integrated approach came from the arrival of Lou Gerstner into the CEO role at the firm. However, immediately prior to Gerstner's arrival, IBM reported what was at that time the largest quarterly loss in US business history, and IBM made the first major layoffs in its corporate history. This dramatically shifted the previous culture of internal focus towards innovation, and many of those laid off were in the R&D organization. When IBM began to adopt a more open approach, it did so at a time when the organization had recognized that the status quo ante was no longer sustainable. A more complete discussion of this transition can be found in (Chesbrough 2003*a*: ch. 5).

P&G's embrace of open innovation also was immediately preceded by a significant layoff in its R&D organization (though a far less severe layoff than the one at IBM). P&G had embarked on a growth campaign in 1990 to double its $20 billion in revenue by 2000. When 2000 came, the organization had only reached $30 billion in revenue, and many spoke of the 'growth gap' for the company. P&G cut expenses significantly, and laid off a large number of people. After these cuts, P&G consciously told its R&D staff that its shift to what it called 'Connect and Develop' would not lead to any further reductions in staff. Instead, the shift in innovation models was positioned to enable P&G to generate more innovation with the (recently reduced) R&D resources on hand (see Sakkab 2002). Again, the effect was to minimize the perception of asymmetric risks among the R&D project leaders and staff. My hypothesis is that this shift would have been received very differently by P&G's internal organization, had Connect and Develop been launched *prior* to the layoffs of P&G's R&D staff. In that context, Connect and Develop might have been viewed as a thinly veiled excuse for downsizing and outsourcing R&D.

This suggests a control variable that could be used in a large sample study of the adoption of external technologies within a company's innovation process. The control variable would be recent changes in R&D staffing levels in the previous period (or, if that were not available, recent changes in R&D spending). If staffing had already declined, perhaps external technologies would be received with less resistance.

2.3 Puzzles in the Limited Offering of Unused Technologies Outside the Firm

A second kind of puzzle emanates from looking at the innovation process from the other end, where companies choose to deploy certain technologies and commercialize them, while leaving a larger set of technologies unutilized. When P&G surveyed all of the patents it owned, it determined that about 10 percent of them were in active use in at least one P&G business, and that many of the remaining 90 percent of patents had no business value of any kind to P&G (Sakkab 2002). Dow Chemical went through an extensive analysis of its patent portfolio starting in 1993, as reported in Davis and Harrison (2001: 146). In that year, about 19 percent of Dow's patents were in use in one of Dow's businesses, while a further 33 percent had some potential defensive use, or future business use. The remaining patents were either being licensed to others (23 percent), or simply not being used in any discernable way (25 percent). In the typical pharmaceutical development process, a company must screen hundreds or even thousands of patented compounds, in order to find a single compound that makes it through the process and gets into the market.[2] From a naïve perspective, it seems wasteful in the extreme to create and develop a large number of technologies, and then only utilize a miniscule fraction of the technologies in any way, shape, or form.

This raises at least two subsidiary questions. First, why do firms develop so many possible technologies, instead of just the most likely ones? a Second, what inhibits firms from making much greater use of unutilized technologies in other ways? These will be considered in turn.

The first question makes an implicit assumption that turns out not to be true in many organizations. That assumption is that the R&D activities of the firm are tightly coupled to the business model of the firm. If one grants this assumption, then it is truly puzzling why so many technologies are so little used. However, the reality is that often the assumption is simply wrong; many firms consciously keep their R&D process only loosely coupled to their business model.

Further, R&D managers often use the number of patents generated by an R&D research or an R&D organization as a metric to judge the productivity of that person or organization. Similarly, some R&D organizations count the number of publications generated by their R&D staff as another measure of productivity.[3] Unsurprisingly, when organizations reward the quantity of patents or papers produced, the R&D organization responds by generating a large number of patents or papers, with little regard as to their eventual business relevance.

To carry this point further, there may be a budgetary disconnect between a R&D group on the one hand, and a business unit on the other. To see this, examine Figure 2.1.

In this figure, the R&D operation produces research results, and operates as a cost center. This is usually how such organizations are funded, since they do not sell their output, and since it is hard to estimate how much money a particular R&D project will need in order to be successful. Instead, companies determine an amount of funding that they can sustain over time, which can be dedicated to R&D tasks. The R&D unit manager must in turn decide how many projects to support with the budgeted funds she has that period. It is bad for her to exceed her budget, since the organization may not be able to sustain the

A Framework for Budgetary Disconnects

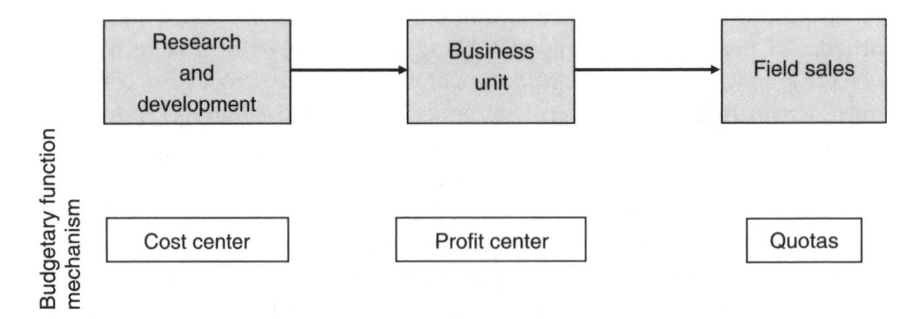

Figure 2.1. A Model of budgetary disconnection between R&D and the business unit

additional expenses. It is also bad for her to come in much under the budget that year, because that may suggest that next year's budget can be reduced as well. So the manager tries to develop as many projects as she can, subject to the budget constraint.

The internal business unit customer, by contrast, is typically managed on a profit-and-loss (P&L) basis. The business unit typically does sell its output to customers, and giving each business unit its own P&L enables that business manager to make the best use of his information to maximize profits for the business. That manager wants to buy low, sell high, and avoid risk. So the business unit manager wants any R&D project coming from his internal 'supplier' to be as fully developed as possible. This reduces any additional costs the manager must incur prior to using the technology in the business. It also reduces any risk to that business's profitability that period.

The stage is now set for the budgetary disconnect between the two functions. The R&D manager wants to push out the project as soon as the publications and patents have been generated. Further development within the R&D budget crowds out other, newer projects that have greater potential for generating still other new patents and publications. So the R&D manager's incentives are to transfer the project sooner rather than later to the business unit. Meanwhile, the business unit manager's incentives are to wait as long as possible before taking over the further funding of the R&D project onto his P&L.

The resolution of this budgetary disconnect is to place a buffer between the R&D operation, and the business unit, as shown in Figure 2.2. This buffer provides temporary storage for the R&D project, until the time when the business unit is ready to invest in its further application within the business. This lets the R&D manager get onto work on her next project, without requiring the business unit manager to commit to further funding on his P&L until he judges it to be beneficial.

While this solves the local problem of each manager, from a system viewpoint, the 'solution' causes many R&D projects to pile up in this buffer. These projects are often termed 'on the shelf', because they are no longer being

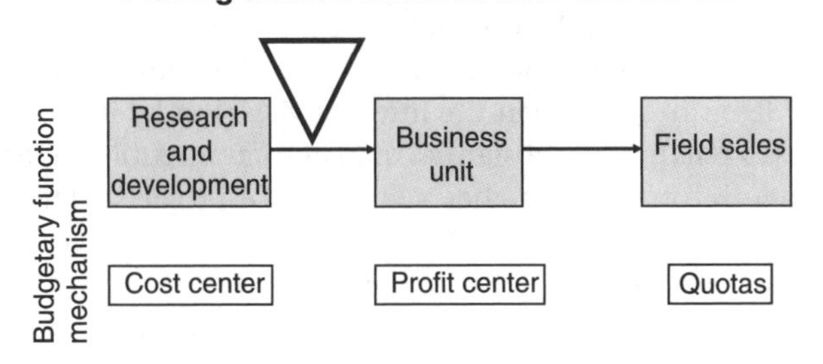

Figure 2.2. Decoupling R&D from the business unit

actively pursued by the R&D organization, nor are they actually being used by the business unit.

The structure of research funding also influences the subsequent utilization of research results within the firm. Some research organizations obtain a significant percentage of their funds from 'research contracts' with their internal business units. These contracts tend to be fairly specific, near term in time frame, and are likely to be utilized by the business units, which pay directly for the output of the work. But other funds for those same research organizations come from a corporate allocation of funds (which is generated from a 'tax' on all of the businesses within the firm). These corporate funds are not tied to any specific business unit objective, and are allocated by research managers to longer term projects whose output may benefit multiple businesses, but may not be immediately relevant to any. Still other research funds come from government research contracts. These funds tend to be academically peer reviewed, and may therefore have little or no relevance to any business unit activity within the firm. So the hypothesis might be that the type of research funding received is correlated to the subsequent business utilization of the research output. Contractual funding with business units would be predicted to lead to higher utilization, while government funded research would be predicted to lead to a much lower level of utilization.

The foregoing analysis suggests that R&D processes are only loosely coupled to the business models of firms (though the tightness of coupling may vary with the type of research funding provided), which may explain why there are a substantial number of technologies that are un- or underutilized within those businesses. This could lead to research on how better to align the incentives of the two units, and potentially how better to manage any buffer that emerges between them. This loose coupling also heightens interest in the second question: What prevents the business from enabling others to utilize those underperforming technologies in their own respective businesses?

There may be parallel forces at work on this question, as well as on the previous one. The expected value of an unused internal technology may be quite low, but there may still be variance in that value. Indeed, the internal view of the technology's potential is likely biased by the business model of the company (Chesbrough and Rosenbloom 2002). This may suggest that an external view of the technology's value may be more unbiased (if less informed, at least initially) than the internal view. But by itself, this analysis would suggest a potentially substantial market for underutilized technologies. After all, when buyers have higher valuations of projects than sellers, it is natural for those parties to find a mutually beneficial transaction that shifts those projects to the party with the higher valuation.

A second concern may be adverse selection. Buyers may worry that the sellers of unutilized technologies will only offer the 'bad' ones (Akerlof 1970). Adverse selection presumes that both parties are rational and unbiased, so the seller (who has more information sooner) will inevitably have an information advantage over the buyer. But the dominant logic of a company's business model would actually suggest countervailing forces that might support the use of external technologies. While companies have significant prior information on a technology project (and therefore might be assumed to enjoy a tremendous information advantage), that information will nonetheless be evaluated within the context of the company's business model. If the buyer has, or can identify, a very different business model, the buyer's evaluation of the project may differ greatly from that of the seller.

To give an example here, consider the experience of Xerox PARC with its many technology spin-off projects (Chesbrough 2002*b*). In that work, I identified thirty-five projects that left Xerox after the further funding for the work had been ended within Xerox. Xerox judged that there was little or no additional value to be gained from continuing this work. In twenty-four of the thirty-five projects, there was little business success after separation. But for eleven of the projects, each of which developed under a very different business model from that of Xerox, there turned out to be substantial value. The collective market value of the companies that emerged from these eleven projects turned out to exceed the total market value of Xerox by a factor of two. I interpret these data to mean that Xerox's estimates of the value of these projects were biased by its business model.

Other barriers to greater utilization of unused technology may lurk inside the budgetary mechanisms of R&D organizations and their business unit customers. There may be a behavioral analogue to NIH that sits within the business units, which I term the Not Sold Here (NSH) virus. NSH is a syndrome that argues that, if we don't sell it, no one should. It is rooted in the surface perception that, if our organization cannot find sufficient value in the technology, it is highly unlikely that anyone else can either (a restatement of adverse selection). At a deeper level, however, the NSH virus seeks to forestall competition with outside entities for accessing internal technology. Most business units enjoy a monopsony position relative to their R&D unit suppliers. Because they have a de facto exclusive right to the technology, they can defer costs and delay commitments to the technology without penalty.

Enabling greater external use of unused technologies alters the business unit's calculation. If a business unit chooses not to incorporate a technology, and that allows others the chance to do so, the business unit now faces a previously latent cost: if it does not use the technology itself, it might 'lose' that technology to an external organization. Typically, internal business units have some defined interval of time during which they can 'claim' the technology. After that interval expires, the technology is then made available to other firms.[4] Depending on who that external firm is, the internal business unit may even have to compete against that technology in the market. Worse (from the business unit's perspective), the external use of the technology might reveal previously unrealized value from the technology, leaving the business unit in the awkward position of explaining why it failed to utilize this now apparently valuable technology. Another asymmetry presents itself: if the technology is licensed externally, the corporation may 'win' through additional licensing revenue, but the business unit may 'lose' through additional competition in its market.

Here, there may be mechanisms that firms can employ to align incentives within the business unit to more closely approximate those of the overall firm. GE and IBM, for example, share any licensing revenues from a technology with the business unit associated with the technology. So the business unit P&L not only bears the risk of competing with the technology in the market (thus negatively impacting the P&L of the unit), but also receives credit for licensing revenue from the technology on its P&L (thus boosting the revenue and profit of the P&L of the unit).

Companies that enable competition for their unutilized technologies might experience more rapid flow of those technologies into the market, both for those taken outside *and* for those that remain inside. The latter implication may require some further explanation to motivate the hypothesis. When business units face external competition for the use of internal technologies, and a defined time limit in which to consider a technology before it is made available to others, it is likely that this limit accelerates the evaluation process within the business unit. It is really a form of buffer management. Technologies get incorporated faster into the business, or else they flow out to other organizations, instead of sitting on the shelf. This increases the flow of ideas from R&D through the business unit, and into the market.

There is a further, more human dimension that could be researched. Companies in which NSH is pronounced likely frustrate many of the R&D staff, because many of the ideas these people work on are never deployed in the market. It is reportedly quite common for a pharmaceutical researcher to never see one of her projects ship into the market, over a thirty-year career, because the attrition rate of compounds is so high. This is an enormous waste of human talent, and must take a toll on any person's initiative. Companies that overcome NSH allow other pathways for internal ideas to get into the market. These other pathways allow the market to provide feedback on those ideas, and lets researchers see their ideas in action in the wider world.[5]

2.4 A Third Puzzle: A Successful Technology without an Apparent Business Model—The Case of Open Source

One of the central tenets of the book *Open Innovation* (Chesbrough 2003*a*) is that business models are essential to unlocking latent value from a technology. On p. xxx of the Introduction, the book asserts:

There is no inherent value in a technology per se. The value is determined instead by the business model used to bring it to market. The same technology taken to market through two different business models will yield different amounts of value. An inferior technology with a better business model will often trump a better technology commercialized through an inferior business model.

This assertion begs an obvious question: What happens when there is no business model being used to commercialize a technology?

This is apparently the case with open source software development. By construction, open source software is created without any one firm owning the technology. No firm can patent the technology, or exclude anyone else from accessing the software code. Enhancements to the code are available to everyone on an equal basis.

Is this simply an exception to the general rule, is this due to a business model of a different kind, or is there something fundamentally wrong by the above claims of *Open Innovation* regarding the importance of business models for the behavior of firms? This is a third puzzle in the context of open innovation.

2.5 Remarks on Each of the Chapters in Part I

While each of these chapters addresses aspects of one or more of the issues above, they go further, introducing additional evidence into the debate. I will briefly highlight some of the insights of each of the authors, and conclude with some synthesis of the material in this section.

2.5.1 *Chapter 3: O'Connor*

Chapter 3 by Gina O'Connor discusses how firms that are pursuing long-term, ambitious, 'breakthrough' innovations incorporate certain aspects of Open Innovation. Building on a fruitful research program on Radical Innovations that has been ongoing at the Rensselaer Polytechnic Institute (RPI) for many years, O'Connor revisits the extensive data collected in the course of this research. The RPI team studied fourteen radical innovation projects in great detail (and some of the investigation is still ongoing). Some of these projects have met with 'success', while others clearly have not. Similarly, a few organizations have attempted to instantiate internal business units to pursue radical innovations, while others have not. And of those who have created a dedicated organizational unit, some have subsequently discontinued the unit. So there is a lot going on here, both in the technology side of the organization, and on the business and strategic side of the organization as well.

In this chapter, O'Connor searches this rich dataset for patterns that illuminate the differing outcomes from these projects. While the sample is too small for any statistical analysis, she presents persuasive evidence that the effective pursuit of radical innovations also appears to benefit from the application of some of the concepts of open innovation. In particular, she reports evidence on the extensive use of external sources of technology in many of the successful projects. Open innovation appears to help not least because it is perceived to have the potential to shorten the time to market for some of the higher impact innovations that otherwise suffer under conventional stage-gate evaluations. These stage-gate processes appear to favor the shorter term projects, and appear to crowd out the longer term, more radical innovation projects. Her concern is not with the evaluation process per se, but rather the overreliance on short term metrics to conduct those evaluations, which have the practical effect of excluding longer term, but higher potential projects. This is quite consistent with the earlier analysis of Clark and Wheelwright (1992). With books like *Execution* (Bossidy et al. 2002), which reinforce a short-term contractual view of meeting commitments selling so well right now, there is a need for a timely response such as this.

Among the new findings she reports in this chapter is the significant degree of 'openness' among her sample firms who are engaged in trying to pursue radical innovation projects. The chapter also points out the complementary relationship between the internal infrastructures to support long-term innovative activity, and the mechanisms created to access external technologies. Instead of seeing Open Innovation as a substitute for radical innovation practices, O'Connor views them as functioning in mutually beneficial ways.

This is helpful on many levels. There is a tendency for some to view Open Innovation as a thinly disguised argument for simply outsourcing R&D to other companies. The RPI research program on radical innovation was motivated precisely to stimulate industrial R&D managers to refrain from cutting out all long-term R&D activity. In O'Connor's chapter, she finds that companies can adopt certain Open Innovation practices without eliminating internal R&D outright. Indeed, a judicious combination of the two appears to be beneficial, and the embrace of Open Innovation may help sustain the pursuit of longer term, more radical innovation.

The chapter breaks other new ground, in its discussion of the discovery/incubation/acceleration stages in the development of a radical innovation. While the path to the development of a radical innovation is tortuous and convoluted, these stages provide a deeper structure within the seemingly ad hoc innovation activities of companies aspiring to radical innovation. In O'Connor's view, Open Innovation is of greater help to companies in the first and second phases of this path. This is likely due to the sample for the RPI work, consisting of very large incumbent firms. Seen from the perspective of smaller firms, Open Innovation may help in the third phase as well, as part of the 'exit strategy' for a firm to partner with or sell to a larger firm in order to finish the commercialization of the technology. This is demonstrated in specialty materials by Robert Kirschbaum (2005), who has created an internal venturing process that spins out new companies, and selectively brings some of them back into the originating organization.[6]

2.5.2 *Chapter 4: Christensen*

In contrast to O'Connor's chapter on how internal innovation processes can be complemented by incorporating more external inputs, Christensen reverses the perspective. He reports on the journey of externally originating technologies into the market, and conditions under which an external technology does or does not get absorbed into the firm. Using a richly detailed study of the transition from analog to digital amplifier circuits in consumer electronics, he finds a variety of firm responses.[7]

This is an important finding in itself. The story of the adoption of digitial amplifier circuitry in consumer electronics reveals different approaches by individual firms to innovation in the consumer electronics industry. It is in part a story about the diffusion of research outside a university. It is also in part a story about an individual entrepreneur, and his attempt to commercialize a new technology. It is also in part a story about Open Innovation at Texas Instruments (TI), and TI's search for an external technology to provide a critical function that they lacked internally. It is even a story about some of the disadvantages faced by a closed innovator, Sony. Sony apparently lacked the processes and perspective to appreciate the value of an external technology that they were being offered (perhaps Sony also overestimated its internal capabilities with regard to digital amplifier circuitry as well).

A second lesson from this chapter is that once the battle for the dominant design is won, the winning firm then faces a new round of choices about how open to be with the winning technology. The desire to appropriate some value from the battle for the design (which was undoubtedly costly) is understandable, but not always advisable. If the design is in the service of supporting a larger system, the choice of how open to be must be taken with the perspective of the system business in mind. Here, for example, Sony's pursuit of an internal version of a digital amplifier (presumably to help it earn higher margins) may have impaired its access to a viable external version (which might have helped increase sales of the overall system).

A third lesson is the illustration of de-verticalization of this portion of the consumer electronics industry. While often described at the industry level, Christensen shows us that the de-verticalization results from the actions of individual firms. Shifts in strategy by some players, while other players remain vertically integrated, and the entry of new participations, act to cause the industry to de-verticalize. This also appears to assist in facilitating entry into the industry, as each of the new entrants appear to enter with some variant of de-verticalization. None chose to enter with a vertically integrated approach.

A fourth lesson, and one with which Christensen is eager to engage the academic community, is the implications of Open Innovation for the core competences theories of the firm in strategy.

Prahalad's emphasis on 'core competence' (Prahalad and Hamel 1990) follows an earlier article he did with Bettis (Prahalad and Bettis 1986) on dominant logic. In the 1986 piece, Prahalad and Bettis were concerned that the dominant logic might filter out important information when that information did not fit with the dominant logic (which, in Christensen's parlance, refers to the specialized but narrow). These concerns vanished by the time of Prahalad and Hamel's 1990 paper on core competences (which now correspond to the more aggregate, integrative competences in Christensen's chapter). In this later incarnation, core competences are viewed as unambiguously good (until qualified by Dorothy Leonard-Barton's article in *Strategic Management Journal* in 1992 on core competences leading to core rigidities). A more recent paper by Prencipe, Brusoni, and Pavitt (2001) situates the knowledge acquisition of firms as being broader than that strictly necessary to make its products, so that competences must be larger than the set of products currently offered.

How to reconcile these different perspectives, in light of Christensen's research on digitial amplifier circuitry? One place to start, from an Open Innovation point of view, is with the business model. TI has chosen to focus its competences upon an OEM business model, whereby TI makes highly complex components, but leaves it to its customers to build system products that utilize those components. It must continually develop new generation component technologies, in order to remain attractive to its systems customers. Sony, by contrast, has developed a business model whereby it makes most of the major componentry in its systems itself, a strategy of vertical integration. Sony makes television displays, DRAM, game consoles, and even uses its own proprietary memory devices (such as Memory sticks) for moving data from its camcorders to its digital cameras to its stereophonic and television equipment. For Sony, vertical integration is a means to capture more value in a highly competitive industry, and also a way to differentiate its products from the competition, as in the case of its proprietary memory formats.

As the Sony episode in Christensen's chapter shows, a vertically integrated business model can influence the firm's care and objectivity in assessing the quality and reliability of external technologies. In the context of core competences, Sony may have overemphasized its component technology competence, when it evaluated the digital amplifier technology from a Danish university, at the risk of its systems or architectural competence. TI, by contrast, though it struggled initially to successfully transfer the technology into its own development organization, has successfully created new systems and chips that benefit from the technology, which exemplifies the increased importance of architectural competence. TI will now profit still further from licensing the technology to other firms, in addition to its own product sales. This is another practice commended in open innovation, as it spreads TI's costs over a larger market, and makes TI's ongoing investment in R&D more sustainable.

2.5.3 *Chapter 5: West and Gallagher*

In Chapter 5, West and Gallagher examine the emergence of open source software in more mature companies. These more traditional companies have begun to craft business models around the open source code base. This is a tricky business to manage, because the founders of open source were well aware of the earlier history of Unix, and how that code base ultimately forked into a variety of incompatible versions. In the construction of the legal structure around open source, they have taken steps to prevent this from occurring this time around. As West and Gallagher note, there has been a schism between the 'open source' software community, and the 'free software' community. This schism derives largely from the fundamental disagreement between the two communities over whether the code base ought to remain free in the public domain, or whether it can be incorporated into other software that becomes proprietary.[8]

What is advancing the embrace of open source in many businesses, West and Gallagher find, is the emergence of what can be termed 'open source business models'. Their analysis is particularly illuminating for the incumbent firms who make products that use other, nonopen source technologies as well. They probe the conditions under which firms will choose to incorporate open source technologies into their overall innovation efforts, and develop 'open innovation business models'. As is consistent with the business model concept, parts of the model are quite open, while other parts are quite proprietary or closed.

As West and Gallagher show, open source is a marked departure from previous 'industrial' models of software development. They are also alert to the unique combination of lower marginal production and distribution costs, with strong network externalities on the demand side of the adoption process. It is not surprising that any strategic moves that enhance these economics will be embraced. And open source, notwithstanding the goals of its initial creators, can sometimes be harnessed for this very purpose. Indeed, West and Gallagher report that some firms now actively choose to sponsor open source projects.

One delight of this chapter is that the two authors go well beyond the label of 'open source', and unpack it into five different types of code, each with distinguishing characteristics. Of particular interest to me is the creation of Mozilla—an open source variant of Netscape Navigator—which started in 1998, languished for a long time with few contributors to advance the code.

However, it had the ability to serve as a browser for Unix workstations like HP, Sun, and IBM (which could not use Microsoft Explorer, since it was tightly integrated with Microsoft Windows, and did not run on each company's respective Unix operation system). These companies made the decision to support Mozilla in the open source domain, in order to continue to sell their (highly profitable and largely proprietary) Unix workstations.

West and Gallagher also consider the use of spin-outs (and later on, spin-ins) from inside the organization to an external body, as another means of harnessing open source to a business model. IBM's strategic placement of its Eclipse technologies into the open source domain was intended to accelerate the adoption of key tools for its overarching WebSphere architecture. Two of its key competitors chose not to join the initiative, but chose instead to create their own open source domains for their technologies. This has not been widely studied in academia yet, and there is likely to be much more of this kind of 'competition'. Perhaps another step along the logic of this chain is the decision to compete by donating one or more technologies into a nonprofit organization of some kind. Indeed, IBM recently donated 500 of its own software patents into the open source domain, to create more activity in this area. Presumably, IBM will find other ways to profit from this activity in other portions of its business model. This is often missed by the advocates of free software or the elimination of all protection for intellectual property (IP): companies will often have motivations to donate or give away IP in the service of their own business model.

These are novel and intriguing ways to create value and to capture a portion of that value from technology, but they are business models nonetheless. As lawsuits around the source code for Linux arise, and as other legal challenges to open source are made, it is likely that some of the most effective defense of open source software will come from decidedly profit-minded organizations who have crafted business models that embrace open source.

2.5.4 *Synthesizing the Chapters*

Chapters 3, 4, and 5 in this volume are quite diverse in content, focus, and method. This makes any synthesis of them quite challenging. From the perspective of Open Innovation, though, one can discern five underlying themes that run through all three chapters:

1. The central role of the business model
2. The role of external technology in advancing the business model
3. The problem of identifying, accessing, and incorporating knowledge
4. The role of start-up firms and new entrants
5. The role of intellectual property

Let us consider each in turn. The concept of the business model is a key construct in open innovation, and figures prominently in these chapters too. A business model has two important functions. It must create value within the value chain; and it must capture a piece of value for the focal firm in that chain. In O'Connor's chapter, the business model appears to constrain firms in the pursuit of longer term, more radical innovations. In Christensen's chapter, the business model appears to influence the innovation approach of different firms (such as that of TI vs. Sony). In West and Gallagher, different variants of business models are emerging to enable the advance of open source software.

External technology. Utilizing external technology can help leverage a firm's business model, both by filling in gaps within the firm's own road map, and by creating complementary products and services that stimulate faster and higher acceptance of the internal technology. In O'Connor's chapter, external technologies potentially can reduce the development time for a radical innovation, making more radical projects more sustainable within the confines of the firm's business model. Firms create new roles, such as idea hunters and idea gatherers, to identify potentially useful external technology. In Christensen's chapter, the digital amplifier circuitry emerges out of a university research program, and struggles to take root inside a commercial entity. In West and Gallagher, open source is the envelope of collectively generated external knowledge around a technology platform, such as Linux, Apache, or Mozilla. A community of contributors emerge through this platform, and supply new technologies to it.

Knowledge. Open Innovation requires an increased emphasis on managing knowledge, both in identifying promising sources of external knowledge (and being able to recognize it as such), and in linking that knowledge together with internal knowledge to create new systems and architectures. In O'Connor, companies often lack the knowledge of how to structure development agreements with outside organizations. This presumably slows down their time to market, suggesting that the firm must going through a learning phase before it truly benefits from a faster development cycle. In Christensen, knowledge exists in many places, and is difficult to transfer from a university setting (via an entrepreneurial spin-out) to a larger company. While TI's discovery of the technology was almost accidental, to its credit, it rapidly developed a working relationship with the inventor and the technology, worked hard over many months to absorb it, and subsequently acquired ownership of the technology. By contrast, Sony lacked the appreciation for the nascent technology's capability, and perhaps overestimated its own internal capabilities to replicate it. Ironically, Sony now may have to negotiate with TI to gain access to this very technology, but at a much higher price than it could have obtained earlier, had Sony worked directly with the Danish entrepreneur. In West and Gallagher, open and transparent parts of knowledge via open source are joined with more proprietary knowledge in the business models of sponsoring companies. Reputations of the individual contributors to the open source code point contributors to those who are contributing the most to the code, and resolving its key issues.

Start-ups. Start-ups play an important role, well beyond that of their share of revenues or employment within the economy. They are carriers of new technologies, and sometimes explorers of new markets. They also often represent experiments with new and different business models. In O'Connor, start-ups provide an initial impetus for radical innovations, and sometimes become important partners in the creation and delivery of those radical innovations. In Christensen, start-ups represent an important source of novel technologies into an industry, even though start-ups do not appear to command much market share in consumer electronics. In West and Gallagher, start-ups experiment with new business models associated with open source. They introduce new variety into the software community or ecosystem, and help that community penetrate into very large enterprises.

IP. As will be explored in Part II of this volume, IP plays an important and nuanced role in Open Innovation. By defining property rights, IP helps to facilitate exchange of ideas and technologies between the many parties who possess useful knowledge. However, property rights that are too strong or too broad might inhibit the flow of ideas and technologies that is necessary for Open Innovation to function well. In O'Connor, IP does not figure prominently. It plays a supporting role to the business strategy of the firms who are pursuing radical innovations. When used, it is primarily to create the design freedom that large company designers require to attack big problems with long term initiatives. In Christensen's account by contrast, the Danish entrepreneur would be completely sunk without IP protection. His discussions with large consumer electronics companies could have resulted in the complete appropriation of the idea by one or more big companies. More subtly, the eventual partner firm, TI, itself had to dedicate considerable time and resources to master the digital technology. That investment would likely not have been forthcoming, if TI could not establish some amount of ownership over the IP. In West and Gallagher, open source has been pressed into service as a marketing complement for decidedly proprietary technologies. If there were no discernable ways to make money from open source software, it might have remained an intriguing curiosity inside university and government laboratories.

A more subtle, and perhaps even more powerful strategy to leverage open source in one's business model is to develop system architectures that build upon it. In a world with lots of useful building blocks, the creation of value shifts from developing yet another building block that is slightly differentiated from the others, to crafting coherent combinations of building blocks into systems that solve real commercial problems. This competition is well underway in Web services (West 2003). Microsoft is trying to establish its. Net architecture as the platform for these services. That architecture will undoubtedly leverage Microsoft's tremendous franchise in its Windows operating system, and the extensive community of developers and other third parties who have based their livelihood upon it. IBM, by contrast, is countering with its WebSphere architecture, which will have to work with Windows, but has the opportunity to leverage open source technologies far more extensively, along with the extensive community that has arisen around those technologies.

2.6 Conclusion

One test of a new paradigm is that extent to which it either identifies new areas of research, or places new emphasis upon previously less salient research areas. This chapter has discussed numerous research areas inside the firm which would benefit from additional scholarly inquiry. The observed variation in utilization of external technologies within the innovation process of a firm raises many interesting questions that were not considered to be of much interest before. The loose coupling between the innovation process of the firm and its business model invites close examination of this coupling, and the ways in which it must be either accommodated or tightened. And the business model construct seems to point the way for very interesting research on the potential for the further adoption of open source development methodologies within industries, and within other sectors of societies.[9]

Notes

1. These data are for the US. Thomas Friedman's new (2005) book, *The World is Flat: A Brief History of the 21st Century*, carries this argument through to an international level. Though his data are largely anecdotal, Friedman's claim that 'the world is flat' is another way of saying that there are fewer economies of scale in R&D globally as well as in the US.
2. For one study among many at the Tufts Center for Study of Drug Development, which conducts period studies on the attrition rate of compounds in the drug development process, see DiMasi JA. Risks in new drug development: approval success rates for investigational drugs. *Clinical Pharmacology & Therapeutics* 2001 May; 69(5): 297–307.
3. See the comment of Rick Rashid, Microsoft's Senior VP for Research, in 2003: 'Our people are judged on peer-reviewed literature, just like they would be in the university environment. And the goal here is to say you have to move the state of the art forward if you're going to be of value to a corporation like Microsoft, and that's what we're trying to do first and foremost.' (*Source*: http://www.microsoft.com/presspass/exec/rick/04-16svalley.mspx)

4. In two cases I have studied, the interval was quite different. In Lucent's New Ventures Group in the late 1990s, the interval was initially nine months, and later condensed to three months, in which the business units had the right of first refusal. In Procter & Gamble, the interval is set at 3 years after a patent is issued to P&G. If the technology is not in use in at least one P&G business by then, the technology is made available to any outside organization (Sakkab 2002).
5. There are further benefits to 'selling' technology and avoiding the NSH. Sales of technology to external parties help companies to control the risk of technological leakages (such as by employees that leave the company). In the case of an unwanted employee departure, the firm has no control over the eventual use of its technology. A controlled spin-out, out-licensing agreement or a nurtured divestment, by comparison, enables companies to control how the technology will be used in future applications, or may provide protections on certain fields of use, or certain time frames, or grantback rights to improvements. In this case, keeping valuable technology on the shelf increases the risks of leakages, and forfeits the ability to direct or control such leakages. I am indebted to my coeditor, Wim vonHaverbeke for this insight.
6. Robert Kirschbaum, 'Open Innovation in Practice', *Research-Technology Management*, July–August, 2005: 24–28.
7. For a richer discussion of the emergence of the digital amplification technology, see the very recent article by Christensen et al. (2005).
8. For a very recent instance of this tension, and the associated risks of co-option, consider the 2005 statements by Jesus Villasante, head of software technologies at the European Community's Information Society and Media Directorate General: 'IBM says to a customer, "Do you want proprietary or open software?" Then [if they want open source] they say, "OK, you want IBM open source." It is [always] IBM or Sun or HP open source.... Companies are using the potential of communities as subcontractors—the open source community today [is a] subcontractor of American multinationals. Open source communities need to take themselves seriously and realize they have contribution to themselves and society. From the moment they realize they are part of the evolution of society and try to influence it, we will be moving in the right direction' (Mason 2005). Villasante's comments capture in a nutshell the tensions within the open source movement between the 'open' and 'free' software camps.
9. To note just a couple of these examples, the Public Library of Science (PLoS) is utilizing an open publishing model to accelerate the dissemination of scientific research to the wider world. And a group in Australia, called BIOS, is developing an alternative approach to genetic engineering that bypasses the strong patents held (many by universities) on the prevailing technology for genetic engineering.

3

Whither Core Competency for the Large Corporation in an Open Innovation World?[1]

Jens Frøslev Christensen

3.1 Introduction

By the early 1990s, it had become broadly acknowledged that large technology-intensive companies, in order to achieve sustainable competitive advantage, should stick to their core technological competencies, exploit these in the development of critical components and product architectures within their core business areas, and pursue opportunities for diversification into technology-related product markets. In several ways this strategic 'wisdom' represented a substantial change in the mental models and practices of corporate strategy that had prevailed during the previous decades. First, it reflected a break away from the tendencies in the 1960s and 1970s towards unrelated business diversification based on portfolio thinking, particularly in the US (Porter 1987). Second, this new strategic perspective generally assumed a more introvert orientation than was the case in the Porterian positioning perspective that had dominated strategic theory and practice during the 1980s. In emphasizing the role of distinctive and company-wide technological competencies as the basic driver for long-term competitiveness, this strategic reorientation represented a clear inspiration from the coherent and technology-based strategies of Japanese companies of the 1970s and 1980s and a critique of the fragmented strategies of divisionalized US companies.

However, recent changes in the context for technological innovation have significantly contributed to undermine the validity of some of the strategy imperatives of the early 1990s. These changes are associated with increasing vertical disintegration, outsourcing, modularization, use of open standards, and the growth of the market for specialized technology. 'Open Innovation' (Chesbrough 2003*a*) can be considered an *organizational* innovation by which large companies seek to adapt to these changes. But what has happened to the core competency perspective?

In order to better answer this question, we shall apply two sets of concepts for understanding technological knowledge. The first set comprises a distinction between narrowly specialized technological capabilities and integrative competencies, including capacities for systems integration and for reconfiguring and building internal and external capabilities to address changing environments. The second set of concepts entails the distinction, suggested by Nelson (1998), between 'bodies of understanding', abstract knowledge in technical fields, and 'bodies of practice', context-specific knowledge associated with the practice of product or process innovation. One important aspect of the changing environment for large companies has been the tendency for new 'bodies of understanding' and specialized technological knowledge to emerge in small firms. This means that small firms often develop new agendas for technology-based business opportunities for large firms, and in order to explore and exploit these opportunities, large innovative firms must put greater emphasis on the dynamic/adaptive, open/extrovert, and systems integration sides of their competencies than what is traditionally associated with the core competency perspective.

This chapter contains three sections. Section 3.2 gives an outline of the dominant logics of corporate strategy and innovation in the late 1980s and early 1990s. This is done through a review of the most important literature on these matters from that period. In particular one stream of literature was associated with the emerging resource- and knowledge-based view of the firm and its strategy and focused on (core) competencies, vertical integration, and technology-related diversification (e.g. Prahalad and Hamel 1990). Introvert modes of innovation were argued to be the standards to be met for large successful companies. However, other research streams did presage the notion of Open Innovation and addressed issues such as absorptive capacity (Cohen and Levinthal 1990; Rosenberg 1990), complementary assets (Teece 1986), and Fifth Generation innovation (Rothwell 1994).

Section 3.3 reviews three interrelated empirical changes in the conditions for technological innovation during the last one or two decades and their likely impacts on the nature of (core) competencies for technological innovation in large companies. First, the general tendency towards vertical disintegration and the 'unbundling' of the vertical corporate structure. Second, the tendency towards more diverse corporate technology profiles and more externally oriented and less cumulative technological competencies; and third, the tendency for increasingly distributed and open modes of organizing research and development (R&D) in large companies, including the increasing requirements for coordinating the innovation processes in and between large and small firms.

To illustrate the central issues raised in this chapter, Section 3.4 presents and analyzes a case study on the current transformation in amplifier technology within the consumer electronics industry.

3.2 Dominant Logics on Corporate Strategy and Innovation in the Late 1980s and Early 1990s

During the early 1990s the dominant perspectives on corporate strategy and innovation underwent significant changes that were reflected in both the management and academic literature as well as in the practices of corporate behavior. In this section, these perspectives are reviewed using the lenses of the most influential papers on these matters from the late 1980s to the early 1990s.

Perhaps most prominently these changes were featured by Prahalad and Hamel in their 1990 article on the role of core competencies for large technology-intensive companies. They maintained that in order for such companies to perform successfully at the longer term, they would have to stick to a limited set of distinctive technological capabilities in which they could obtain specialization and synergistic economies and through which they would be able to deliver an ongoing flow of innovations to multiple product markets. The paper had a powerful impact on corporate managers' (and their consultants') general conception of what constituted the foundation for sustainable competitive advantage in large corporations. It was part of a broader wave of strategy literature that surfaced in the late 1980s and early 1990s, under the common term Resource-Based View (RBV). This literature provided theoretical and empirical support for the basic idea that competitive advantage stems from imperfectly imitable, imperfectly substitutable and imperfectly tradeable, and valuable assets (Barney 1986, 1991; Dierickx and Cool 1989; Grant 1991; Peteraf 1993). It formed a comprehensive critique of the two hitherto dominant perspectives in corporate strategy, portfolio-based strategy that flourished in the 1960s and 1970s, and the Porterian positioning view that prospered during the 1980s. The RBV was inconsistent with unrelated diversification strategies while providing support for competence-based strategies associated with related diversification strategies (Markides and Williamson 1994). Likewise, much of the RBV literature criticized the predominant multidivisional mode of organizing the large company (the M-form), especially in its decentralized (Williamsonian) version, which was argued to lead to corporate fragmentation and short-termism and to undermine the capacity for developing core competencies and radical innovations (Prahalad and Hamel 1990; Chandler 1991; Hedlund 1994; Teece et al. 1994; Christensen and Foss 1997). Some level of central planning was needed to identify and build company-wide core competencies and to overcome the 'tyranny of the SBU' (strategic business unit) (Prahalad and Hamel 1990).

As Porter already in 1991 observed, much of the RBV literature shared an introvert inclination: *Your* company is, or should be, the best in what *your* company is doing, an inclination stimulating not only a high achievement spirit but also (potentially) a Not Invented Here arrogance.[2] That also accounted for the way technological innovation in large corporation was generally perceived, although Prahalad and Hamel explicitly referred to the effectiveness with which Japanese firms during the 1970s and 1980s acquired external knowledge as an important means of building core competencies.

While Prahalad and Hamel (1990) and much of the other RBV literature made strategic arguments for nurturing core competencies in order to leverage long-term innovative, hence competitive, performance, Henderson and Clark's paper on Architectural Innovation from the same year (1990) addressed the more downstream issues of managing and organizing innovation in large companies. The paper obtained lasting impacts on the theory and practice of management of innovation, and in particular brought the issue of modularity and systems integration out of the narrow confines of design and engineering disciplines and into the strategy and management fields.

Henderson and Clark (1990) proposed a distinction between two levels of innovation; the component (or module) level and the architectural (or systemic level). This allowed them to specify the well-established distinction between incremental and radical innovation by adding two new categories: modular and architectural innovation. Their case study pointed to particular difficulties for large companies in dealing with not only radical innovation (which is not surprising) but also architectural innovations that involve substantial systemic changes but no dramatic technical changes. Their explanation for the difficulties in managing architectural innovation was that existing product architectures tend to become ingrained in organizational routines and division of labor, the inertia of which provides a barrier to architectural innovation—even when the cognitive barrier associated with the technological change is low. Accordingly, large companies would have to explicitly engage in organizational adaptations in dealing with such innovations, and this would require some element of centralized planning.

When scrutinizing Prahalad and Hamel's paper (1990), a duality emerges in their use of the term core competencies. Sometimes, core competencies are associated with company-wide and integrative competencies needed for developing architectural and radical innovations. Sometimes, they are associated with deep and narrowly specialized technological capabilities needed to develop core components. Henderson and Clark's analysis makes it clear that there is no identity between the two categories. The distinction between the two corresponds to Christensen's distinction (1996, 2000) between, on the one hand, a specialized, technical *capability* that reflects a team-based capacity to mobilize resources for particular productive activities, and, on the other hand, an (integrative) *competence* that reflect a higher-order managerial capacity to mobilize, harmonize, and develop a diverse set of (tradeable) resources and capabilities to create value and competitive advantage at the systems level (e.g. in systemic products). In the following, we shall apply this analytical distinction as signifying two qualitatively different types of (potentially core) capacities.

Prior to Prahalad and Hamel's 'embracing' of the integrative core competency perspective, Prahalad and Bettis (1986) wrote a paper on the dominant logics of companies. Here they raised the concern that the dominant logic might filter out important knowledge when that knowledge is not well-integrated into the corporate logic. In my parlance above, such knowledge would exactly be the new, specialized and narrow capabilities emerging under the radar of the existing dominant logic, and eventually emerging to become a critical technology that will feed into existing or new integrative competencies. While these concerns vanished in Prahalad and Hamel's later notion of (integrative) core competency, which was considered an unambiguously positive asset, they were later qualified by Leonard-Barton (1992) who argued that core competencies may turn into core rigidities.[3]

Other concerns have later been raised by Williamson (1999) who states that the concept of core competency is expansive, elastic, and tends to be identified as an *ex post* 'good' asset: 'There being no apparatus by which to advise firms on when and how to reconfigure their core competencies, the arguments relies on ex post rationalization: show me a success story and I will show you (uncover) a core competence.' (1093)

Despite variations, and some critical concerns among scholars in corporate strategy and management of innovation as well as among business consultants and analysts, there was a broadly shared view in the late 1980s and the early 1990s that the ideal large R&D-intensive company should incarnate a core competency view, control both the systemic and the most critical parts of the component level of innovation (the more simple parts should be outsourced), and be occupied with the need for ongoing organizational adaptation. That would imply a more coherent and synergistic organization than the one accounted for in the strictly multidivisional structure (Christensen 2000; Hedlund 1994; Markides and Williamson 1994). Pavitt (2003) precisely points to the continuing importance of '... [d]ealing with an inevitably imperfect M-form organization, given the impossibility of neatly decomposing technological activities with pervasive applications into specific product divisions ...' (2003: 105). A focus on core competencies, technology-related diversification, and fairly introvert modes of innovation were the standards to be met for the large company, and among the successful benchmark cases frequently mentioned in the literature at the time were companies such as IBM (prior to the crisis and turnaround in the early 1990s), Intel, Texas Instruments, Ericsson, 3M, Philips, Siemens, and large Japanese players such as Canon, Casio, Honda, NEC, Matsushita, Sharp, and Sony.

This is not to say that there was no sense of the need for external relations in corporate innovation. Two seminal papers, Teece's 1986 paper on complementary assets, and Cohen and Levinthal's 1990 paper on absorptive capacity, clearly precipitated later more open innovation perspectives. Teece (1986) made a distinction between technological innovation and the complementary assets required to commercialize the innovation, and he developed a contingency framework, combining insights from resource-based and transaction cost theory, for determining whether complementary assets should be outsourced, accessed through alliances or licensing agreements, or developed in-house. He argued that pioneers in technological innovation often overrate the strength of the appropriability regimes surrounding their innovations and underestimate the importance of complementary assets. Even if Teece gives examples of owners of complementary assets (mostly large companies) capturing the major rents from innovations pioneered by other firms, he takes the view of the pioneer, whether small or large, and doesn't expand his framework into an analysis of large owners of complementary assets in search of (possibly) external innovation ideas, projects, and technology entrepreneurs. This latter perspective has only more recently become a central part of a more open innovation perspective.[4]

In their opening statement, Cohen and Levinthal (1990) placed Open Innovation (without using the term) as an upcoming agenda: 'Outside sources of knowledge are often critical to the innovation process, whatever the organizational level at which the innovating unit is defined' (p. 128). The central idea in their paper is that internal R&D investment plays two functions; to provide improved and new technologies and innovations and to provide a capacity to absorb relevant knowledge emerging in the external environment. Hence, absorptive capacity is primarily seen as a by-product of a firm's R&D investment. In the same vein, Rosenberg (1990) argues that an important reason why (some) large firms spend their own money on basic research, despite it having no or very little value as direct input to ongoing innovation, is that it positively impacts their capacity to integrate relevant, external science-based knowledge. Basic research may be thought of as 'a ticket of admission to an information network' (p. 170) and '. . . a basic research capability is often indispensable in order to monitor and to evaluate research being conducted elsewhere' (p. 171). Both Cohen and Levinthal's and Rosenberg's arguments are grounded in the fundamental insight that R&D processes are inevitably associated with spillovers, and to build absorptive capacity through in-house R&D is one way of capturing spillovers from external R&D. Moreover, as argued by Rosenberg, large multibusiness companies can better than small firms make internal use of spillovers from in-house research. Both papers show a certain bias towards internal mechanisms that influence an organization's absorptive capacity, and Cohen and Levinthal articulate a skepticism towards more 'open' forms of absorptive capacity:

The discussion thus far has focused on internal mechanisms that influence the organization's absorptive capacity. A question remains as to whether absorptive capacity needs to be internally developed or to what extent a firm may simply buy it via, for example, hiring new personnel, contracting for consulting services, or even through corporate acquisitions. We suggest that the effectiveness of such options is somewhat limited when the absorptive capacity in question is to be integrated with the firm's other activities. A critical component of the requisite absorptive capacity for certain types of information, such as those associated with product and process innovation, is often firm-specific and therefore cannot be bought and quickly integrated into the firm. (135)

From the secure position of the hindsight, it is clear that Cohen and Levinthal underestimated the extent to which such more 'open' mechanisms would come to penetrate many companies' mode of innovating and developing their absorptive capacity. Thus, for example, Lane and Lubatkin (1998) find that alliances can also develop absorptive capacity, and Mayer and Kenney (2004) show how Cisco since the early 1990s has successfully used acquisitions as a form of absorptive capacity, and, at least partially, a substitute for internal R&D. Cohen and Levinthal's concept of absorptive capacity is also limited to cover only knowledge areas related to or overlapping with those targeted by the firm's general R&D investments. If the firm wishes to acquire and use external knowledge that is unrelated to its current R&D activities, it must dedicate efforts exclusively to creating absorptive capacity, and Cohen and Levinthal state that firms are likely to underinvest in such areas (1990: 149–50). In somewhat contrast to this position, however, they also predict a need for companies in the future to expand the diversity of their absorptive capacity:

We also suggest... that as the fields underlying technical advance within an industry become more diverse, we may expect firms to increase their R&D as they develop absorptive capacity in each of the relevant fields. For example, as automobile manufacturing comes to draw more heavily on newer fields such as microelectronics and ceramics, we expect that manufacturers will expand their basic and applied research efforts to better evaluate and exploit new findings in these areas. (1990: 148)

As we shall see in Section 3.3.2, this prediction has later been verified by empirical research.

More explicitly Open Innovation perspectives that treat spillovers as potential resources to be managed either by bringing in external spillovers (in the Cohen and Levinthal mode) or by fostering external utilization of internal spillovers through licensing, spin-offs, and so on, had to await yet another decade.

That external relations are needed in technological innovation, has for long been reflected in both the practice and theory of management of innovation (dealing with innovation at the project level and in the context of an R&D organization). Since the 1970s, much of the management of innovation literature has addressed the interactive, cross-disciplinary, and (mostly) inter-organizational nature of innovative learning and searching (Rothwell et al. 1974; Rosenberg 1982; von Hippel 1988; Lundvall 1992; Pavitt 1998), and in his excellent review of generations of (somewhat different) modes of managing innovation, Rothwell (1994) clearly presages the notion of Open Innovation when, in the early 1990s, seeking to identify prevalent features in current streams of innovation practices (termed Fifth-Generation Innovation Process).

However, even if the importance of external relations were acknowledged, the predominant logic of innovation in large high-tech companies was introvert and proprietary (the technologically complex parts of innovation should be done in-house, while the simpler parts could be outsourced). In Section 3.3, I shall argue that the emergence of increasingly open modes of managing technological innovation in large companies reflects substantial changes in the external conditions for conducting technological innovation.

3.3 Empirical Insights on Innovative Dynamics Since Early 1990s

In the years since the papers reviewed above appeared, much seems to have changed. Below, we shall address three interrelated aspects of these changes: First, the general tendency towards vertical disintegration (Section 3.3.1), second, the tendencies towards more diverse technology profiles of large R&D-intensive companies (Section 3.3.2), and thirdly, the tendencies towards more distributed modes of organizing R&D in large companies (Section 3.3.3).

3.3.1 *The General Tendency Towards Vertical Disintegration*

> It has become exceedingly clear that the late twentieth (and now early twenty-first) centuries are witnessing a revolution at least as important as, but quite different from, the one Chandler described. Strikingly, the animating principle of this new revolution is precisely the *unmaking* of Chandler's revolution. Rather than seeing the continued dominance of multi-unit firms in which managerial control spans a large number of vertical stages, we are seeing a dramatic increase in vertical specialization—a thorough-going 'de-verticalization' that is affecting the traditional Chandlerian industries as much as the high-tech firms of the late twentieth century. (Langlois 2003: 352)

Likewise, Sturgeon (2002) argues that a new mode of industrial organization, characterized by increasing modularity, specialization, outsourcing, and networking, has been driving American capitalism (and probably most other parts of modern capitalism) since the 1990s. Two interrelated economic and institutional dynamics seem to underly this change: First, the world has seen dramatic increases in population and income as well as reductions of barriers to trade implying increasing division of labor and increased coordination through 'the market'.[5] Second, an important aspect of this development has been the emergence of market-supporting institutions (North 1990) reducing the costs of coordinating through the market. One case of illustration is the powerful trend in favor of open market standards (Steinmueller 2003). The rise and diffusion of the venture capital institution to promote technological entrepreneurship represent another important case. In combination with the increasing scope for secure and alienable intellectual property rights, these institutional dynamics have been critical drivers in the enhanced effectiveness of markets for specialized technological knowledge, whether this knowledge takes the form of a patent, an intangible asset (e.g. a software program), or a component to fit into a module or an end-product (Arora et al. 2001*a*). The shaping of (much more) well-functioning markets for technology has fuelled the generation of small technology entrepreneurs dedicated to the development of and commercial exploitation of highly specialized technological capabilities. Their 'core competency' (cf. previous discussion in Section 3.2) thus only reflects the specialized and deep side of Prahalad and Hamel's double-sided concept of core competency.

Also the very nature of technological change seems to have reinforced vertical disintegration in the sense, as argued by Langlois (2003), that technical change generally tends to reduce (minimum efficient) scale, making it possible and profitable for small firms to drive technological innovations in many areas and thereby 'unbundle' the vertical corporate structure.[6]

The tendency towards vertical disintegration, modularization, outsourcing, and networking gives rise to more open innovation models: 'Rather than being limited to the internal capabilities of even the most capable Chandlerian corporation, a modular system can benefit from the *external capabilities* of the entire economy' (Langlois 2003: 375). It can generate external *economies* of scope (Langlois and Robertson 1995), thus allow more entry points for innovation.

These tendencies have implied that large companies have had to give way to specialized suppliers (often independent start-ups, sometimes later to be acquired by large companies) at the level of component-based innovation and beyond (subcomponents or knowledge or service inputs in intangible form). If this also implies giving up front positions in an increasing array of relevant technological specialty fields, one can ask whether large incumbents may still be able to maintain the other side of the classical notion of core competency, those stemming from interaction and interfaces across components and their underlying capabilities? Or to use the concepts of Henderson and Clark, can incumbents maintain superior abilities to innovate at the architectural level when they, at least partially, have had to surrender at the component level? In order to come closer to answering this question, we shall take a look at what we know about the proliferation of corporate technology bases.

3.3.2 *Tendencies in the Proliferation of Corporate Technology Bases*

As stated above, the nature of technological change in recent decades seems to have favored vertical disintegration and market dynamics. But two other aspects in the accumulation of technological knowledge have, in combination, given rise to nontrivial challenges in the technology strategies of large firms. The growth in global R&D investment (Kodamoa 1992) leads to an increasing number of technical fields providing new opportunities for problem-solving, and moreover, a tendency for specialized knowledge in each field to deepen leading to ongoing enhancement of the opportunities for performance improvements in problem-solving. Altogether, we witness an expansion in the global technological opportunity set, an expansion most likely to be exponential in times of global market expansion and improved effectiveness of markets for specialized technology, as witnessed since the 1980s as the Asian Tigers, China and Eastern Europe have become strongly enrolled in the global market economy, and as institutions for technology markets have been strengthened.

However, companies can generally not (at least not on an enduring basis) expand their R&D investments at the same rate due to budgetary constraints and limited organizational capacity of firms to absorb and integrate new knowledge. With R&D funding in large incumbents being constant (or slowly growing) and the global technology base rapidly expanding, incumbents must acknowledge that an increasing share of relevant technological knowledge is being accumulated externally, and they will have to choose between (at the extremes) whether they strive for world-leading positions in one or a few fields or wish to obtain some (more superficial) level of knowledge in many areas.

How have large R&D-intensive companies responded to these strategic dilemmas? Are they sticking to a few interrelated core areas as would be expected by Prahalad and Hamel (1990), or do they try to follow suite into a broader array of technologies more in accordance with an architectural view (Henderson and Clark 1990) and the proposition of increasing diversity of R&D investments, as predicted by Cohen and Levinthal (1990) (cf. Section 3.2). Several empirical studies based primarily on patent data (covering especially the 1980s and early 1990s) have shown that in large companies, technology diversification has been more pronounced than product diversification (Granstrand 1982; Pavitt et al. 1989; Granstrand and Sjölander 1990; Granstrand et al. 1997; Patel and Pavitt 1997; Gambardella and Torrisi 1998). While their technological diversity has tended to increase, their product range has typically not expanded to the same degree, or become narrower. Among the world's largest technology-intensive companies, by far the most had expanded the number of technical fields in which they are active from the early 1970s to the late 1980s and have developed significant capabilities outside their distinctive technologies (Granstrand et al. 1997).

Granstrand et al. (1997: 13) make the following interpretation of their empirical data (both patent statistics and case studies):

> Large firms built up and maintained a broad technology base in order to explore and experiment with new technologies for possible deployment in the future. The creation of corporate competencies in new fields was a dynamic process of learning, often requiring a combination of external technology acquisition and in-house technological activities and usually resulting in an increase in R&D expenditures. While technology sourcing was rarely a substitute for in-house R&D, it was an important complement to it.

Large companies clearly also had a focus on a number of 'core' technological capabilities,[7] as recommended by Prahalad and Hamel, but in addition they sustained an increasing and broader (if less deep) set of technological capabilities, what Granstrand et al. (1997) term background competence enabling the company to coordinate and benefit from technical change (and exchange) in its supply chain, and moreover explored new opportunities emerging from scientific and technological breakthroughs. In short, they had become multi-technology firms (Granstrand et al. 1997; Patel and Pavitt 1997).

The studies furthermore show that firms producing similar products tended to master similar technologies. These results are contemplated by Patel and Pavitt (1997) as follows: 'Given that some technologies underpin a range of competing and differentiated product configurations, product variety in an industry is compatible with technological homogeneity' (p. 154). This interpretation '... is compatible in the sphere of product development with variety, experimentation, social shaping, and trade-offs at the margin, but in the sphere of technology, it is underpinned by quite rigid one-to-one technological imperatives' (p. 155). Pavitt (1998) elaborates on this interpretation by applying Nelson's (1998) distinction between two complementary forms of knowledge, 'bodies of understanding', abstract knowledge underlying technological fields and giving rise to patenting and publishing, and 'bodies of practice', context-specific knowledge related to engineers' experience and firms' practices in product and process development. The former is reflected in the technology profiles as indicated by the patent studies, while the latter is interpreted as 'organizational knowledge', and Pavitt concludes that competitive advantage is primarily based on *organizational* characteristics of the firm (e.g. interactions between different functional departments) rather than on distinctive technological competencies. This interpretation is contested by Nesta and Dibiaggio (2003) who make an empirical account of Nelson's analytical distinction in their study of the dynamics of technology profiles in biotech firms. They find that even if these firms also tend to develop similar profiles in terms of technical disciplines (bodies of understanding), they diverge in terms of the particularities of their technology combinations which are used as indicators of application- and experience-based competencies (bodies of practice). While this analysis specifies the role of (hence saves *some* role for) technology as a source of competitive advantage, it does not contest the proposition that organizational characteristics are also important, and there is indeed a key element of organization to 'bodies of practice'.

Generally, the results of the studies discussed above do not support the proposition that successful firms primarily tend to focus on few distinctive 'core technologies' as would be expected following the more narrow conception of core competency (cf. the discussion in Section 3.2). 'Core technologies' play a significant but relatively decreasing role in the technology profiles (bodies of understanding) of large companies, while they show increasing involvement in noncore technology areas, 'background competencies' and emerging areas of knowledge. Most of these studies, however, deal with industry averages and mask inter-industry differences between firms.[8] Moreover, they cannot say much about the possible role of (core) competencies in the broad sense of being (more or less) company-wide integrative competencies.

A richer picture of innovative and technological competencies of large firms has been emerging from a number of detailed field studies (Prencipe 1997, 2000; Iansiti 1998; Brusoni et al. 2001; Gawer and Cusumano 2002; Chesbrough 2003*a*; Ernst 2003). Generally, these studies have addressed the increasingly important role of large companies as system integrators, innovation architects, platform leaders, standards creators, or in short, market coordinators of increasingly distributed and vertically disintegrated value chains. Prencipe (1997, 2000) finds that aircraft engine manufacturers retain knowledge about components whose production is outsourced. Thus, one engine maker developed capabilities to specify and test externally produced components, and to coordinate the integration of new technologies. Brusoni et al. (2001), who further explore the development of the aircraft engine control systems, find evidence that such development requires the mobilization and maintenance of a loosely coupled network organization:

> A key characteristic of a loosely coupled network organization is the presence of a systems integrator firm that outsources detailed design and manufacturing to specialized suppliers while maintaining in house concept design and systems integration capabilities to coordinate the work (R&D, design, and manufacturing) of suppliers (pp. 617–18).

The nature of this 'modern' concept of integrative competencies differs in two respects from that of Prahalad and Hamel's (1990) company-wide core competencies. First, from the technology side, integrative competencies are not as strongly associated with particular areas of technological knowledge ('bodies of understanding') as the case is with Prahalad and Hamel's core competencies. Integrative competencies rather relate to application-specific knowledge ('bodies of practice') engaged in product design (both of components and architectures), including the processes by which firms synthesize and acquire knowledge resources and transform these resources into applications

(Kogut and Zander 1992). Second, from the managerial side, the integrative competencies need to be responsive and adaptive to changing external contingencies (e.g. changes in component markets, the emergence of new external technologies), while 'core competencies' are usually assumed to be subject to long-term strategies for cumulative competence building and improvement. While features relating to the technology side are reflected in recent research into systems integration competencies (Prencipe et al. 2003), the managerial side is much closer to the concept of dynamic capabilities (Teece et al. 1997, Eisenhardt and Martin 2000) by which firm managers 'integrate, build, and reconfigure internal and external competencies to address rapidly changing environments' (Teece et al. 1997: 516). And more generally, this notion of integrative competencies is more consistent with open modes of innovation than the 'old' notion of core competency is.

3.3.3 *The Organization of Corporate R&D and the Coordination with Technology Specialists*

'Open Innovation' (Chesbrough 2003*a*) can be conceived as an *organizational* innovation in the way large companies try to come to grips with the changes in the context for technological innovation that have been outlined above.[9] This organizational innovation is overlapping with and extending the scope of earlier organizational changes since the 1980s from the 'central R&D lab' mode that became prevalent in large high-tech companies after World War II to an increasingly distributed mode through a wave of downsizing of central laboratories and delegation of responsibility for technical innovation to product divisions and subsidiaries (Coombs and Richards 1993; Christensen 2002). A particular feature of this transformation has been the tendency, although somewhat reluctantly, towards internationalization of corporate R&D (Kuemmerle 1998; Gerybadze and Reger 1999; Boutellier et al. 2000; Kim et al. 2003).

Neither in the case of increasingly distributed corporate innovation nor the case of increasingly open innovation, are we dealing with one paradigm replacing another. While the overall trend in the 1980s seems to have involved a predominant process of decentralization of R&D to lower levels in the corporate structure, hence a weakening, at times a full elimination, of the previously dominant position of the central laboratory, there are no evidence that this trend has continued to create a dominant model of fully decentralized and distributed R&D. Rather, according to two surveys of R&D-intensive companies in 1994 and 2001 by Industrial Research Institute (here referred from Argyres and Silverman 2004), the largest group of the surveyed companies (about 60 percent) in both years reported hybrid structures, while only a small minority (about 10 percent) reported a decentralized structure and a larger group (about 30 percent) a centralized structure. Thus, many corporations still maintain quite powerful central laboratories and experiment with different ways of coordinating R&D at the central and decentral levels (Coombs and Richards 1993; Argyres 1995; Christensen 2002; Argyres and Silverman 2004; Tidd et al. 2005).

Likewise, corporations do not externalize all research and innovation in the transition from relatively more closed to more open innovation. A recent study by Laursen and Salter (2006) indicates that while external relations are critical for successful management of innovation, there are limits to the scope of external relations that companies can effectively manage in innovation projects. Furthermore, neither the distributed nor the open mode of innovation should lead to the interpretation that all companies act according to herd behavior and practice identical or very similar modes of governing innovation. Huge variation exists across as well as within industries and companies are not only moving in a one-way direction towards delegation and externalization, but may, under various contingencies, also change the direction and partially recentralize and internalize.

An important premise for large high-tech companies in an increasingly open innovation world is that superior technological capabilities are increasingly emerging outside the boundaries of large companies. As markets for technology have improved, we increasingly witness a division of labor between, on the one hand, technology entrepreneurs, often in collaboration with universities and other research institutions, providing emergent, deep technological capabilities, and, on the other hand, large companies providing integrative and dynamic competencies. While the advanced technology entrepreneurs develop the technologies in their more abstract form ('bodies of understanding') and experiment with early adaptation of the knowledge to practical applications (e.g. prototypes, early products/components for high-end markets), the large companies further transform the technologies into application-specific use ('bodies of practice') which, among other things, imply the use of modularity tools for systems integration and the experience-based maturing of the technology for large-scale throughput. The strength of large firms, however, often extends beyond the scope of their innovative assets (Christensen 1995, 1996) and capacities for systems integration. Large companies also tend to be endowed with powerful complementary assets for large-scale commercialization of innovation (Teece 1986), even if these more operational types of assets (in particularly manufacturing assets) are also increasingly being subject to 'de-verticalization'. Thus, from an innovative asset perspective, large companies will have to look out for external (as well as internal) innovative ideas, new technologies, concepts, or IPs to align with and integrate into new or improved product architectures. And from an operational asset perspective, large companies will have to look out for external (and internal) innovations in search of, and sometimes in exchange for, complementary assets.[10]

3.4. Open Innovation: The Case of the Digital Amplifier in Consumer Electronics[11]

The industrial and strategic dynamics underlying the recent breakthrough of a new amplification technology, termed class D or switched amplification, can provide us with an improved empirical understanding of the critical issues discussed in this chapter. More specifically, the case can illustrate:

- the way knowledge for leveraging a new, complex technology can be decomposed into a set of specialized and deep capabilities, on the one hand, and particular forms of integrative competencies, on the other hand;
- how the division of knowledge between small technology-based firms and large incumbents involve a division of labor in terms of the roles in developing, maturing, and commercializing the new technology; and
- the diversity of more or less open innovation strategies conducted by large incumbents engaged in the development of the same new technology.

3.4.1 *Specifics About Class D Technology and its Market Prospects*

Since the mid-1990s a radically different approach to amplification, class D or switched amplification, has been subject to a major scientific, technological, and commercial breakthrough.[12] *'It marks a clear break with tradition, and incidentally demands an almost entirely different set of design skills than those we are used to seeing in analog electronics generally'* (Sweeney 2004: 5). While known at least in conceptual form for more than forty years, class D amplifiers had never been successfully applied in an audio context. Even if early class D amplifiers offered big advantages as compared to conventional class A/B amplifiers in terms of space efficiency, energy efficiency, and low heat dissipation, they also suffered from severe fidelity and reliability problems and tended to burn up due to overload or radiate unacceptable amounts of interference (Sweeney 2004: 7). However, as these problems have recently been overcome, we are currently witnessing a technological transformation comparable with the solid state revolution in amplification some fifty years ago. In less than ten years, since the mid-1990s, this technology has undergone a condensed cycle from a stage of embryonic experimentation pioneered by university scientists and small start-ups, to a fairly mature stage characterized by chips-based technology and mass production controlled, to a great extent, by large incumbents (Christensen et al. 2005).

Class D amplifiers can be embedded in either discrete modules (based on discrete standard components) or in chip-based modules (based on integrated components). The former are high-cost/performance amplifiers which have since the late 1990s penetrated parts of the high-end niche markets, while the latter have gained increasing positions in the mid-level mass markets, in particular the DVD receiver market, and increasingly are moving down towards the lower-end markets. The big audio markets are still dominated by conventional technology. Rodman & Renshaw Equity Research estimates the size of the analog amplifier market between $2.1 billion to $3.0 billion as of 2003 and the size of the switched amplifier market between $80 million to $100 million, or only 2–3 percent of the total amplifier market (Rodman and Renshaw 2003). This level is expected to increase to $515 million, or 15 percent of the total amplifier market by 2006. Forward Concept (Sweeney 2004) estimates the total class D amplifier 2003-market at $84 million, and forecasts steep growth rates as cell phones, automotive audio, and other markets are expected to kick in. By 2008, the market is expected to exceed $800 million.

3.4.2 *Competence Requirements for Class D Innovation*

Even if the traditional class A/B amplifiers and the new class D amplifiers share some components, such as power supplies, filters, and semiconductors, the knowledge underlying their respective core components and systemic interdependencies differ in fundamental ways. Thus, despite some technological heredity (Metcalfe and Gibbons 1989) in peripheral parts of the amplifier, this new technology reflects a radical competence-destroying discontinuity signifying substantial cognitive barriers (Tushman and Anderson 1986) to overcome for incumbents.

During the embryonic stage of this technology (mid- to late 1990s), successful innovation in class D technology required the alignment not only of three complementary types of innovative knowledge assets: science-based assets, product design assets, and lead-user assets (von Hippel 1988; Christensen 1995), but also the alignment of operational (complementary) assets. The knowledge base necessary for leveraging the functionalities of class D technology to acceptable performance standards was (and still is) highly complex. To design a full amplifier system, including integrating a class D amplifier chip with high-power transistors and other components, requires capabilities in signal modulation, electro magnetic compatibility (EMC), error correction and electric power engineering, chip design as well as competencies in optimizing and integrating the components associated with the new technology into a complete amplifier module, and the integration of this module into the particular end-product system (Lammers and Ohr 2003). These requirements thus involve both deep, specialized capabilities in numerous technical fields with a bias towards 'bodies of understanding', and complex system integration competencies with a bias towards 'bodies of practice'—and both are very different from those at work in traditional amplification technology. Hence, the digital amplifier represented an engineering challenge beyond the existing capacities of most amplifier incumbents.

3.4.3 *The Pioneering Role of Technology Entrepreneurs*

The breakthrough in class D amplification occurred as a result of basic university research, and especially the achievements of a research community lead by Professor Michael A.E. Andersen at Technical University of Denmark where the research culminated in two spin-off ventures: Toccata Technology and ICEpower, now owned by, respectively, Texas Instruments and Bang & Olufsen. Both ventures were founded on a strong IP base of patents reflecting the technical novelties obtained through the founders' previous PhD-projects. Together with the US-based start-up, Tripath, and Dutch Philips, Toccata and ICEpower were the early pioneers of class D amplifiers, launching products in 1998 and 1999.

Figure 3.1 shows the cumulative number of firms' first launches over the period 1997–2004. By early 2004, twenty-four firms with at least some activity in the area have been registered. They can be divided into three groups: First, a number of small start-up ventures, including, beyond the previously mentioned early pioneers, Apogee (US), JAM Technologies (US), and NeoFidelity (Korea); Second, a group of large vendors of semiconductors and digital signal processing chips, for example National Semiconductor, STMicroelectronics, and Texas Instruments; and thirdly, a few large-scale Audio-Visual (AV) OEMs, including first of all Philips and Sony.

Small technology-based firms set the agenda for this upcoming technological innovation founded on a core of highly specialized and deep technical knowledge. Several of the start-ups did not provide any amplifier products but only IP assets covering only part of the class D value chain. Hence, in order to become technologically mature and commercially viable, the innovation process required complementary contributions from different types of players. In the early stage of the technology cycle, the major challenge to small high-tech

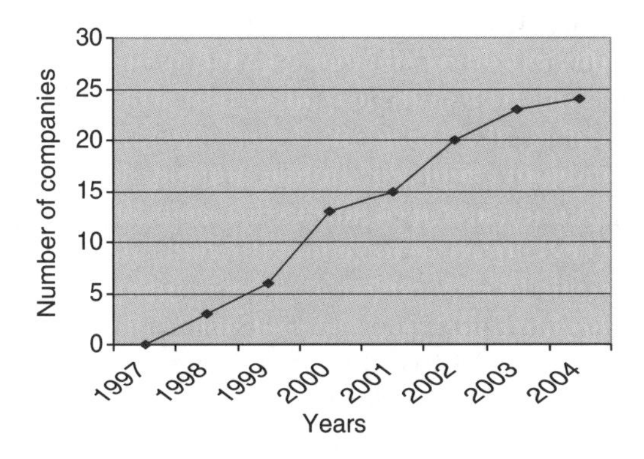

Figure 3.1. Accumulated number of firms' first product launches within class D amplification

start-ups was twofold. First, to establish a deep technology base that could be well-protected from quick imitation. Second, through codification, documentation, and communication to make this technology base attractive in the eyes of one or more complementary players and try to persuade them to engage in cooperative efforts to create functional solutions and to test market potentials. Such partnership could form the beginning of an evolving and interactive learning process based on a mutual recognition of the opportunities for innovative synergies between the two parties. This is exemplified by Apogee's partnership with STMicroelectronics. Or the partnership could be the first step towards a takeover of the technology entrepreneur by a larger incumbent as the case was with Texas Instrument's takeover of Toccata. The most successful of these technology entrepreneurs were able to establish a fairly strong regime of appropriability around their technological knowledge due to a combination of patents, a high level of complexity of the knowledge base, and the fact that this knowledge was generally unrelated to the knowledge bases of the large complementors and prospective competitors. They were moreover able to access complementary assets (both innovative and operational) through partnerships with (or eventually takeovers by) large incumbents.

Similar early-stage dynamics dominated by technology entrepreneurs have been well documented in the literature (for a recent case, see Giarratana 2004), but less attention has been addressed to the particularities of the 'core competencies' of these firms and especially the fact that their innovative practices not only require deep and specialized technological capabilities, but *also* managerial and organizational capabilities to link up with owners of critical complementary assets *without* loosing out of their capacity to capture rents from their technological knowledge. In other words, small high-tech start-ups are bound to embrace some form of Open Innovation (for an extended analysis, see Christensen et al. 2005).

Next we shall more closely address the particularities of the 'core competencies' of the large incumbents that engaged in innovative endeavors in class D amplification, and how these competencies were associated with (more or less) Open Innovation practices.

3.4.4 *Modes of (More or Less) Open Innovation Response from Large Incumbents*

For the large players with strong engagements in class A/B amplification, there were good reasons to expect that they would aggressively try to take control over this new technology. The conventional amplifier represented a critical module in any AV product,[13] hence to give up on the new amplifier paradigm would not only imply the loss of control over a critical module, but also the loss of a potentially large source of revenue and profits.

Table 3.1. The Response of categories of incumbents to the challenge of switched amplification technology

Category of firms	Firms	Response to class D amplification technology
Semiconductor companies with a strong position in A/B amplifier technology	National Semiconductor	Limited response—few products
	STMicroelectronics	Strong position with Apogee and Tripath in chip-based amplifiers
	Texas Instruments	Strong position in chip-based amplifiers
AV OEMs with a strong position in AB amplifier technology	Philips	Internal technology—few products
	Sanyo	Slow response/partnership with ICEpower
	Toshiba	No digital amplification technology
Large AV OEMs without a strong position in A/B amplifier technology	LG-Electronics	External technology—e.g. Pulsus
	Matsushita	External technology—e.g. Tripath
	Samsung	External technology—e.g. NeoFidelity
	Sharp	Limited response (1-bit technology)
	Sony	Internal module—external chips

Source: Websites of the individual companies, www.classd.com; www.puredigitalaudio.org, Daniel Sweeney, technology expert and author of Forward Concept's report on the emerging class D market (Sweeney 2004).

Table 3.1 shows the response (registered by mid-2004) to the new amplifier technology from three categories of large incumbents: Large-scale semiconductor firms with strong positions in conventional A/B amplifiers, large-scale AV OEMs likewise with strong positions in A/B amplifiers, and finally large-scale AV OEMs with no or weak positions in A/B amplifiers.

The two former categories comprise those firms with the strongest incentives to jump unto the new paradigm and indeed, with the exception of Toshiba, they have all engaged in the development of class D amplifiers. Toshiba's reluctance may be explained by the fact that Toshiba is operating in the low-price AV markets which have not yet faced any competitive threat from class D technology. The firms in the last category had already (at least to a large extent) outsourced traditional A/B amplifiers, and have, with the exception of Sony, so far also primarily been using external class D technology.

Table 3.2 shows the substantial variety of innovation strategies pursued by the five AV and semiconductor incumbents involved in class D innovation. In terms of their timing, we can identify three early, dedicated movers (Texas Instruments, STMicroelectronics, and Sony), one early but slow mover (Philips), and one late mover (Sanyo). In terms of their external/internal orientation, three of the firms (Texas Instruments, STMicroelectronics, and Sanyo) have demonstrated a strong external orientation (acquisition-based, partnership, or licensing oriented), while two firms (Sony and Philips) have exerted more internal approaches. Below, the particular strategies of each of these firms will be addressed.

Table 3.2. Innovation strategies of incumbents engaging in class D development

	External focus		Internal focus	
Timing	Acquisition-based	Partnership/Licensing-based	Tight system integration	Closed style
Early/dedicated movers	Texas Instruments	STMicroelectronics/Apogee	Sony	
Early/slow mover				Philips
Late mover		Sanyo		

As of early 2005, the commercial leaders are Texas Instruments (henceforth TI) and STMicroelectronics, the semiconductor firms that were early movers *and* strongly externally focused. Prior to entering the class D market, both companies were heavily embedded in the old solid state amplification paradigm and witnessed small high-tech frontrunners such as ICEpower and Toccata leverage the new technology and offer class D IP and early products to high-end market niches. They were early/dedicated movers in the sense that they fully engaged in catch-up efforts *as soon* as the small pioneers had demonstrated the viability of the new technology by the end of the 1990s and *before* any substantial market inroads had been obtained. Both companies used Open Innovation strategies, STMicroelectronics based its strategy around a long-term alliance with the technology specialist Apogee, while TI demonstrated a concerted set of actions to get access to complementary, innovative assets through acquisitions (it already possessed innovative assets in chip design and the necessary operational complementary assets). First, TI acquired Unitrode, a major supplier of power management components and thereby obtained a strong position in catalog analog semiconductors for power management. Second, TI acquired Power Trends, a leading supplier in the fast-growing market for point-of-use power solutions. In part through these acquisitions, TI had obtained key components and knowledge necessary for transferring fully digital class D amplification into chip design. Both in the area of chip design and chip manufacturing, TI was recognized as one of the world's leading companies, but it lacked key knowledge associated with digital/class D amplification. This knowledge was initially acquired through a licensing contract in 1999 with the technology entrepreneur, Toccata (one year exclusivity and to IC manufacturing only). However, in March 2000, following a mutual recognition that the technology transfer and the related chip design project was proving more complex than expected, TI came up with an acquisition offer and, after some negotiations, acquired Toccata. Through the acquisition, TI reduced the vulnerability and uncertainty of being dependent on critical capabilities located in an independent firm, and eliminated further contracting issues as well as royalty outlays. TI moved quickly to integrate all R&D activities in and related to digital amplification in order to ensure a more effective design process, and later in 2000, TI was able to launch its first generation of digital amplifier chips. By late 2003, TI was producing its fourth generation chipsets in millions.

TI has clearly exercised dynamic capabilities (Teece et al. 1997; Eisenhardt and Martin 2000), that is, organizational and strategic capacity to alter its resource base, through combining in-house R&D with timely licensing and acquisition policies. In particular, TI managed to orchestrate various sets of complementary innovative assets (Christensen 1995) through a succession of three acquisitions followed by organizational integration of the class D-relevant R&D of the various parties. This made it possible for TI to take the lead in transforming the technology into amplifier chips and to use its powerful operational complementary assets (in manufacturing, marketing, distribution) to create a first-mover spearhead for mass-produced digital amplifier chips in the expansive market for DVD receivers.

At about the same time during the late 1990s, Sony engaged in establishing a proprietary module system, the S-master technology, which seems dedicated to its captive product markets in order to seek differentiation gains. However, the core component of this module, the amplifier chipset, was from the early start provided by Mitsubishi and more recently by other class D chips vendors. Sony has shown a strong commitment to S-Master as a brand and an in-house technology that has become an integrated part of many Sony products and the system is offered on a licensing basis to other AV OEMs. Sony was an early-mover incumbent as were TI and STMicroelectronics but decided for a system integration strategy allowing for external chips suppliers willing to adapt to the particular system requirements of Sony's S-master system. Hence, even if Sony has exerted a more internal and proprietary systems orientation than TI and STMicroelectronics, with respect to the core component, the amplifier chipset, Sony has been using external suppliers. Sony's strategy can be seen as an attempt to set the standards for a dominant design in digital amplifiers and has been pushing the system into many of its audio products. It is too early to judge the broader success of this strategy, but so far, Sony has refrained from in-house development of the heart of the digital amplifier, the amplifier chipset.

Philips was probably the only company which already by the late 1990s possessed a fairly complete endowment of innovative assets and specialized technological capabilities (in both power and front-end technologies) necessary for leveraging class D amplifiers as well as the complementary assets for commercialization. Philips could therefore pursue a more introvert mode of innovation and indeed provided one of the first class D products to the market. What seems more surprising is that Philips, despite its in-house technological strengths and early product launch, has not so far (by early 2005) demonstrated a capacity (or willingness) to commercialize class D products more broadly into its products. This may be due to missing corporate commitment to do what Sony has apparently done, namely to force the end-product divisions to adopt the new technology. Hence, Philips may be a case of not only fairly closed innovation but also of the 'tyranny of divisions' in decentralized multidivisional companies (Prahalad and Hamel 1990).

Sanyo combines a strong position in traditional analog amplifier and chip production with a position as large AV OEM. Sanyo has shown a strong commitment to ongoing optimization of its conventional module technology and manufacturing capabilities, and when the paradigmatic shift in sound amplification emerged, Sanyo was ill prepared. Around the time when TI and STMicroelectronics launched their first amplifier chips (2002), Sanyo established a royalty-based licensing contract with the technology entrepreneur ICEpower to develop its own amplifier chipsets. In this way Sanyo seeks to combine ICEpower's technology with Sanyo's chip manufacturing and miniaturization capabilities and distribution network. The downside for Sanyo is that ICEpower controls the new technology and that Sanyo will have to pay a royalty for each amplifier chip sold. By the end of 2004, as TI and STMicroelectronics seem to have consolidated their leading market position in amplifier chips, Sanyo has just begun to ship its first licensed amplifier chipsets. Sanyo combines a slow response with an active catch-up effort based on external technology.

The large incumbents have, with the possible exception of Philips, applied elements of open innovation, and since small technology start-ups pioneered the embryonic stage of the technology cycle, even the strategies of the early moving incumbents have implied some kind of reactive rather than proactive response to the challenge of the new technology. An interesting inherent paradox of a strongly acquisition-based way of practicing open innovation, the case of TI, is that it leads to vertical integration. Hence, after an innovative entry involving a highly extrovert strategy that is considered necessary for managing and controlling a technological discontinuity, the company can internalize the next rounds of follow-up innovations, much more in accordance with the closed model. This points to the significance in some cases of nonregular cyclical changes from a relatively more open style of innovation associated with a company's attempt to realign a company's resource base in the face of radical (competence-destroying) and architectural innovation, and subsequently to a more closed style of innovation as the technology matures and incremental change and technical upgrading come to prevail. This cycle may eventually, as suggested by Chesbrough and Kusunoki (2001) create the basis for reexternalization of increasing parts of the components as the technology and the associated interfaces become commoditized and standardized. Thus, we do not necessarily see a once-and-for-all replacement of closed innovation by open innovation. This shows that companies cannot freeze their modes of managing innovation into one particular set of routines.

The case has illustrated the key analytical perspectives and issues discussed in this chapter. First, it has shown that the development of new complex technologies can fruitfully be analyzed using the distinction between specialized technical capabilities, possibly, but not necessarily, with a bias towards 'bodies of understanding', and integrative competencies, mostly with a bias towards 'bodies of practice'. Second, the case has illustrated the new division of knowledge and labor between small technology specialists and large incumbents emanating from the dynamics of vertical disintegration. Accordingly, technology-based start-ups will tend to have an advantage in the embryonic stages of a radically new technology requiring deep and specialized knowledge unrelated to the knowledge of conventional technology possessed by incumbents. By contrast, incumbents with strong incentives to capture commercial value from the new technology, will be better situated to mobilize the integrative competencies needed to provide the appropriate systemic/architectural innovation and large-scale commercialization. Critical features of such integrative competencies are the capacities for technical systems integration, for coordination with technology-based specialists, for reconfiguring the knowledge base (dynamic capability), and for mobilizing complementary assets.

Finally, the case has demonstrated that different incumbents engaging in the development of the same technology and associated products apply different innovation strategies. The most successful incumbents involved in class D development have been the two early movers dedicated to Open Innovation strategies (TI and STMicroelectronics). The outline of TI's strategy has indicated that system integration competencies may have to be closely aligned with capacities to reconfigure the existing knowledge base—in this case through acquisitions followed by R&D integration—*and* the mobilization of critical complementary innovative assets (such as chip design) and complementary operational assets (such as manufacturing and marketing). The more closed strategies of Sony and Philips have not so far proven as successful. Sony has followed a tight system integration strategy trying to establish a dominant systems architecture based on external amplifier chips, while Philips seems to have possessed the in-house knowledge assets to provide most elements of a class D amplifier, including the chipsets, but not the corporate commitment to commercialize the amplifier at large scale. This seems to indicate that the core competency strategy with a strong introvert orientation cannot adequately meet the challenges of increasing vertical disintegration and improved markets for technology.

3.5 Concluding Remarks

What has happened to the core competency perspective in corporate strategy and innovation that in the early 1990s was generally praised as *the* strategy for achieving sustainable competitive advantage among large technology-intensive companies? To what extent is this perspective at odds with the empirical tendencies towards vertical disintegration, enhanced markets for advanced technology, and increasingly open innovation? No doubt, companies will still have to develop or maintain in-house core competencies and innovative assets that are unique, complex, and difficult-to-imitate in order to obtain competitive advantage. However, in a world of increasing vertical disintegration and expanding technological opportunities in many industries, large incumbents have had to accept, more or less voluntarily, to give up full control and ownership over increasing parts of the value chain within their product markets and instead leave the provision of these parts to external suppliers with highly specialized expertise. This development has had numerous implications for the conception of 'core competencies' and associated innovation strategies in large companies. First, as has been demonstrated in several empirical studies, large companies have expanded the diversity of their technology profiles (technical fields in which they have at least a fairly deep level of generic/abstract knowledge—'bodies of understanding') putting relatively increasing emphasis on developing 'background competencies', a sort of absorptive capacity enabling the firm to coordinate and benefit from external technical development in the supply chain, and to explore new opportunities emerging from scientific and technological breakthroughs outside the firm. This is not to say that large firms (should) give up developing deep core technologies, but these seem to play a relatively decreasing role in the overall technology profiles. Second, as large firms increasingly take on the role as innovation architect and market coordinator of increasingly distributed value chains, they have to develop integrative competencies for systems integration involving experience-based and firm-specific architectural knowledge ('bodies of practice'). Third, as an increasing share of relevant innovative knowledge and component development takes place outside the large firm, dynamic capability, the capacity for reconfiguring the firm's knowledge and resource base, becomes a central asset which is strongly related to, but not identic with, competencies for systems integration. The diverse and highly dynamic nature of integrative competencies cannot in a stable way be contained in either a central laboratory or 'imprisoned' in isolated business units—but must be reflected in ongoing (sometimes erratic) changes in the organization and delegation of tasks and the mobilization of external relations (Brown and Eisenhardt 1997; Galunic and Eisenhardt 2001).

Open Innovation is premised on the presence of widespread useful knowledge, such that even the biggest and most knowledgeable companies cannot develop all of the important technologies they require on their own. This has been illustrated in the case of the new amplifier technology. It emerged from a university context, not from any of the leading consumer electronics firms. And it has diffused very unevenly into the consumer electronics market, as different firms with different innovation strategies varied in their competence to absorb this innovation into their own systems. While TI's discovery of the digital amplifier technology was almost accidental, to its credit, it rapidly developed a working relationship with the inventor, and though it struggled initially to successfully transfer the technology into its own development organization, it has successfully created new systems and chips that benefit from the technology. This exemplifies the increased importance of architectural competence build through dynamic reconfiguring of the parts of the firm's knowledge base. By contrast, Sony perhaps overestimated its own *ex ante* systems integration competence and underemphasized the requirements for the development of the core component technology. This may have left Sony with a difficult bargaining position vis-à-vis the increasingly strong suppliers of amplifier chipsets.

In a world of widely available knowledge, there are virtues in seeking external technologies, and hazards in ignoring them in favor of one's own technologies. Indeed, Chesbrough (2003*a*) argues that architectural knowledge will be increasingly important when knowledge is widely available. We appear to see that supported in this instance. But that doesn't mean that architectural knowledge is the only asset that matters for large firms. 'Old style' core competencies will most likely still be needed, but the dark side of core competencies, when they turn into core rigidities, has become increasingly prevalent as the technological opportunity set expands rapidly and as the external knowledge expands more rapidly than the internal knowledge. Hence, companies cannot any longer base themselves on a few deep core competencies that are cumulated over decades.

Notes

1. I gratefully acknowledge the constructive and insightful comments on earlier versions of this chapter from Henry Chesbrough, Constance Helfat, Joel West, and Volker Mahnke. The usual disclaimer applies. I also acknowledge the CISTEMA funding from the Danish Social Science Research Council.
2. The relative introvertness in much of the RBV-literature can be ascribed to its emphasis on criticizing the strong extrovert bias of the positioning strategy school. This latter extrovertness, however, was predominantly occupied with the external *competitive* environment, not the (potential) *cooperative* environment which later came strongly onto the agenda of both the strategy and innovation literature.
3. Thanks to Henry Chesbrough for suggesting these points.
4. Christensen (1995, 1996) takes this discussion one step further by arguing that technological innovation, per se, is not just the outcome of some unitary R&D function, but the outcome of the mobilization of a specific constellation of *innovative* assets. Four generic innovative assets are delineated from Pavitt's taxonomy (1984) of firm-based technological trajectories: scientific research assets, process

innovative assets, product innovative application assets, and aesthetic design assets. Most innovations involve more than one type of innovative assets (just like most innovations require more than one type of complementary asset for their commercialization), and firms may access some of these innovative assets externally. Likewise, most innovation requires the mobilization and integration of various specialized technological capabilities. Although the main focus of the papers is on firms' innovative asset profiles, they precipitate a more explicitly Open Innovation perspective.

5. Langlois (2003) uses the term market in a broad sense encompassing '. . . a wide range of forms many of which are not anonymous spot contracts but rather have "firm-like" characteristics of duration, trust, and the transfer of rich information' (2003: 351).
6. Moreover, some have argued that while coordination technologies (associated with information processing, communication, and transportation) previously tended to favor internal organization, they have more recently favored market dynamics (Malone and Laubacher 1998): 'The coordination technologies of the industrial area—the train and the telegraph, the automobile and the telephone, the mainframe computer—made internal transactions not only possible but advantageous' (p. 147). With more recent information and communication technologies, most notably the Internet, the value of centralized decision-making has decreased. While these arguments are intriguing, only a profound comparative analysis could discern whether external markets are indeed favored over internal coordination by these technologies.
7. Indicated as being technical fields in which the firm has a relatively high share of its patenting plus a relatively high share of total (global) patenting.
8. Thanks to Connie Helfat for pointing to this issue.
9. Organizational innovation is here defined in a broad sense as comprising a general set of organizational features that emerges as a response, among an increasing and eventually substantial part of a given population of organizations, to the emergence of either external or internal incongruities (Christensen 2002). A well-researched other example of the same kind of organizational innovation, is the innovation, in the early twentieth century, of the multidivisional organization (Chandler 1962). In other words, I do not by the term organizational innovation adhere to more specific organizational changes such as those taking place on a regular basis within, for instance, an open innovation model, or a multidivisional form.
10. According to Teece (1986), complementary assets are required in weak appropriablity regimes, when strong they are not required. Most technology entrepreneurs have limited access to complementary assets and limited resources (financially and competence-wise) for building complementary assets. In cases of tight appropriability regimes, which are rare, the technology provider may gain a good return on its technology from a licensing contract with little risk of having the technology expropriated by the licensee (or by others). Still a good return may also accrue to the licensee, due to synergistic economies from integrating the technology into a complex system of other technologies and complementary assets.
11. This case is based on a more detailed account of the innovation dynamics of switched/digital amplification technology in Christensen et al. (2005).

12. Class D amplifiers produce a power output by modulating a carrier frequency with an audio signal through a technical principle termed Pulse Width Modulation (PWM). A conventional class D amplifier is not digital, because the width of the pulses is continuously variable rather than variable according to some given number of discrete values. However, through various modifications, it is possible to make class D amplifiers truly digital. In the final stage of the audio signal path, a passive low-pass filter transforms the PWM signal into an analog power signal that can drive a speaker.
13. The amplifier module would typically account for about 20–30 percent of the total sales price of a traditional home stereo system.

4

Open, Radical Innovation: Toward an Integrated Model in Large Established Firms

Gina Colarelli O'Connor

4.1 Introduction: The Problem of Radical Innovation in Large Established Firms

Organizational growth and renewal are fundamental to any firm's long-term survival (Jelinek and Schoonhoven 1990; Morone 1993). Firms pursue multiple approaches to renewal. One path is to gain new capabilities via acquisition of or merger with companies that offer technologies or market entrée that the focal firm may lack. Another approach is organic, generative growth, meaning growth through the development of new lines of business based primarily on technical competencies nurtured from within the organization. When the promise of the opportunity is very large, and the concomitant risk and uncertainty of the opportunity are high, the technology and innovation management literature refers to that phenomenon as radical innovation (Morone 1993; Leifer et al. 2000).

Whether or not large established companies can develop and commercialize radical innovations (RI) is a moot point. The fact is, they need to. Mature firms depend on radical, breakthrough innovation to provide the next platform for growth as mature businesses become commoditized and loyal markets become saturated. But even though big firms rely on breakthroughs, they have not built the supportive infrastructure necessary to enable breakthroughs to be commercialized. Instead, large firms have tended to rely on maverick champions with a connection to a supportive senior management sponsors to push the project through a system that's tuned for incremental innovation (Leifer et al. 2000). Depending on these 'one-off' RI projects to be successful every ten years is not enough to fuel the organizational renewal necessary for the established firm.

While RI is widely viewed as one approach to generative growth available to large, established organizations, the evidence suggests that forces operate within such organizations to impede RI success (Cyert and March 1963; Gilbert et al. 1984; Dougherty 1992; Dougherty and Heller 1994; Teece et al. 1997; Leifer et al. 2000; Hill and Rothaermel 2003). Organizations grow by gaining efficiencies of scale and scope in specific core competency areas that, ultimately, become core rigidities (Leonard-Barton 1992), or core incompetencies (Doughtery 1995). They lack patience in terms of converting investment of time and resources into profits due to the pressures of equity markets, yet radical innovation can require more than a decade of investment before financial returns are seen (Gilbert et al. 1984; Quinn 1985). Some scholars believe that, in fact, large established firms are incapable of meeting the demands of current stakeholders and simultaneously being proactive regarding future disruptive technologies (Christensen 1997).

A host of other scholars argue, however, that organizations are capable of developing appropriate management systems for radical innovation, but are simply underdeveloped in this regard (Jelinek and Schoonhoven 1990; Morone 1993; Leifer et al. 2000; Ahuja and Lampert 2001; Hill and Rothaermel 2003). Schumpeter's (1950) early observations of the 'process of creative destruction' describing the ability of new companies to commercialize radical technology at the expense of incumbent firms, has been validated by many scholars (Rosenbloom and Cusumano 1987; Utterback 1994). The challenge has been for such groups to build their competencies before senior leadership loses patience. It has been documented that most new ventures groups and radical innovation hubs last, on average, 4–5 years (Fast 1978). Just as they are coming up to speed on the appropriate tools and mechanisms to use, they are defunded due to changes in the organization's growth strategy or because they have not 'delivered enough'. A generation later, they are resurrected, but the learning has dissipated.

Thus large established firms are seeking ways to develop RI competencies that can be sustained over time. The Open Innovation model offers firms an enormous help. If discoveries can be sourced from external parties as well as internal groups, and the innovation required to nurture those discoveries into business opportunities becomes more interactive with market and technology partners sooner, the life cycle of RI can be substantially shortened. As I reviewed our research program on large established companies' attempts to build RI competencies and infrastructures, I came to understand how companies' innovation programs have incorporated an increased orientation toward Open Innovation, and to observe how it is manifesting itself across the commercialization spectrum. Our participating companies are partnering and leveraging universities and other companies as a way to (*a*) learn quickly and inexpensively; (*b*) develop or co-opt new capabilities that radical innovation spaces require; and (*c*) actually begin to create new markets.

4.1.1 *Defining Radical Innovation and RI Competency*

We define Radical Innovation as the ability for an organization to commercialize products and technologies that have (*a*) high impact on the market in terms of offering wholly new benefits, and (*b*) high impact on the firm in terms of their ability to spawn whole new lines of business. We operationalized these impact levels as projects with the potential to offer either (*a*) new to the world performance features; (*b*) significant (e.g. 5–10x) improvement in known features, or (*c*) significant (e.g. 30–50%) reduction in cost[1] (Leifer et al. 2000; McDermott and O'Connor 2002). RI's often require the use of advanced technology and can enable applications in markets unfamiliar to the firm (Hage 1980; Meyers and Tucker 1989; Morone 1993). They may result in dramatically modified consumption patterns and business models in existing markets (Roberts 1977; Kozmetsky 1993; Dhebar 1995) or the creation of entirely new markets (Roberts 1977; Betz 1993). All of this is reflected in the high levels of market, technical, resource, and organizational uncertainty (Galbraith 1982; Maidique and Zirger 1985; Day 1994; Utterback 1994; Rice et al. 2002) that the project teams experience, which translates into long project maturity durations, unpredictability (Schon 1967) and nonlinear project development (Cooper et al. 2002). Such uncertainty makes conventional project management approaches inappropriate and requires the firm to develop new, situation specific competencies in technology, market, resource management, and organizational domains (Vanhaverbeke and Peeters 2005).

A RI Competency, then, is the ability for a firm to commercialize RIs repeatedly. The working hypothesis that drove our research program beginning in 1995 was that large established firms had become highly capable at managing incremental innovation using stage-gate like processes, but that the processes and evaluative criteria used to fulfill a stage-gate approach, if applied to the high uncertainty regime of RI, would kill potential breakthroughs before they could mature enough to impact the market or the company. Because established companies excel based on high volume operational efficiencies, the management system in place is oriented toward efficiency. Stage-gate processes align with those objectives, and ensure that firms work in familiar markets and technology domains where they are leveraging current know-how and relationships. RI, almost by definition, stretches firms into new market, technical, and business model territory. The result is that the management system that works so well for incremental innovation is mismatched with the requirements of RI.

4.1.2 *The Importance of Open Innovation to RI Competency*

We observed twelve potential RI projects in ten firms from 1995–2000 (Leifer et al. 2000) and developed timelines for each project to capture the uncertainties, discontinuities and to analyze how the project teams have dealt with them. Figure 4.1 depicts the chronology of Texas Instruments' development of the Digital Micromirror Device (DMD®). The solid horizontal lines represent applications pursued, the thickness of the lines indicates level of commitment of human and financial resources, and the short vertical lines mark project discontinuities. The figure reflects the long years in the laboratory 'experimenting' with the technology, but the fact is that, once the team had a direction in terms of a potential application to pursue, the project gained momentum. New technical directions were pursued (though not always successfully) and new market partners were engaged. Eventually TI's Digital Imaging business emerged from this effort. It is now part of the Semiconductor Business group. There are four product platforms, and TI commands 70–90 percent market share in several of those, with new applications continuing to emerge.

One wonders how long that early experimentation work would have gone on if TI had engaged early on in considering the market possibilities, or, in fact, what the benefit was of TI supporting this work fully in-house rather than working through a university or other laboratory to support it. Certainly the timeline of this ultimately extremely successful RI project would not have been twenty years as it is currently depicted.

Similar issues arise in all of the projects we observed. The RI life cycle is so rife with uncertainty, stochasm, starts, and stops, that it is difficult for large established organizations, who thrive on operational excellence, to tolerate them from beginning to end. Given the length of the RI life cycle, the Open Innovation concept offers great promise for helping enable RI in large established firms. While expectations for its contributions to business growth and profitability are high, management's patience for investing in the scientific discovery and invention is quite thin, as evidenced by the reduction in R&D investment over the past twenty years in US Corporations (O'Connor and Ayers 2005). In addition, while large established firms are highly adept at managing markets that currently exist, their skill sets, operating models,

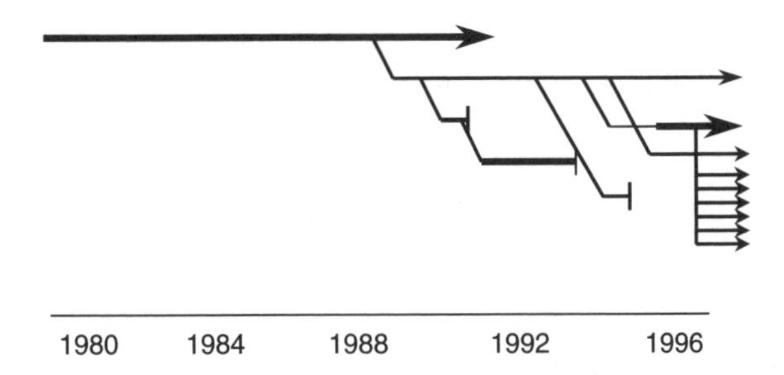

Figure 4.1. Texas Instruments' digital micromirror device (DMD®) timeline

performance measurement systems, and organizational structures severely infringe on their ability to create wholly new markets (O'Connor and Rice 2005). Any new conceptual mode of operating that can help speed any part of the process is welcome. The Open Innovation model does just that by helping companies leverage their vast resources and market power to identify and partner appropriately but also to provide the context for potentially game-changing innovations. As Open Innovation emphasizes, and as our research also concludes, the value from innovation lies more in identifying the context and applying the necessary business resources to commercialize the technology, than in having the initial idea originate in one's own laboratory.

Most expectations are that start-up organizations are more appropriate as engines of RI (Leifer et al. 2000). Start-ups arise frequently on the basis of a RI that holds promise of offering wholly new performance features. In fact, that is the promise that the venture capital community seeks to fund. In addition, start-ups do not suffer the organizational bureaucracy of large established firms, and so can be flexible in terms of reading market signals, structuring appropriate business models, and accepting smaller orders initially. We have learned that, in fact, markets for RIs emerge in just this manner (Lynn et al. 1996; O'Connor 1998; O'Connor and Rice 2005).

However, start-ups face numerous disadvantages in commercializing RIs, including a lack of resources. They do not have an identifiable company brand name and therefore lack credibility with partners and the market. In addition, start-ups do not have a broad base of knowledge assets to draw upon. They typically lack the complementary assets needed to scale the innovation. Large firms, we have seen, depend heavily on rich, powerful internal networks to answer questions, gain contacts, and get technical and market related questions answered (Kelley et al. 2005).

In the past, large established companies have operated on the assumption that they must develop everything internally to maintain competitive advantage. Open Innovation can help alleviate this and enable large companies to contribute in ways that leverage their richest capabilities. What is needed is to understand the balance of Open Innovation and internal competency development that best enables the large organization to constantly renew itself through game-changing innovation.

4.2 The Radical Innovation Research Program at Rensselaer

Overview. The Lally School of Management and Technology at Rensselaer Polytechnic Institute has been home to the Radical Innovation Research Program since 1995. The research program has occurred in two phases and it has been sponsored throughout by the Industrial Research Institute, a professional organization of research and development (R&D) Directors and CTO's of Fortune 1000 US-based companies.

Phase I, conducted from 1995–2000, tracked projects that were ongoing in large established companies that senior technical leadership identified as having the potential to be breakthroughs, should they succeed. So the unit of analysis for the first five years of study was the individual project. We believed that, by tracking projects identified by senior leaders as having breakthrough potential as they were being nurtured within their companies, we could at least describe what companies are doing, the extent to which those practices differ from incremental innovation practices, and begin to arrive at some theories for prescription that could be tested more conventionally. Projects were qualified into the study if they had an identified team and a budget, and had the potential to offer either (*a*) new to the world performance features; (*b*) significant (5–10x) improvement in known features; or (*c*) significant (30–50%) reduction in cost.

Key Learning, Phase I. Tracking the projects over five years led to a number of important insights regarding the challenges that RI project teams face in large established companies. We provided our findings in the book *Radical Innovation: How Mature Firms Can Outsmart Upstarts* (HBS Press 2000) and the series of papers listed on our website (www.lallyschool.rpi.edu/programs). We identified four dimensions of uncertainty and seven challenges that companies faced in maturing RIs. We noted that, although RI was in fact occurring, appropriate management practices were ad hoc, unsystematically applied, and occurred on an exception basis. They were not recognized as legitimate practices or treated as part of the routines of the business. These findings led us to ask the higher level question of how companies could build a capability to enable RI to happen over and over rather than relying on singular strong-willed highly gifted individual project leaders who had access to a senior executive sponsor.

In Phase II, carried out from 2001–2005 we have studied companies who have a declared strategic intent to evolve a RI capability. The unit of analysis for phase II was not the project, but the Corporate Radical Innovation initiative, that is, the building of a competency to do RI over and over. Companies have tried and failed to build organic growth and renewal engines. Sometimes called incubators, sometimes called corporate venturing organizations, and sometimes called Radical Innovation hubs, these are organizational entities charged with finding the new, 'really big' growth opportunities for large, established, sometimes stagnant companies. Yet history shows that very few of these internal organic growth organizations (*a*) have lasted very long and (*b*) have had real impact on their companies' growth and renewal patterns.

The formal objectives of Phase II are to understand how organizations can systematically develop and sustain their RI capabilities. Our conceptual framework has been that RI cannot be managed as a process like incremental innovation can, but rather requires a management approach consisting of multiple elements aligned as a system (O'Connor 2005).

Sample. Participating firms are large industrial North American based companies. Phase I firms are termed Cohort I, and Phase II is comprised of two sets of companies, Cohort II and Cohort III. Ten firms participated in Phase I, including Air Products, Analog Devices, Dupont, GE, General Motors, IBM, Nortel Networks, Polaroid, Texas Instruments, and United Technologies. In Phase II, a total of twenty-one companies participated across the two cohort groups, including four from Phase I. Examples of Phase II companies include 3M, Corning, GE, Dupont, Intel, and Hewlett-Packard.

Our methodological approach is a longitudinal, cross-case approach (Eisenhardt 1989; Yin 1994), with the added component of a multidisciplinary research team, as documented in O'Connor et al. (2003). The approach for Phase II is the same. Sample and methodological details can be found in O'Connor and DeMartino (2005). A total of 143 interviews were conducted for Phase II's initial round of data collection, between 9 and 14 managers per company. Four rounds of follow-up interviews have been completed to date, for a total of 224 interviews.

4.3 Results and Insights

4.3.1 *Firms are Investing in Building a Radical Innovation Capability Much More Today Than They Were Ten Years Ago*

Based on the project level data in Phase I and, later, the company level data gained in Phase II, it is clear that firms are becoming increasingly sophisticated in building a RI capability. They recognize the need for it, are investing in improving that capability, and recognize it as more than a process, but in fact a complex system. Of the ten companies in Cohort I, only two had a programmatic approach to managing RI. By 2000, four of the companies that were cohort I companies had evolved a more sophisticated strategic intent and programmatic approach such that they opted to participate with us in Phase II. In addition, the number of companies that have a recognized, identifiable RI group or program or strategic intent was overwhelming. Cohort III was formed, in fact, on the basis of companies contacting us to learn from the research and to network with other companies that were doing this well. So we observe that the trend for finding new paths to growth via highly innovative products and businesses is of keen interest and importance to large companies, much more so today than it appeared to be in 1995 when the research program began.

4.3.2 *Radical Innovation is Not a Single Capability. Rather, it is Comprised of at Least 3 Distinctive Sets of Competencies: The Discovery-Incubation-Acceleration model*

We have traced the organizational structures of the twelve cases in Phase II, and their evolution as they confronted particular challenges over time (O'Connor and DeMartino 2005). This exercise provided insight into the competencies required to develop a mature RI capability. We identify three such competencies—discovery, incubation, and acceleration—each of which requires distinctive types of expertise and processes (Figure 4.2).

Discovery. A discovery capability involves activities that create, recognize, elaborate, and articulate RI opportunities. The skills needed are exploratory, conceptualization skills, both in terms of technical, scientific discovery, and external hunting for opportunities. One of our Cohort III firms distinguishes between invention and discovery. Invention is the creation of something that was previously unknown. Discovery is becoming aware of something that may be known in other venues but was not known to the company. RI activities can include invention, but need not always, according to our companies. This implies that a mature discovery capability includes not only internally focused laboratory research that industrial R&D laboratory scientists perform (witnessed in the vast majority of our sample companies), but also activities that embrace the Open Innovation concept.

Incubation. The analysis also suggests that an *incubation* capability is necessary for RI. Whereas discovery competencies generate or recognize RI opportunities, the *incubation competency* involves activity that matures radical opportunities into business proposals. A business proposal is a working hypothesis about

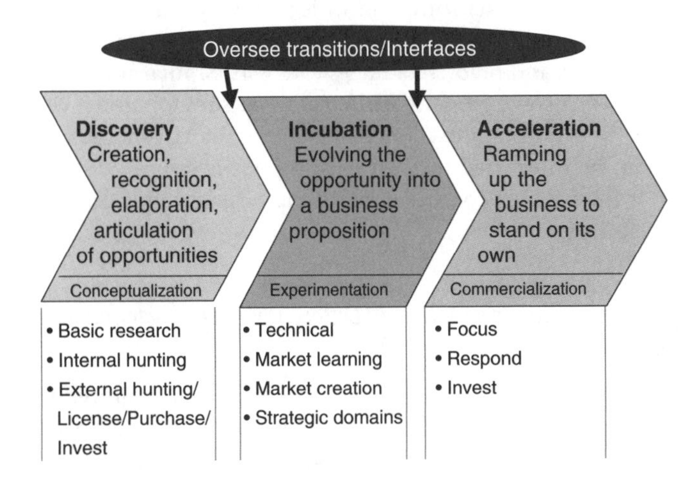

Figure 4.2. RI requires three sets of competencies

what the technology platform could enable in the market, what the market space will ultimately look like, and what the business model will be. Incubation is not complete until that proposal (or, more likely, a number of proposals, based on the initial discovery) has been tested in the market, with a working prototype.

The skills needed for incubation are experimentation and interaction skills. Experiments are conducted not only on the technical front, but, simultaneously for market learning, market creation, and for testing the match of the business proposal against the company's strategic intent.

Acceleration. Acceleration activities ramp up the fledgling business to a point where it can stand on its own relative to other business platforms in the ultimate receiving unit. Whereas incubation reduces market and technical uncertainty through experimentation and learning, acceleration focuses on building a business to a level of some predictability in terms of sales and operations. As one Radical Innovation Director notes:

> I need a landing zone for projects that the business unit does not feel comfortable with. If I transfer these projects too early, the business unit leadership lets them die. I need a place to grow them until they can compete with ongoing businesses in the current operating units for resources and attention.

The skills needed are those required for managing high growth businesses. Acceleration involves exploitation rather than either exploration (which Discovery requires) or experimentation (which Incubation requires). The activities of acceleration include investing to build the business and its necessary infrastructure, focusing and responding to market leads and opportunities, and beginning to institute repeatable processes for typical business processes such as manufacturing and order delivery, customer contact, and support. Acceleration involves turning early customer leads into a set of qualified customers and predictable sales forecasts. Similar to an independent start-up firm in first stage growth, acceleration pursues top line revenue rather than bottom line profitability. Once a RI program is generating profitable returns, it can be integrated into an existing business unit with less chance of neglect. It may also become a stand alone business unit or spin-out with P&L responsibilities.

4.3.3 *Open Innovation is Manifested Differently in Discovery, Incubation, and Acceleration*

While we note that each of these three competencies is required to enable a sustainable RI capability, we also note that Open Innovation is apparent across our companies in each of these aspects, and, we believe as do our participating companies, offers the possibility of speeding the arduous life cycle of RI. As mentioned in Chapter 1, it is an open question as to whether OI will further the RI cause or in fact drive companies to seek technologies they can quickly commercialize. The participating companies in this study do not perceive this trade-off, but rather are experimenting with ways to engage in Open Innovation that further their RI capability development efforts. In other words, a major part of many of the companies' efforts to develop an RI capability involves developing a capability to use the OI model appropriately for that cause. However, they are not yet completely successful. Documented in this section are the ways in which they are engaging in OI to further RI objectives, and the challenges they face in doing so.

Discovery and Open Innovation: What's happening and what's missing? As mentioned, Discovery activities observed in the sample companies include not only an investment in basic R&D, but also hunting inside and outside the company for ideas and opportunities and licensing technologies or placing equity investments in small firms that hold promise. Ten of the twelve Cohort II companies are involved in each of these activities simultaneously, to increase the opportunity space for RI. Nine of our twelve firms noted external programs to locate outside opportunities through universities, venture capital investments, or strategic alliances. One of the cohort III firms describes placing 'eyes and ears' investments in small companies to maintain a seat on the board so the large firm can understand the novel technology domain that the start-up is exploring, to help reduce the company's risk through quick, cheap learning.

New Roles Emerging. A number of formal organizational roles and structures are emerging to create, recognize, or elaborate RI opportunities in the discovery phase. Four companies relied heavily on a relatively large number of dedicated research staff personally responsible for developing radically innovative ideas, which were generated from technical interest primarily. However, each one of these companies (and most of the remaining sample companies) were members of research centers at universities so they could stay abreast of new discoveries that could be leveraged into their innovation plans. Five sample firms used dedicated RI 'hunters' and/or 'gatherers' responsible for identifying RIs within internal and/or external environments. An idea hunter is an individual who actively seeks out RI ideas. They may conduct 'idea generation workshops' in the business units, or visit laboratory scientists working on exploratory research and identify opportunities that are raised as potential business proposals.

In one company, for example, the RI group became the home for the founders of small companies that the larger organization had acquired. The RI group found these people very valuable as idea hunters because (*a*) they had rich external networks due to their stature and previous activities as the founder of an organization; and (*b*) their skill at opportunity recognition given their entrepreneurial experience in starting up and running a company. Another company, formed an 'externalization' team devoted to the development of future trend analyses based on visits to universities, and built a 'hunters' network' of creative individuals throughout the company as well. In another company, a permanent team of technical and business development middle managers comprised the 'Technology Identification Process' team, challenged with finding new opportunities to help fuel R&D projects.

Three companies are experimenting with 'exploratory marketing groups' which serve as a mechanism to proactively discover RI opportunities at the technology/market nexus. Finally, one sample company relied upon an informal network of external contractors to generate and develop wild ideas and inventions. This network was maintained and funded by a senior executive who elected not to bring them within the company for fear that their creativity would be stifled.

In contrast to idea hunters, a 'gatherer' is a central locus for idea generators to turn to for help. It is a more passive role than the idea hunter. Thus, 3M, Dupont, and Kodak have websites for inventors from outside the company to propose ideas. Nearly all the companies in the sample had an idea tracker system for employees within the companies to contribute ideas. These are screened and evaluated. From an internal perspective, employees or researchers who have ideas but do not know where to turn for funding or help articulating them or even guidance in what to do next use idea gatherers for coaching. Idea gatherers were present in six companies.

Several firms used a modified Open Innovation model in that they focused heavily on sharing with and borrowing ideas across divisions of the firm, so the openness was within the company so that company resources could be well leveraged. In one case, a technology board of senior leaders across the divisions met to share information and ideas on a monthly basis. Another divisional CEO within a large diversified company, whose businesses were primarily in low margin consumer product categories began pirating researchers and product managers from the company's pharmaceutical division to help germinate cross-industry ideas.

Thus we observe that ideas come not just from the scientist's bench, but from groups of creative people within the organization, from idea hunters who uncover ideas inside and outside the organization, from formal relationships with universities and venture capital funds, from efforts to cross-fertilize within an organization across divisional and industry boundaries, and from single creative individuals who may be maintained outside the organization but whose efforts are dedicated to the organization's needs. A broad spectrum of structural mechanisms exist to ensure a rich Discovery competency for the company. When asked to describe the rationale behind the various Discovery mechanisms the companies were using, they mentioned using exploratory marketing and external idea hunters to find 'system level problems that we can contribute to solving,' and 'to talk with potential customer/partners that we currently don't know'. Venture investing, university liaisons, and targeted search for small companies were described as 'eyes and ears investments to find promising new technologies that are emerging,' or 'to solve a competency gap'. Finally, one company mentioned hiring external contract inventors for the sole purpose of ensuring that creativity was maintained without being hampered by corporate norms and bureaucratic burdens.

Observed Challenges with Open Innovation and the Discovery Capability. While nearly every one of the sample companies recognized the value of the Open Innovation model, they are struggling in the discovery phase with several issues. Firms currently lack the capability to create 'development partnerships', that is, relationships with other firms that are neither hands-off licensing agreements or subcontract relationships nor highly integrated joint ventures. In the RI domain, the outcome of any discovery activity is highly unpredictable. In addition, since firms are pushing the boundaries of their capabilities and expertise, partnerships for joint development are particularly appropriate means for accessing new knowledge and expertise quickly. But companies are confused about how to structure such agreements given the vast array of potential outcomes of the development project. In one company, a partnership was needed for manufacturing of the product. The manufacturing process itself required major innovation, given that the technology was heretofore not manufactured in large scale and was rather different from previous approaches. The manufacturing partner was never able to generate the process innovations necessary; these were ultimately developed in the principal firm's R&D laboratory. The partnership was never actively leveraged and the manufacturing firm became a subcontractor. In another case, a large firm relied on a small contract R&D company for much of the ceramics-based innovation that was required in a particular project but that was only resident in a few of the large firm's key scientists. The two companies became tightly partnered in the development work. The smaller firm wanted to be acquired ultimately, but actually had no production capability, which was what the larger firm ultimately needed, and so the smaller firm was disappointed about missing out on the value of the innovation they had contributed to so intensively.

Finally, the Open Innovation model, as it invades the Discovery process, may be interpreted in ways that threaten the role of R&D in large established companies. Will R&D now be evaluators and 'assemblers' of technology, much as the large automotive companies became large-scale assemblers of cars, but created almost none of the technology in house? How much component level expertise is required, in order to be an effective systems integrator of technology? How much resident R&D expertise in general is required to enable RI in a large established company? While Open Innovation exemplars like IBM, Intel, and Procter & Gamble maintain deep ongoing internal R&D programs, many other companies may be tempted to utilize the Open Innovation concept as a pretext for hollowing out their internal technical capabilities.

Incubation and Open Innovation: What's happening and what's missing? Incubation requires experimenting with the opportunity such that, ultimately, a new market can be created. It cannot occur, even by luck, within the confines of the company. Interaction in an extensive manner with the market is critical to understanding the aspects of the discovery that are valued, by whom, and the mechanisms by which those can best be delivered to the market (Lynn et al. 1996, O'Connor 1998). Preliminary answers to those questions then drive technical development. A case example helps clarify this critical issue.

Intelligent Packaging. Intelligent Packaging is a methodology for unique identification of a package through wireless technology. It is like a bar code that need not be scanned, or even seen by the naked eye. That ability allows for inventory control to be handled from the raw material stage through to final assembly in a radically new, more specific manner. It reduces uncertainty about what stock levels are at each step in the value chain. To an ERP system this is worth 1–2 percent of sales per year in reduced inventory shrinkage and pilfering. Another application may be for security purposes in airports. The possibilities are endless, but, as is typical, the costs presently are too high to justify adoption in many major industries.

In one of our participating companies, the inventor of this technology, Paul, described his early experiences in the discovery phase. He knew the invention was exciting, but, he had reached a stage in his technical exploration at which he was stalled; he didn't know what to do next. He was actually receiving negative performance evaluations because he was not providing immediate, direct support to business unit related products. He stated:

I could've done what most research scientists do in that situation . . . ask for more money and a bigger lab. However, I realized I still wouldn't know what to do in that lab on Monday morning. I got forced out of the research lab because I didn't know who my customer was {so which business unit would take this on] . . . The market didn't exist . . . so I couldn't develop the technology without studying the market. But if I wrote papers about the market that wasn't really a technical research problem. So I wandered down the hall and described my discovery to Linda, [who worked in the group responsible for nurturing radical innovations in the company]. She asked me a lot of questions but convinced me that we had to start talking about this in the market. We found and joined a university center that had 60 member companies whose purpose was to discuss standards in the intelligent packaging domain.

I could've made up some stories and gotten a lab and a team. But it would only be because I didn't know what to do. The competitive intelligence we've gathered has convinced me that the places we thought this would work would not have worked out.

The champion of the project (Ron) told us:

This could have remained a 'research project,' but it wouldn't have worked because it would've been treated as a technical problem. Really, though, it's a business problem.

To study the markets that did not yet exist, Paul worked with Linda to conduct a gap analysis on the technology. The small team started looking for companies to buy to catapult them forward in the technology. The inventor (Paul) later stated:

> I didn't realize that we were getting a lot of competitive intelligence in the industry because of the due diligence we were doing on these companies. We tried that for 9 months or so... but didn't get anywhere because we talked to small company startup founders who kept trying to hose us on the price. We asked 'How will this pay itself off?' This caused me, a technical person to see how the financial side is important to me too.

Ron ultimately challenged them to stop looking for acquisition candidates to help them drive the business, and decide that they could evolve the opportunity in a manner that leveraged the company's core competencies. This reorientation would have dramatic influence on the business model, but that could not have happened without their early market interactions. Paul described this clarification:

> We had become experts on the technical side by joining that university center and conducting those due diligence interviews. We knew where the technology was falling short of the market... what it needed to do. A lot of people were talking about characteristics of the technology that weren't really important. We concluded that because we weren't an RFID company proper, but rather a commodities manufacturing company, we might see something a little different here. We could take the technology that exists today and apply it in a way that makes its weaknesses irrelevant. We could do this because that's how a commodities company thinks. We couldn't have done it if we weren't in the university center... because we could see all the mistakes the other members were making in how they were pitching it. Being from a different industry helped us see the problem differently.

The team began to work with market partners that distributed lower volume high value added goods rather than high volume bulk goods, so the cost could be justified. The project continues and is regarded as an impending major success for the firm at this writing.

We note that incubation was neither recognized as a necessary activity nor systematically engaged in across the companies. Lynn et al. (1996) have documented the critical importance of this sort of 'probe and learn' activity, but few companies recognize the critical interconnection and simultaneity between technology development and new market creation. Of the twelve companies in Cohort II, only one had a mature incubation capability at the outset of their RI initiative. Ultimately, however, nine companies recognized the need for this activity and attempted to build it in some way. Only a small proportion of the cases ever achieved a high level of incubation competency, though many of the companies expressed lack of business acumen and inability to build businesses linked to the company's strategic intent as challenges they faced (O'Connor and DeMartino 2005). Incubation becomes increasingly important as the opportunity poses challenges to the company's current business model and therefore to finding an ultimate home for the new business opportunity. When companies do not engage in incubation, they risk leveraging the full value of the opportunity because they typically 'force-fit' it into a current business and adopt the business model of that SBU (Rice et al. 2002).

One very interesting practice we have noted that stimulates incubation opportunities is presented in Figure 4.3. Dupont's biodegradable polyester, Biomax® was one of the projects studied in Phase I. The New Business Development director, situated in the R&D laboratory, acted as the role of idea hunter and gatherer described earlier. But he also helped nurture those ideas into business proposals with his team. The advertisement pictured in Figure 4.3 was run when Biomax® was struggling to find application niches to help guide its continued development. The ad does not provide the secrets to the material, but offers enough information that potential codevelopment partners' interest would be stimulated. The ad was run in trade and professional journals such as *Chemical Weekly* and *Scientific American*. It generated over 100 inquiries and resulted in more than 30 initial trials of the material in various application domains. The NBD director used this technique more than once, and it was becoming a useful practice. However, when he left the firm, the practice was forgotten.

Observed Challenges with Open Innovation and Incubation. It is clear that getting involved with the market early and often is of critical importance to learning about the value of the potential business and to help guide the direction of technology development. Two issues emerge. The first is how to define which companies to work with. If left to the current organizational structures and roles, marketing people tend to take on the responsibility of identifying customer partners. But in creating new markets, current customers are frequently the worst possible choices (Christensen 1997). Greenfield methods for finding interested parties, like the advertising mechanism Dupont used, are important.

Second, the probe and learn methodology for market learning and market creation has been viewed increasingly as a legitimate, accepted process in the literature. However, we note that questions remain about how to investigate applications and business models optimally using this approach. One question that arises is whether probes should be conducted serially (to allow learning to occur between each probe) or in parallel, working with multiple customer partners in multiple application domains. Participating companies related that having too many partners at the same time requires too many resources. However, it appears that this could substantially shorten the time involved to understand the boundaries of the technology's value to the market, and thus speed commercialization. A possible line of future inquiry is how to optimize the number, sequencing, and selection of partners to work with to maximize learning output.

Figure 4.3. Dupont's Biomax® Ad

Acceleration and Open Innovation: What's Happening and What's Missing? Acceleration is not a major issue for RI opportunities that fall completely in line with a particular business unit. So long as there is high level communication and, typically, shared funding between the receiving unit and R&D, these projects can be commercialized within prescribed application spaces, business models, and infrastructures that the organization typically uses. However, when the opportunity requires any stretching of the customer set, the business model, the manufacturing processes, or the value proposition that a business unit is used to, transition problems occur (Rice et al. 2002). The fact is that most RIs do not fall neatly within the scope of the current businesses. Acceleration is the set of activities needed to build the business to a point where it can compete within its eventual home (should it be one of the existing business units) for resources, attention and priority with the ongoing lines of business.

In our Cohort II sample of companies, six firms created permanent organizations to accelerate businesses, some of which were located within a division but were given senior management level oversight (4 cases) and others (2 cases) located outside the current divisions, reporting to the Corporate Officers directly.

Open innovation is not as pertinent in this phase, since fledgling businesses are generally free to establish their own partnerships and customer base. In fact, that is the purpose of acceleration. However, two examples arise in the data that should be mentioned.

First, a brand new technology platform was developed within R&D, and the company wanted that platform to spawn a number of opportunities that would be useful to many different business units. Because the technology was so new, the businesses hesitated to invest in project ideas emanating from the platform group. The group became frustrated and hired Opportunity Brokers to shop the technology to external companies for potential codevelopment partnerships. Setting up the external competition quickly caught the attention of the company's senior leadership, who had invested in this development effort specifically because they viewed it as a strategically important growth path for the company.[2] This is the use of an Open Innovation model for the purposes of expediting focused attention to the project, and, in this case, it worked. The platform is now its own business unit with a general manager and is on the path to significant revenue streams.

Second, we and others note that new businesses do not get built quickly. Sometimes the expectations on magnitude of the RI business are so heightened that early market entries appear disappointing. The theme of interacting with the market so that the market learns about the technology is again key in the acceleration phase. None of the projects in our first study achieved a 'killer application' early in their commercialization phase. Figure 4.4 shows the case of Analog Devices' accelerometer. The company's vision of the killer application was the automotive market, first for airbag detonation and then in other applications. The company sacrificed early profits to gain those volumes and manufacturing learning curve advantages, but the market actually evolved way beyond that, and, in fact, brought profitable applications to Analog because they saw the potential as Analog's accelerometer technology became understood. The point of Figure 4.4 is that application migration occurs, meaning that an early entry application may be in a niche market, but others arise and, in fact, seek out the innovating company to learn more. So being too driven towards any single application space and expecting a killer application is not in alignment with reality. It is critical to be open to inquiries from fields far removed from those originally envisioned.

4.4 Concluding Thoughts: Radical Innovation Must be Open Innovation

It is clear that Open Innovation, if managed in a balanced way with internal capability development, can help speed RI through its emphasis on interaction and networks. In fact, RI efforts in large companies have not been sustained, and Open Innovation is quickly becoming viewed as a critical aspect to helping gain the efficiencies of learning necessary to make RI sustainable. Any company choosing to develop RIs is, by definition, stretching the boundary of what is already known, certainly within its own domain. Accessing technologies, market partners, and expertise in arenas that are dramatically different from the company's core enables creativity, opportunity

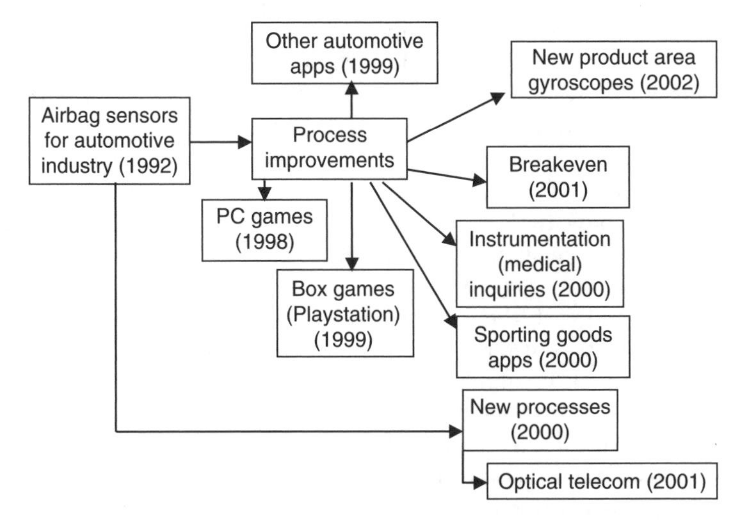

Figure 4.4. Application migration: Analog Devices' accelerometer MEMS device

recognition, and connectivity into new domains. However, a number of questions remain for research in this unchartered space.

From the perspective of Discovery capabilities within a firm, we need a better understanding of appropriate agreement terms with partners who contribute sources of technology and development know-how. Agency theory and balance of power issues between firms are stalling progress in this regard. Firms continue to hesitate to engage in Open Innovation because they are concerned about intellectual property ownership issues. In addition, the Open Innovation model raises the issue of the role of R&D in the large established company. How does this impact the core competency of an organization that has heretofore prided itself on its discovery capability? What, in fact, is the core competence of a company who competes on the basis of technological innovation, given the challenges posed by the Open Innovation model? What are the boundaries of Open Innovation? How can we think about RI as a dynamic capability (Teece et al. 1997), and Open innovation's role in this?

In terms of incubation, a critical theme that comes through in our data is to engage in early market participation, with usable prototypes. But most of these prototypes are clumsy and unrefined. How open can we be, how early on, with 'klugey' technology? And how is this relationship with potential new markets best managed, given that customer-partners' expectations will never be met?

In terms of Acceleration, how does the OI model extend to customer-partners? How can it help us understand the appropriate pace of a RI's impact on the market? Can it help speed that? Can it help set realistic expectations for senior management, so that appropriate metrics are used to gauge a RI's success over time? These issues plague large companies today.

Finally, some overall questions arise regarding the interaction between Open Innovation and the RI competency model defined in this chapter. First, what are the relative risks and rewards of using an Open Innovation model across Discovery, Incubation, and Acceleration? So far, attention has primarily been paid to OI in the Discovery aspect of RI. Developing the model further with regard to Incubation and Acceleration seems paramount for successful commercialization and new market creation.

Additionally, how is Open Innovation leveraged differentially across RI opportunities that are aligned with the firm's current business models versus those that do not fit? It is fairly clear that attempts to move into white spaces, that are unaligned with the firm's current business models, require partnerships and openness. But territoriality, the desire for economies of scale and scope, and competitive dominance are dynamics that exist in the aligned opportunity space. This issue deserves further investigation.

Finally, how critical is leveraging internal strengths and networks compared to those outside the organization? The use of internal networks is a critical success factor that arises in our RI research data, and that effect appears initially to be stronger than engaging in external networking, which Open

Innovation prescribes. Again, a systematic exploration of the relative impacts of internal networks versus external partnerships in large companies is warranted.

We have an important research agenda ahead. It will certainly have enormous influence on management practice, and hopefully can help enable successful RI in large established firms.

Notes

1. We use the word *potential* because the study's methodological approach was a longitudinal one rather than a case approach of data collection post hoc. We therefore did not know if the projects would be successful when they were first qualified into the study.
2. This was similar in spirit to the effect Chesbrough (2003*a*) observed in Lucent's Bell Labs, when an internal venture capital arm was established to commercialize technologies that otherwise would not have been used by the internal businesses. The presence of a second path to market created internal competition that prompted faster and more thorough consideration of technology projects within Lucent.

5

Patterns of Open Innovation in Open Source Software

Joel West
Scott Gallagher

5.1 Introduction

Models of Open Innovation offer the promise that firms can achieve a greater return on their innovative activities and their resulting intellectual property (IP). Open Innovation models stress the importance of using a broad range of sources for a firm's innovation and invention activities, including customers, rivals, academics, and firms in unrelated industries while simultaneously using creative methods to exploit a firm's resulting IP (Chesbrough 2003*a*). While the use of external sources of innovation is nothing new, recently some of the most successful high-tech firms have been those that utilized Open Innovation rather than more traditional 'vertically integrated' innovation approaches.

The Open Innovation paradigm is often contrasted to the traditional vertical integration model where internal R&D activities lead to internally developed products that are then distributed by the firm. Consistent with Chapter 1, we define Open Innovation as systematically encouraging and exploring a wide range of internal and external sources for innovation opportunities, consciously integrating that exploration with firm capabilities and resources, and broadly exploiting those opportunities through multiple channels. Therefore, the Open Innovation paradigm goes beyond simply the externalization of research and development (R&D) as identified by von Hippel (1988). Rather than just a shift in the technical production of IP, Open Innovation reflects a transformation of how firms use and manage their IP.

Over the past twenty years, an increasingly popular example of Open Innovation has been open source software, exemplified by the Linux operating system. Open source software involves collaboration between firms, suppliers, customers, or makers of related products to pool software R&D to produce a shared technology. At the same time, open source IP policies mean that this shared technology is available to potential buyers at little or no cost.

Together, the shared production and low cost of open source software has forced firms to reconsider the proprietary business models used by commercial software companies for the past twenty-five years. The essential issue for a firm's business model is: 'How does the firm create value for the customer while simultaneously extracting some of that value for itself?' The rise of open source software has enabled a wave of experimentation in software business models that is ongoing even today.

In this chapter, we consider how open source addresses what we identify as three management challenges of open innovation: maximizing the use of internal innovation; incorporating external innovation into the firm; and motivating a supply of such external innovation to support the firm. We then classify the strategies taken by companies selling open source software based on their business models, and suggest how this fits into broader issues of Open Innovation.

5.2 Challenges of Open Innovation

The pace of technological advance has often been subdivided into two phases: invention (a scientific breakthrough) and innovation (commercialization of the invention)—a distinction Nelson and Winter (1982: 263) attribute to Schumpeter (1934). This split parallels a similar bifurcation between research and development, where inventions come from basic research and innovations from the development group. Many organizations, however, define additional phases between the two extremes, as with Intel's 'advanced development' step (Tennenhouse 2003). Others have attempted to subdivid innovation into 'radical' and 'incremental', where the former more closely resembles the 'invention' concept (e.g. Leifer et al. 2000).

Like Nelson and Winter, we use 'innovation' in its broadest sense to refer to the entire process by which technological change is deployed in commercial products. Such innovation may incorporate formally protected IP (such as patents or copyrights) that is difficult to imitate, or it may reflect codified knowledge that is readily imitated and at best provides a transient competitive advantage.

In contrast to earlier models and 'fully integrated innovators' like AT&T (now Lucent) Bell Laboratories and IBM which do basic research through commercial products, Open Innovation celebrates success stories like Cisco, Intel, and Microsoft, which succeed by leveraging the basic research of others (Chesbrough 2003*a*). Under this paradigm, firms exploit both internal and external sources of innovation, while maximizing the returns that accrue from both sources (Table 5.1).

Table 5.1. Models of innovation and resulting managerial Issues

Innovation model	Management challenges	Resulting management techniques
Proprietary (or internal or 'closed')	1. Attracting 'best & brightest' 2. Moving research results to development	1. Provide excellent compensation, resources, and freedom. 2. Provide dedicated development functions to exploit research and link it to market knowledge.
External	1. Exploring a wide range of sources for innovation 2. Integrate external knowledge with firm resources & capabilities	1. Careful environmental scanning 2. Developing absorptive capacity, and/or using alliances, networks, and related consortia.
Open	1. **Motivating** the generation & contribution of external knowledge 2. **Incorporating** external sources with firm resources & capabilities 3. **Maximizing** the exploitation of diverse IP resources	1. Provide intrinsic rewards (e.g. recognition) and structure (instrumentality) for contributions. 2. As above. 3. Share or give away IP to maximize returns from entire innovation portfolio.

Tactics that embody an Open Innovation approach include exploiting knowledge spillovers and consulting with venture capitalists, while also using both inbound and outbound licensing of key technologies. Although earlier frameworks acknowledged the role of external knowledge and 'accidental' internal discoveries, it is the systematic encouragement and integration of these issues coupled with creative exploitation of IP that distinguishes Open Innovation from earlier innovation models.

Firms practicing open innovation face three inherent managerial challenges:

- *Maximization*. Firms need a wide range of approaches to maximize the returns to internal innovation—not just feeding the company's product pipeline, but also outbound licensing of IP, patent pooling, and even giving away technology to stimulate demand for other products.
- *Incorporation*. The existence of external knowledge provides no benefits to the firm if the firm cannot identify the relevant knowledge and incorporate it into its innovation activities faster than its rivals and new entrants. This requires scanning, recognition, absorption, and also the political willingness to incorporate external innovation.
- *Motivation*. Open Innovation assumes an ongoing stream of external innovation, but this raises the question of who will continue to generate IP externally that can be used by the firm? We suggest that over the long term, firms must cultivate ways to assure continued supply of relevant external technologies and IP.

We discuss each of these challenges below.

5.2.1 *Maximizing Returns to Internal Innovation*

A central concern to Open Innovation is how to best use the internal R&D capabilities of the firm to maximum advantage. Those capabilities can be used for:

- generating innovations to be internally commercialized (the traditional model);
- building absorptive capacity and using that capacity to identify IP laying beyond the boundaries of the organization, i.e. external innovation;
- generating innovations that generate returns through external commercialization (e.g. licensable patent portfolios or spin-offs); and
- generating IP that does not produce direct economic benefit, but indirectly generates a return through spillovers or sale of related goods and products.

Successful approaches will often combine a variety of these approaches. For example, to identify promising technologies, Intel establishes research laboratories near top university research groups, with open flows of information in both directions. If an innovation proves promising, Intel recruits the top academic researchers to help commercialize the technology and see it through to production (Tennenhouse 2003).

This approach can be used cooperatively as well, as with the GSM patent pool assembled by European telephone makers in the early 1990s. While the patents were often the result of basic research, contribution of a patent to the patent pool allowed firms to have favorable access to all of the IPR of the GSM standard, creating a cost advantage for European pool participants over potential Asian rivals (Bekkers et al. 2002).

5.2.2 *Incorporating External Innovations*

While firms may generate considerable internal knowledge to support their innovation activities, von Hippel (1988) identified four external sources: (*a*) suppliers and customers; (*b*) university, government, and private laboratories; (*c*) competitors; and (*d*) other nations. Various models have been developed to explain how firms can exploit external knowledge. Perhaps the simplest method is to imitate a competitor: such free riding on the product and market investments of rivals is a common way for firms to overcome a first-mover strategy (Lieberman and Montgomery 1998). Consulting with customers can provide firms ideas about discovering, developing, and refining innovations (von Hippel 1988). Public sources are also an important source of knowledge, for example government R&D spending was identified almost fifty years ago as an important stimulus for private R&D (David et al. 2000). University research is one key source of external innovation for some industries (see Chapter 7).

The managerial challenges of utilizing external knowledge then center around identifying useful external knowledge, and then integrating that knowledge with the firm. For example, for new products there are significant trade-offs between innovation speed, development costs, and competitive advantage for relying on external rather than internal learning (Kessler et al. 2000). Environmental scanning, competitive intelligence, sponsored research, and membership in relevant trade organizations is a way to uncover external knowledge opportunities. Developing absorptive capacity, via internal R&D investments appears to be an important prerequisite for converting external knowledge into internal innovation (Cohen and Levinthal 1990).

Even if external innovations are identified, that does not mean they will be incorporated into the firm's product strategies. A firm that was once highly successful at the vertically integrated innovation model will tend to believe its innovations superior to any competing ideas from outsiders. For example, flush from its successful user interface innovations of the 1980s, engineers at Apple Computer rejected external ideas in areas such as handheld computers, adopting the phrase 'not invented here' to describe such rejection (Kaplan 1996: 156).

5.2.3 *Motivating Spillovers*

With external innovation, there is often an unstated assumption that the supply will continue. But what happens if everyone tries to use others' basic research? Will 'innovation benefactors'—such as government and nonprofit research sponsors—continue to provide as fertile a field (Chesbrough 2003*b*)? If commercial firms do not realize a return on their innovative activities, they will tend to underinvest in innovative activities that are either highly risky (e.g. basic research) or that are easily imitated by free-riding competitors. Therefore, we consider the incentives for generating the knowledge spillovers at two levels—the individual and the organizational. This also has important societal implications, such as the funding of basic R&D by national governments. At the simplest level, such incentives can reflect direct financial payment to the innovators, but innovation recipients can and do exploit a wide range of alternatives.

Motivating individuals to generate and contribute their IP in the absence of financial returns is a significant management challenge for an Open Innovation approach. One of the simplest models of motivation is expectancy theory that posits that individuals are motivated when both valence, the attractiveness of a reward, which can be either intrinsic (e.g. happiness) or extrinsic (e.g. fame or fortune) and instrumentality, the path to that reward, are present (Lawler 1971). The integrated innovation model solved this challenge though extrinsic compensation from the firm coupled with adherence to traditional professional scientific norms, e.g. scientific freedom, support for publishing. The external model doesn't formally address individuals but appears to rely upon others, e.g. universities, to partially or wholly provide motivation.

The incentives for organizations to contribute spillovers fall into two categories. In the one case, the innovation benefits the innovator and there is no loss from sharing that benefit with others. For example, customers often share their innovations with their vendors if it means improved products in the future (von Hippel 1988). And of course suppliers invest in innovations to sell more products, as when Intel increases the performance of microprocessors that it sells to Dell.

Spillovers to a direct competitor are more problematic, but still are economically rational under conditions of 'co-opetition'. Firms in the same industry complement each other in creating markets but compete in dividing up markets (Brandenburger and Nalebuff 1996: 34). So if a firm stands to benefit from an innovation that grows the market, it will accept spillovers if the return from its share of market growth is attractive enough. For example, Intel's venture capital arm makes investments to grow the ecosystem around its microprocessors, even though a small portion of that investment accrues to AMD, the second-place maker of microprocessors (Chesbrough 2003*a*: 125–31).

5.3 Research Design

Our three challenges led to three related research questions:

- What circumstances motivate firms to embrace Open Innovation approaches as part of their R&D efforts?
- Why would for-profit firms commit their IP as well as *ongoing* human resources to an effort that they know will benefit others, including competitors?
- Why do individuals contribute their IP to a project that benefits firms without receiving financial remuneration?

We chose to study the use of external innovation in the software industry, in particular the 'open source' movement. Open source and other collaborative development techniques in the software industry offer examples of how the three key challenges of Open Innovation can and have been addressed by commercial firms. Open source also offers an approach to address what West (2003) refers to as an 'essential tension' in information technology (IT) innovation: appropriating the returns from an innovation versus winning adoption of that innovation.

5.3.1 *Traditional Software Production Models*

Modern software engineering techniques are based on an abstraction of the software design to minimize interdependencies between individual components of a complex system. This enables both specialization in the software development effort and reuse of existing technology to facilitate cumulative innovation (Krueger 1992). As such, software development efforts are highly modular within or between firms (Sanchez and Mahoney 1996). This specialization in software mirrors the specialization of other forms of industrial production.

The output of this software production effort is an information good, marketed to individual consumers or businesses. Even if software is distributed on a physical medium such as a magnetic tape, floppy, or optical CD-ROM, the value of the software is tied to the information on those tapes or discs, not the value of the physical medium or the cost of reproduction. As with any other good, firms price their software to appropriate some but not all of the utility derived by consumers.

While the marginal cost of each copy of a software application is thus very low, initial development costs are quite high. Commercial firms spend millions to hundreds of millions of dollars to make a commercial software release, and in most cases are forced to reinvest comparable amounts every one to three years to continue their ongoing revenue stream. This gives software firms huge supply side economies of scale: the first copy of Windows is very expensive, but additional copies cost almost nothing.

Interestingly, many software firms also benefit from not only supply side economies of scale but demand side economies as well. Most software users would face significant switching costs in using some other software package, due to some combination of retraining user skills and converting data stored in proprietary file formats. As Arthur (1996) observes, software thus has tremendous positive returns to scale, generally allowing only one (or a small number) of winners to emerge. These winners are tempted to extract rents from their customers by increasing prices and creating additional switching costs to protect those rents (Shapiro and Varian 1999). From these production economics, commercial software firms seek to build complete systems to meet a broad range of needs, in hopes of forestalling potential competitors and protecting high gross profit margins.

These supply and demand side economies of scale have fueled ongoing consolidation in the software industry. Microsoft is the most obvious beneficiary of these scale economies of having exploited both supply and demand side economies of scale, having captured not only the desktop operating system market but the most common office applications as well. However, consolidation has also occurred in other segments, such as in enterprise software as with Oracle's 2005 acquisition of PeopleSoft.

5.3.2 *Open Source Software*

Open source software emerged as a reaction to the proprietary software model, with two direct antecedents. One was the open systems movement—centered around Unix and its variants—reflecting an attempt by customers to reduce their dependence on proprietary software vendors. Another antecedent was the creation of university research software during the 1980s, including the BSD variant of Unix from UC Berkeley. However, the open source (and related free software) movements differed from the open systems and university initiatives by focusing on user rights, especially establishing a series of principles to enable shared development and perpetual use rights. The rapid rise of open source in the late 1990s was enabled by the dissemination of software tools and the Internet to facilitate the shared distribution production of software (West and Dedrick 2001, 2005).

Open source software differs from proprietary software in two ways: in its IP philosophy and how it is produced (West and O'Mahony 2005). Researchers have considered each of these in turn:

IP Philosophy. Differences between the open source movement and proprietary software are most dramatic over the treatment of the source code of software. The definition of 'open source' requires free redistribution of software in source code form and the right to modify the software. An allied but distinct group, the 'free software' movement, also requires that software remain perpetually 'free' by compelling users to return all modifications, enhancements, and extensions (West and Dedrick 2005). These conditions are enforced by a wide range of software licenses approved by one or both factions (Rosen 2004). In comparison, proprietary firms aggressviely protect their software source code.

Production via Collaboration. Another important difference between open source and proprietary efforts is the collaborative open source production process. In examining two projects, Mockus et al. (2002) concluded that the development was controlled by a small group, but received occasional error correction from a much larger group of developer-users. Considering a broader group of projects, Healy and Schussman (2003) showed that participation was highly skewed according to a power-law (log-log) distribution: a handful of projects attract most of the attention, and participation in these projects is heavily skewed towards a small number of contributors.

What motivates individuals to contribute to open source projects? Consistent with expectancy theory, empirical research on more successful projects

(Hars and Ou 2002; Hertel et al. 2003; Lakhani and von Hippel 2003) found three general categories of contributor motivations: *direct utility,* either to the individual or to one's employer; *intrinsic benefit* from the work, such as learning a skill or personal fulfillment; and *signaling* one's capabilities to gain respect from one's peers or interest from prospective employers.

Firm Participation. Despite these key differences between open source and proprietary software, for-profit IT producers have gotten involved in open source software. Why? West (2003) showed that in a positive returns environment, such firms made limited use of openness to win adoption of their technologies, either by opening portions of the their technology or providing partly open access to key technologies. O'Mahony (2003) demonstrated that while open source projects employ IP licenses that grant use rights to a broad class of users, they also used a combination of legal and rhetorical tactics to aggressively safeguard the independence and permanence of their development efforts, particularly when negotiating with firm seeking to advance potentially conflicting proprietary interests.

But in addition to these independent, organically grown open source projects, firms have also sponsored their own open source projects. Such sponsored projects differ from the organic ones in both production and governance, as the bulk of the resources are provided by the sponsor to achieve its goals (O'Mahony and West 2005). Firms have also experimented with sponsoring collaborative software production that is similar to open source in most but not all dimensions, such as 'gated' communities in which collaboration is available to some but not all (Shah 2004).

Here we consider the strategic motivations and business models of firm involvement in open source software projects, both those projects that begun autonomously and those created by sponsors to directly advance firm goals.

5.3.3 *Research Methods*

Given the comparative newness of the Open Innovation paradigm, we chose to use a theory-building approach grounded in the context of rich data. This draws on established procedures for generating theory from qualitative data (Glaser and Strauss 1967), as well as management studies that employ the inductive method to draw theory from a set of case studies (Harris and Sutton 1986; Eisenhardt 1989).

Our research efforts included both primary and secondary sources. From 2002 through mid-2005, one author conducted fifty-six interviews with forty-six informants representing thirty organizations. Of the thirty organizations, eighteen were for-profit companies in the IT industry. The interviews also included seven major open source projects, as well as other professionals indirectly involved in the industry. Interviews typically ranged from 45–90 minutes, and most were tape recorded for later consultation. This was supplemented by observation of (and, in some cases, participation in) six Silicon Valley industry conferences and seminars from August 2003 through March 2005 that focused solely on open source software. This primary data were complemented by a secondary data sources, we also incorporated secondary data. During the observation period, one of the authors reviewed approximately 800 news articles from trade journals, business press, and websites related to Linux and other open source topics. We also reviewed prior research from technology management, sociology, and computer science on open source collaboration.

From our data, we sought to identify regular patterns of Open Innovation among IT firms actively participating in open source projects, and (where possible) to identify the goals and motivations for such innovation strategies. Once we identified key patterns, we shared preliminary conclusions with a subset of informants and used their feedback to refine our conclusions.

5.4 Open Source as A Manifestation of Open Innovation

Open source as an Open Innovation strategy has two key elements: shared rights to use the technology, and collaborative development of that technology. Unlike many individual participants, firms must also consider a third issue: capturing an economic return to justify their investment.

Here we consider four approaches for external innovation in software, where firms have both invested in open innovations and benefited from them (Table 5.2).

5.4.1 *Pooled R&D: Linux, Mozilla*

A familiar model of Open Innovation is that of pooled R&D. While cooperative research often occurs to save costs, prior research also suggests that firms

Table 5.2. Open source projects as examples of Open Innovation

Project	Product Category	Approach
Apache	Web server	Shared R&D
Darwin	Operating system	Selling complements
Berkeley DB	Database	Spin-in, then dual license
Eclipse	Programming environment	Spin-out, then shared R&D
Jikes	Java compiler	Spin-out
Linux	Operating system	Shared R&D
Mozilla	Web browser	Spinout, then shared R&D
MySQL	Database	Dual license
OpenOffice	Business productivity	Selling complements
Sendmail	Mail router	Spin-in, then dual license

cooperate in cases where they cannot appropriate spillovers from their research (Ouchi and Bolton 1988), in areas that are highly risky or for industries most dependent on advanced science (Miotti and Sachwald 2003). They also tend to collaborate in industries with strong vertical relationships (Sakakibara 2001), with firms that share overlapping technological capabilities and are in the same industry or sector (Mowery et al. 1998).

Two highly visible open source examples are support for the Linux operating system through the Open Source Development Labs (OSDL), and the Mozilla web browser. For both, firms donate R&D to the open source project while exploiting the pooled R&D of all contributors to facilitate the sale of related products.

A simple example is the Mozilla open source project, a descendant of the Netscape Navigator browser offered for a wide range of systems—Windows, Macintosh, Linux, and at least seven Unix variants. This browser was among the first commercial browser products ('Netscape Navigator' 2004). Navigator held more than two-thirds of the browser market until late 1997—surpassing Microsoft's Internet Explorer—but only two years later the shares were reversed due to IE's bundling with Microsoft Windows (Bresnahan and Yin 2004).

Netscape created the Mozilla open source project in 1998 and terminated all internal development of it in July 2003—deferring further work to the open source community. At this point, Unix system vendors such as IBM, HP, and Sun were left without a supported browser, which they needed to sell Internet-connected workstations. Thus, each of them assigned software engineers to work with the Mozilla project, both to help keep the project moving forward and to assure that new releases are compatible with their respective systems.[1]

The R&D cooperation in the OSDL for Linux is more complex. Founded in 2000, the OSDL takes as its mission 'To be the recognized center-of-gravity for the Linux industry; a central body dedicated to accelerating the use of Linux for enterprise computing' ('Corporate Overview' 2004). In its first five years, the consortium began work on three projects: data center Linux, carrier grade Linux, and desktop Linux.

The founders, sponsors, and other members of OSDL and their motivations for supporting OSDL could be grouped into four broad categories: vendors of computer and telecommunications systems, producers of microprocessors, Linux distributors and support organizations, and developers of complementary software products (Table 5.3).

How do such projects address the three open innovation challenges? For firms participating in Mozilla, the quid pro quo is straightforward: systems vendors *maximize* the returns of their innovation by concentrating on their own needs (such as platform-specific customization), and then *incorporate* the shared browser technology into their integrated systems. However, the *motivation* challenge is not completely solved, in that the systems vendors assume a

Table 5.3. Members of the Open Source Development Laboratories

Category	Companies	Motivation
Computer systems vendor	Dell, Fujitsu, *Hitachi, HP, IBM, NEC,* Sun	These vendors spent the late 1980s and 1990s fighting the 'Unix wars' with mutually incompatible Unix implementations for their workstations and servers. In the late 1990s, they began shifting resources from their proprietary Unix implementations towards adapting and extended a shared implementation of Linux.
Telecommunications vendor	Alcatel, Cisco, Ericsson, NEC, Nokia, NTT, Toshiba	These vendors used Unix to run their switching systems but began shifting to Linux. As with systems vendors, interested in assuring that Linux evolved to work with their respective hardware and customers.
Microprocessor producer	AMD, *Intel,* Transmeta	Makers of Intel-compatible processors wanted to speed the shift of enterprise applications from proprietary RISC processors to their commodity processors.
Linux distributor (server and desktop)	Miracle, Linux, NEC, Soft, Novell, Red Hat, SuSE, Turbolinux	Distributors have a clear interest both in free riding off the work of others in developing Linux, and making sure that the software met the specific needs of their customers.
Embedded Linux distributor	LynuxWorks, MontaVista, TimeSys, Wind River	Similar to motivations of desktop and server Linux distributors, but need to support more heterogeneous customer needs for use with custom system configurations.
Linux support company	VA Software, Linuxcare, LynuxWorks	Without development capabilities, the firms both want to leverage the work of others and understand how it met customer needs.
Software developers	Computer Associates, Trolltech	Want to make the operating system more reliable for running their specific applications and libraries.

Founding member in **bold**
Source: OSDL and company websites (as of May 2004).

pool of individual open source contributors that sustain innovation in the core product.

For the OSDL, firms contributed their specialized knowledge (e.g. in telecommunications operations or microprocessor architecture) to build a common platform. OSDL resembles other self-supporting industrial research consortia, where firms pool interests towards a common goal, and assume they can both cooperate in supporting that goal and compete in selling their respective products.

However, both Mozilla and OSDL differ from typical consortia in two ways:

- *Spillovers are not controllable.* Many consortia reward members by limiting direct access to the consortium's research output to member-participants, limiting access to indirect spillovers. Open source licenses typically make it impossible to limit even direct access, allowing nonmembers to accrue many of the same benefits as members.
- *Contributions from nonparticipants.* The engineering contributions to these open source projects extend beyond the sponsoring companies to include user organizations, academics, individual hobbyists, and other interested parties. Unless the corporate contributions eventually dwarf the individual ones, the projects must continue to motivate such contributions to survive.

Given these factors, an open source innovation model is inherently more 'open' than a typical R&D consortium, both in terms of exploiting information from outside the consortium, and sharing that information back out to nonmember organizations and individuals.

5.4.2 *Spin-outs: Jikes, Eclipse, Beehive*

Open Innovation can release the potential of technologies within the firm that are not creating value. In some cases, the technologies are no longer strategic, as with AOL Time Warner's decision to spin off Mozilla into a stand-alone open source project after firing its Netscape development team (Hansen 2003).

But in addition to spin-off (and, frequently, abandonment), firms also have opportunities to release more value from their technologies by situating them outside the firm, but at the same time maintaining an ongoing corporate involvement. Here we use the term 'spin-out' to refer to all cases where firms transform internal development projects to externally visible open source projects.

If a firm gives away its IP, how can such spinouts create value? One way is that the donated IP generates demand for other products and services that the donor continues to sell. Two examples of this come from IBM and its efforts to promote the Java programming language developed by Sun Microsystems, that was widely embraced by firms competing with Microsoft in web-based technologies. As Java become more widely adopted, IBM would generate increased revenue from sales of its own hardware and supporting services, especially its consulting services which have become an increasingly important component of IBM's overall revenues.

IBM came to realize that it made sense to take other software projects into the open domain, as part of its overall strategy. IBM's first open source spinout came from a preproduction R&D project. In response to IBM's growing interest in Java, in early 1996 two IBM researchers began work on an experimental Java compiler, which they named 'Jikes'. They quickly developed a prototype that was more efficient than Sun's industry standard compiler. After customer requests for a better Java compiler, in December 1998 IBM announced the release of Jikes in open source form to allow external programmers to extend and improve the compiler. Since 2000, development has been led by non-IBM engineers.[2] Jikes has been widely adopted, and is now bundled with several distributions of open source operating systems.

A second IBM spinout came with Java development tools. In 1996, IBM purchased a Canadian software company that created such tools for its WebSphere application server product. IBM released much of this technology in open source form when it founded the Eclipse project in 2001 and these efforts were further boosted by its acquisition of Rational (Brody 2001). Other software companies involved in web application development, including Borland, Red Hat, SAP, and SuSE, and well as hardware makers HP, Fujitsu, and Intel joined the Java development tools effort. In 2004 the project became an independent nonprofit corporation ('Eclipse Forms Independent Organization' 2004), although IBM engineers retained technical leadership of key projects. As an IBM executive later explained, 'It is not that we are looking to make more money off the platform. It is just that we are looking to accelerate the adoption of Java and the building up of it for all of us' (Southwick 2004). Very recently, IBM even donated 500 software patents to the open source community, while it continues to license other parts of its patent portfolio for significant revenues.

But despite this openness, BEA and Sun—IBM's two major rivals in Java applications servers—chose not to join IBM's coalition, instead promoting the rival Java Tools Community (Taft 2003). During 2004 BEA also created a 'Beehive' open source project to release key application libraries from its WebLogic product for use with other development systems; it also helped a third party development of a 'Pollinate' library to link Beehive with Eclipse. Finally, in March 2005 BEA officially joined the Eclipse project.

These spin-outs also differ in the ongoing participation of their respective firm sponsors. IBM continues to provide hundreds of programmer-years of software development for Eclipse but limited support for Jikes, while Netscape has cut all sponsorship ties to Mozilla. The conditions under which a 'foundling' project can become immediately self-sufficient are likely to be much narrower than one which enjoys ongoing support from its original sponsor.

The spinout thus makes sense for technologies that either are not yet commercialized (as with Jikes), or that will eventually become commoditized and thus of limited commercial value (as many predicted for Java development tools). Both IBM and BEA donated internal innovations to create open source projects, which were intended to fuel adoption of the innovations. As with other organizations that sponsor open source projects, the benefits included:

- helping establish their technology as de facto standards, which, at a minimum, reduces the likelihood of having to reimplement their technology to conform to competing standards;
- attracting improvements and complements that make the technology more attractive;
- together, the innovation and complements enable the sale of related products (such as other components of WebSphere and WebLogic);
- generating mindshare and goodwill with the same audience that includes the potential customers for these related products; and
- Lowering or eliminating the ongoing costs of supporting projects, while providing customers of the project some possible source of ongoing support.

These motivations for open source spin-outs go beyond those identified for one of the most cited firm sponsor of spin-outs. Xerox PARC spun out technologies that no longer aligned with the company's strategy (Chesbrough and Rosenbloom 2002), which is consistent with Netscape's exit through creation of the Mozilla project. However, for the Jikes, Eclipse, and Beehive projects, the sponsoring firms spun out open source projects that were closely aligned with the firm's ongoing strategies. As West (2003) observed for other open source projects, relinquishing some level of control was essential to win adoption.

5.4.3 *Selling Complements: Apache, KDE, Darwin*

Many goods in computers and electronics fall into what Katz & Shapiro (1985) term the 'hardware-software paradigm'. As Teece (1986) notes, the base innovation ('hardware') requires an investment in producing complementary goods ('software') specialized for that innovation, in order to make the entire system useful. In many cases, these complements are more valuable than the core innovation. For example, makers of videogame consoles deliberately lose money or break even on the hardware so that they can make money from software royalties (Gallagher and Park 2002).

In other cases, a system architecture will consist of various components. Some mature (or highly competitive) components may be highly commoditized, while other pieces are more rapidly changing or otherwise difficult to imitate and thus offer opportunities for capturing economic value. Two open source examples are the IBM's WebSphere and Apple's Safari browser.

Customers access the WebSphere e-commerce software using standard web browsers, so IBM originally developed a proprietary httpd (web page) server. IBM later abandoned its server for the Apache httpd server, recognizing that it would be wasting resources trying to catch up to the better quality and larger market share enjoyed by Apache (West 2003). Today, IBM engineers are involved in the ongoing Apache innovation, both for the httpd server and also related projects hosted by the Apache Software Foundation (Apache.org website).

Similarly, in 2002 Apple Computer decided to build a new web browser called Safari, to guarantee one would be available for buyers of its computer systems. The browser built upon libraries from the Konqueror open source web browser, which in turn were developed to support the KDE desktop interface for Linux users (Searls 2003). The move paralleled Apple's earlier use of BSD Unix as a foundation for its OS X operating system, in which it created a new open source project (Darwin) to share all modifications of the BSD code (West 2003). For both Safari and OS X, Apple used open source and contributed back its changes, but the company did not release the remainder of the proprietary code for its browser and OS, respectively (Brockmeier 2003).

In the case of the Apache, Konqueror and Darwin open source projects, the firms adopting open source components had four common characteristics:

- there was preexisting open source code being developed without the intervention of the focal firms;
- the 'buy vs. build' decision to use external innovation was made easier because the code was 'free';[3]
- the firms were willing to contribute back to the existing projects on an ongoing basis, both to assure that the technology continued to meet their respective needs and to maintain absorptive capacity;
- the firms could continue to yield returns for internal innovation by combining the internal and external technologies to make a product offering that was not directly available through open source.

Another model for selling complements is the 'dual license' strategy, where a firm develops code and releases it both as an open source project and a commercial product. Buyers who want free software get no support and restrictions on source code distribution in exchange for development feedback. Less price sensitive buyers (e.g. corporations) pay the sponsoring firm a license fee to receive full features and support (Välimäki 2003; West and O'Mahony 2005). However, the ongoing proprietary control of such sponsored projects mean that they have trouble attracting external innovation, and open source thus becomes a marketing technique to attract adopters and build network effects through a large community of adopters.

5.4.4 *Donated Complements: Avalanche, PC Game 'Mods'*

In other cases, firms make their money off of the core innovation but seek donated labor for valuable complements.

For decades, IT companies have encouraged their users to collaborate and share user-developed software that filled in the gaps for their proprietary offerings. This has been particularly relevant for medium and large buyer organizations (companies, universities, and government) with large internal IT organizations. In the 1960s, IBM sponsored its SHARE user group, while in the 1970s Digital Equipment Corporation had its DECUS.

More recently, firms have indirectly or directly supported user collaboration that is coordinated using open source techniques. One example is the Avalanche Technology Cooperative, a Minneapolis-based nonprofit founded in 2001 to pool IT customizations developed by local enterprise IT users. This would allow companies to integrate disparate packages such as PeopleSoft and SAP that do not provide their own integration modules.[4]

Another example of donated complements is the use of 'mods' for PC video games. The PC gaming industry competes with lower priced dedicated gaming platforms such as Sony's PlayStation2 or Microsoft's Xbox. The commercial publishers of PC games thus have decided to exploit the one key advantage they have versus the consoles: the ability for PC users to update and modify their games. To do this, publishers release editing tools for their games to encourage user mods that create different environments, scenarios, or even total rebuilds of the game; the users then freely distribute these mods on the Internet. A few of the mods (such as Battle Grounds) are developed as open source, but most are developed as closed source.

While mods do not directly generate publisher revenues, the novelty of the mods extends the otherwise relatively short demand period for most computer games. Meanwhile, the mods keep the name of the game in front of consumers for additional months, while the publishers need years to prepare follow-on products. This external innovation keeps the product current without tying up internal innovation resources. In rare cases, the publisher serves the need identified by the mod by creating its own game or even buys the mod outright.

As with open source, a key issue for mods is motivating the contributors. The motivations parallel those for open source: direct utility, intrinsic reward, or external signaling. Individuals (or virtual teams) contribute mods because of their creative nature, love of either the computer game they modified or the milieu they recreated via their mod (Todd 2004). Students are also frequent contributors, increasing their enjoyment of a favorite game (direct utility) as well as signaling their value to potential employers.

The computer game industry highlights three key ideas for attracting external innovation that are similar to those for open source:

- *minimizing technical obstacles*. Contributors develop mods because they can build upon the publisher's proprietary innovation to make a compelling game experience. As with other software development platforms (such as operating systems or databases), third party developers are attracted by platform capabilities and the prompt availability of development tools.

- *creating an infrastructure* that encourages participation and collaboration. For open source, this is a project website and email lists, but for mods this would be a distribution site that highlights the mods. Modern technologies make the cost of such infrastructure quite low.
- *recognition for contributors,* including added visibility for the most popular creators. For example, since 2002 Apple has given annual awards for the best use of open source related to its OS X operating system.

However, the mods also help address one problem that's very different from those of business-oriented source projects. As with other entertainment products, novelty-seeking consumers eventually grow bored with a PC game; by combining the core game engine with new externally generated game scenarios, the external innovation extends the life of the core (internally developed) innovation. Of course, this novelty comes at the potential expense of loss of some control over the software product. For example, recently 'modders' unlocked some features in the PC version of Grand Theft Auto: San Andreas (GTA: SA) that enabled a pornographic 'mini-game.' As a result of this activity, the rating on all versions of GTA: SA (not just the PC version) were changed to 'Adults Only' thereby severely curbing the availability of the title.

5.5 Discussion

5.5.1 *Open Source as Open Innovation*

How did firms in open source software efforts effectively exploit Open Innovation? We identified four strategies of Open Innovation in software that addressed the unique combination and exploitation of innovation from multiple sources. Table 5.4 cross references these four strategies with the key Open Innovation challenges we identified.

Despite the wide popularity of open source collaboration, software firms appear to embrace Open Innovation only when there is no alternative—specifically a broad dispersal of both production knowledge and market share that forces vertically integrated producers to admit that they no longer can 'do it all'. The use of open source by firms typically begins in ways that do not disrupt their fundamental business strategy (e.g. selling complements), or comes at a time when their existing strategy is so threatened that they are forced then to make drastic changes.

At the same time, firms faced (and often acknowledged) the risks of collaborating, risks that have not yet been fully realized. For example, encouraging users to develop complements could reduce the availability of vendors

Table 5.4. Open source strategies as solutions to open innovation challenges

Open source strategy	Maximizing returns of internal innovation	Role of external innovation	Motivating external innovation	Challenges
Pooled R&D	Participants jointly contribute to shared effort	Pooled contributions available to all	Ongoing institutions establish legitimacy and continuity	Coordinating and aligning shared interests
Spin-outs	Seed noncommercial technology to support other goals	Supplants internal innovation as basis of ongoing innovation	Free access to valuable technology	Sustaining third party interest
Selling Complements	Target highest value part of whole product solution	External components provide basis for internal development	Firms coordinate ongoing supply of components	Maintaining differentiation as shared components add capabilities
Donated Complements	Provide an extensible platform for external contributors	Adding variety and novelty to established products	Recognition and other nonmonetary rewards	Third parties can control the user experience

to achieve proprietary lock-in, an explicit long-term goal of the Avalanche project.

Especially significant is the role of open source in enabling pooled R&D, which has been seen in other industries but rarely in software. This scarcity is hardly surprising, given that up-front R&D forms one of the major barriers to entry in the software industry (Arthur 1996; Cusumano 2004). Even for products for which software is only part of the value creation—such as computer systems—the shared R&D available through open source software such as Linux can significantly reduce the barrier to new entrants (West and Dedrick 2001).

Interviews suggested that firms were willing to share software R&D—and thus eliminate potential differentiation for the component—in areas that were necessary prerequisites for selling a complete system (e.g. 'infrastructure layer') but offered few opportunities for differentiation. Removing duplicative investment reduced costs for this type of commodity component, consistent with Miotti and Sachwald's (2003) finding that R&D collaboration with more direct competitors was associated with reducing costs.

Is such pooled R&D an example of Open Innovation? As with other aspects of Open Innovation, it depends on the ability of firms to capture value for their investment (See Chapter 1). So if the firms use open source to provide a common base, but have a way to sell complements, then investing in an open source project to provide pooled R&D is consistent with Open Innovation.

To what degree are the lessons of pooling open source R&D applicable to other industries? If cost savings are a major goal of the collaborative R&D, then we would presume that the transaction costs of creating and maintaining such collaboration would have to be limited to make collaboration more attractive than a go-it-alone strategy. Two factors tend to reduce these costs, at least for open source projects. First, such virtually dispersed, decentralized production takes advantage of software and communication technologies that facilitate joint software development (West and Dedrick 2001; Lerner and Tirole 2002). Second, considerable effort has gone into developing mechanisms for coordinating and governing such efforts (O'Mahony 2003; Shah 2004; O'Mahony and West 2005). While we would not suggest that such collaboration and governance advances are necessary for other forms of pooled Open Innovation, we predict that adapting such practices to other industries would make such pooled R&D more likely.

5.5.2 *Where Open Source and Open Innovation Part Ways*

For the firms and projects in our sample, we concluded that most firm involvement in open source fits the Chesbrough (2003*a*) definition of Open Innovation, in which firms both use a broad range of external sources for innovation and seek a broad range of commercialization alternatives for internal innovation. However, we would not mean to suggest that all open source software is an example of Open Innovation—or for that matter, that all Open Innovation in the IT industry relates to open source software.

In fact, some form of Open Innovation has become the norm in the computer industry for decades. For the reasons identified by Teece (1986), Shapiro and Varian (1999) and others, computer vendors have long relied upon third-party suppliers of complementary software products; this has been true of even the most proprietary of systems, such as Apple's Macintosh (West 2005). However, what we now refer to as Open Innovation was extended by IBM's 1980 decision to source its PC CPU and operating system from Intel and Microsoft instead of its traditional vertical integration.

Examples outside the overlap of open source and Open Innovation include the following (Figure 5.1):

- *Open Source but not Open Innovation.* Open Source is only Open Innovation if it has a business model. There are tens of thousands of Open Source projects created and run for nonpecuniary motivations—such as the work done within Project GNU, which is motivated by a strong ideology (West and Dedrick 2005). Also this category could be AOL's exit strategy with Mozilla, which (like many of the Xerox PARC spin-offs) reflected the failure of the sponsor to create a viable business model, leaving the foundling innovation abandoned to whoever is willing to nurture it.
- *Open Innovation but not Open Source.* The 'Wintel' PC using Windows and Intel components represents a powerful embrace of what we now term Open

	Open Innovation	*not Open Innovation*
Open Source	Apple: Darwin BEA: Beehive IBM: Apache, Eclipse, Jikes, OSDL	Project GNU
Not Open Source	PC makers: CPU Windows Game Mods	Microsoft: applications † Intuit: Quicken †

Figure 5.1. Overlap of Open Source and Open Innovation
† Increasing use of vertical integration, with some reliance on external innovation

Innovation. On the one hand, it lowered barriers to entry, enabling the rise of numerous makers of PC 'clones' (Moschella 1997). On the other hand, IBM's failure to appropriate adequate returns from its PC systems integration activities led to its 2004 exit from the PC industry.

- *Neither Open Source nor Open Innovation.* As noted above, the use of independent software vendors (ISVs) for external innovations is the norm in the computer industry. However, some firms are heading in the other direction, becoming increasingly integrated. For example, Microsoft has integrated downstream from operating systems into applications such as Windows, Money, and SQL Server—decreasing the relative importance of third party application providers. Intuit is adding additional services (such as loans) to extend its Quicken financial management software. In video games, console makers supply both hardware and software, with Nintendo particularly dependent on its internally generated game franchises like Pokemon and Mario Brothers rather than ISV-supplied titles.

In a self-interested prediction, Grove (1996) argued that the IT industry had been irrevocably transformed by component-based systems integration model, in which component suppliers using horizontal specialization achieved insurmountable economies of scale over vertically integrated innovators. We believe that open source is also having a profound impact on IT value creation and capture, but it is too soon to say what effect open source open innovation will have upon proprietary alternatives.

5.6 Open Source Collaboration Within a Value Network

The chapters in Part III show Open Innovation both enables and builds upon interorganizational collaboration. Such collaboration has been variously referred to as a network form of organization (Powell 1990), value network

(Christensen and Rosenbloom 1995) or an ecosystem (Iansiti and Levien 2004*a*). Open innovation in the IT industry—and particularly the commercial support of open source software—certainly would fit such a classification.

Due in large part to IBM's aforementioned shift from vertical integration to externally sourced components, the IT industry has moved to a separation of technical activities based on a modular subdivision of labor between firms that parallels the technical modularity of the overall systems architecture (Grove 1996; Langlois 2003*b*). This form of Open Innovation enables firms that provide a component for a complex system to use various appropriability mechanisms to capture a return from their portion of the system's value creation (see Chapter 6). When open source software is used as part of a complex system, a firm still faces the fundamental issues of coordinating the systemic innovation, assuring overall value creation and capturing the firm's portion of that value. These issues are common to any Open Innovation value network, as noted by Maula, Keil, and Salmenkaita (Chapter 12) and Vanhaverbeke and Cloodt (Chapter 13). But the use of open source software changes this ecosystem management in at least two ways.

The task is made easier where open source software reflects the commoditization of some portion of the overall complex system. Because open source is (by definition) free,[5] the use of open source makes value capture less important for that component: pooled R&D to develop a shared component reflects an acknowledgement of the commoditization of that component. Value capture is still a factor if the firm merely shifts its value capture from selling software licenses to selling support services for that software.

However, economic coordination of the systemic innovation is made more difficult if participants are motivated not by pecuniary goals such as value capture, but instead some combination of intrinsic and extrinsic factors. This chapter has identified some of the ways that firms can motivate the supply of external innovations from individuals, while the work of O'Mahony (2003) O'Mahony and West (2005) highlights the motivations of open source communities in their collaboration with firms. But such coordination poses difficulties, both in practice and also for our theories of Open Innovation.

One thing that is clear from studying open source innovation—both through direct observation and from prior accounts of the process (e.g. DiBona et al. 1999; O'Mahony 2003; Shah 2004)—is the distinct set of attitudes and norms that set open source software production apart from commercial IT innovation. Shared organizational culture is long-identified form of governance within organizations, particularly for creative, individualistic workers and when direct monitoring of performance is difficult (cf. Kunda 1992). But little is known about how such culture is created and maintained across a network organization, particularly if that culture is used to facilitate the process of Open Innovation.

5.7 Limitations

There are key limitations to the generalizability of our findings. While we were informed by broader secondary data on the open source movement, the framework was constructed using inductive theory-building from a small number of cases, anchored to specific firms in US IT industry. As such, they may not generalize to other firms, let alone other industries.

More seriously, there are fundamental questions about drawing conclusions regarding an emergent phenomenon: open source software as part of corporate open innovation strategies is still a comparatively recent phenomenon, and there are many key questions regarding the sustainability of this model.

The open source movement built on a confluence of ideology, professional norms, and enthusiasm; some question the long-term sustainability of such motivations. Also, many projects have been created as challenges to an entrenched incumbent (e.g. Microsoft), and if such challenges are largely unsuccessful, vendor interest in sponsoring future open source efforts could wane. In addition, as measured by traditional profitability standards, many open source projects have had problems. For example, currently JBoss and MySQL appear to be unprofitable and Linux vendor Red Hat is reportedly operating at break-even or slight profit levels (Lyons 2005).

Also, open source has yet to fully resolve the IP issues of accepting donations from a wide community of unknown contributors, as reflected by SCO's legal attacks against Linux. While such potential infringement has been attributed to ignorance, others have suggested that infringing 'stealth' IP could be deliberately donated to projects to sabotage their success (Cargill and Bolin 2004).

Finally, many have accused open source software as being more about reimplementing and commoditizing prior technologies than creating new innovations. The adoption of Linux as a low-cost Unix is the best known example of such commoditization, but other examples include MySQL and OpenOffice. To rebut such criticisms, open source supporters point to the mainly Internet innovations that were first implemented using open source projects (such as X, BSD, sendmail, or Apache) and to ongoing university-led research projects such as the Globus Alliance. Even without this, prior researchers have shown how radical innovations disrupt existing market structure if they merely deliver similar capabilities at significantly lower cost (Christensen 1997; Leifer et al. 2000; See Chapter 4); open source software certainly would qualify as a radical (or 'disruptive') innovation under such definitions.

5.8 Future Research

There are many factors that enable Open Innovation in open source. Future research could consider whether these characteristics of open source are prerequisites for other forms of open innovation:

- *feasibility of virtual teams* as a way to organize innovation, enabling pooled R&D and other collaboration between organizations;
- *a culture of open innovation* throughout such teams that spans organizations, vanquishing both a 'not invented here' attitude towards external innovation and a 'crown jewels' attitude of controlling internal innovations;
- *modularization* of technologies and products, to allow the external production of components or complements;
- *formal IP mechanisms* that encourage collaboration;
- *economic prerequisites* for effective open source collaboration; and
- *abandonment* of open source projects: how and when do they terminate?

Our attempts to define open innovation uncovered questions beyond those specific to software. Two relate to the availability of external spillovers:

- *Commercialization of public research.* Universities have gotten increasingly sophisticated about profiting from their research spillovers, a trend encouraged in the US by the Bayh-Dole Act (Colyvas et al. 2002; see Chapter 7). Will this restrict the flow of external innovations or provide an ongoing incentive for greater supply?
- *Increasing conflict over patents.* The increasing scope and commercial value of patents has spawned various concerns that patents will inhibit traditional closed innovation (e.g. Jaffe and Lerner 2004); the threat to external spillovers is likely greater.

Other questions relate to potential patterns for leveraging external knowledge:

- *Boundaries of the firm.* While firms are making increasing use of virtual teams, collaborative R&D consortia and other shared fora, the root cause is far from clear. Is this evidence that R&D is no longer necessary to internalize in firms? Or are these or merely examples of specific innovations that cannot be appropriated by firms, symptomatic of industry segments that have become commoditized and thus where R&D produces little competitive advantage?
- *Role of process innovations.* Companies like Dell combine external product innovations with internal process innovations. Research on Open Innovation has focused on innovation to produce products, so would the process of Open Innovation be fundamentally different when it incorporates process innovations?

- *Low R&D intensity firms*. Many firms have low R&D intensity, either due to size (e.g. small businesses) or industry characteristics (low tech). Are they pursuing 'external innovation', 'Open Innovation,' or (as commonly assumed) 'no innovation' strategies?

5.9 Conclusions

Open source software offers a significant example of how Open Innovation can transform an industry. While producers of complements and add-ons have occurred since the IBM 360, the rise of open source has increased attentions on alternate forms of organizing to exploit firms' IP. The use of external innovation is not a wholly new idea, however the activities of firms surrounding open source software highlights ways firms can reap returns by 'giving away' their IP and related firm resources. The transformation appears hastened by the nature of the good, the available tools, and previous trends in the sector away from vertical integration.

Open source software provides a powerful example of how firms can manage a complex ecosystem to combine external and internal innovations, creating an architecture for the whole product solution that both creates and captures value. Some of the sociological and legal characteristics may seem particularistic to open source, particularly with the culture of 'free software' (West and Dedrick 2005). However, the conflicts around open source—between those who want to share value and others who want to capture value from shared innovation—anticipate parallel concerns in other industries, as will be discussed in Chapters 6, 7, and 8.

Notes

1. Interview, Asa Dotzler, Mozilla.org, March 8, 2004.
2. Interview with David Shields, IBM Corporation, May 24 and June 10, 2004, as well as news coverage including Gonsalves and Coffee (1998).
3. Both Apache and BSD packages were open without restriction in the typology of West (2003), while KDE contained the compulsory sharing restrictions of the GPL.
4. Interview with Scott Lien and Andrew Black, Avalanche Technology Cooperative, March 16, 2004; see also Gomes (2004).
5. The ongoing debate over open source as 'free speech' versus 'free beer' is beyond our scope; see, for example, West and Dedrick (2005).

Part II

Institutions Governing Open Innovation

6

Does Appropriability Enable or Retard Open Innovation?

Joel West

6.1 Introduction

Open Innovation reflects the ability of firms to profitably access external sources of innovations, and for the firms creating those external innovations to create a business model to capture the value from such innovations. Contrasted to the vertically integrated model, Open Innovation includes the use by firms of external sources of innovation and the ability of firms to monetize their innovations without having to build the complete solution themselves.

But as Teece (1986) noted some twenty years ago, the ability of firms to pursue this latter course (and thus create a supply of external innovation) depends on appropriability. Absent appropriability, imitators will commercialize the ideal and the innovating firm will lack the incentive (and possibly the funds) to ever innovate again.

Formal appropriability by and large depends on intellectual property (IP) laws, and certain types of Open Innovation are only possible through such IP protection. Thus, the remaining chapters in this section consider the relationship of IP policies (whether at the nation-state or organizational level) to the practice of Open Innovation. At the same time, from their studies of biotechnology and information technology (IT) innovations, the authors suggest cases under which too much appropriability is also bad for Open Innovation.

Implicit in these and other studies—but explicit in West (2003)—is that firms can voluntarily surrender appropriability to achieve other firm goals, such as seeking adoption in the presence of demand-side economies of scale. Appropriability decisions are thus not just those of infrequent changes in national policy, but also the ongoing strategies of individual firms for specific technologies.

In this chapter, I first review the role of IP in providing appropriability, and from that its role in enabling Open Innovation. I then discuss how strong IP can also hinder the flow of innovation, using a discussion of the remaining chapters of the section to contrast Open Innovation (as defined in Chapter 1) with three other uses of 'open' in the context of innovation: open science, open standards, and open source software. From this, I review a brief case study of the effect of IP on innovation in mobile telecommunications, and then conclude with observations and questions about Open Innovation and appropriability.

6.2 IP Enables Appropriability

6.2.1 *Importance of Appropriability*

Nearly two decades ago, David Teece (1986: 285) wrote:

> It is quite common for innovators—those firms which are first to commercialize a new product or process in the market—to lament the fact that competitors/imitators have profited more than the firm first to commercialize it!

Teece's observation anticipated a subsequent burst of research that showed that technological pioneers have as many advantages as disadvantages. Pioneer investments are highly risky due to technological, market, and financial uncertainty, and their efforts to create a new market usually benefit imitators, particularly fast followers. Meanwhile, imitators have lower costs if they can wait for the pioneer to identify a winning strategy rather than having to make their own investment in technological and market experimentation (Aaker and Day 1986; Lieberman and Montgomery 1988; Golder and Tellis 1993; Schnaars 1994).

Teece's strategic recommendations were contingent upon the level of appropriability available to the firm. If the level of appropriability is high, firms will have time to develop the idea, experiment in search of a dominant design, and enjoy the fruits of any eventual success of the technology. If not, the innovative firm must vertically integrate to build a complete solution or, barring that, hope to create an enforceable contract with suppliers of complementary products and capabilities necessary to commercialize the innovation (Teece 1986).

If firms are unable to lock up key strategic resources to assure competitive advantage, then the path to profiting from innovation is more tenuous. More recent research suggests firms must change rapidly to be able to exploit new opportunities and achieve at least temporary competitive advantage (Tushman and O'Reilly 1986; Teece et al. 1997; Rindova and Kotha 2001). But such dynamic or transient competitive advantage is both hard to achieve and often fleeting even if won, thus raising considerable doubt about the likely payoffs to innovators. Without returns to innovation, the temptation is for all firms to free ride on the innovation of others, with none willing to invest in creating their own innovations.

6.2.2 *Potential Role of IP*

Avoiding this problem of underinvestment in innovation is exactly the point of granting temporary monopolies[1] through intellectual property rights. As Besen and Raskind observe:

> The objective of intellectual property protection is to create incentives that maximize the difference between the value of the intellectual property that is created and used and the social cost of its creation. (Besen and Raskind 1991: 5)

In the US, such a policy objective dates back to the Constitution (1787), which calls on Congress 'To promote the Progress of Science and useful Arts, by securing for limited Times to Authors and Inventors the exclusive Right to their respective Writings and Discoveries' (Article I, Section 8). Two centuries later, responding to a fear that Europe was lagging in innovation due to its IP system, a European Commission 'green paper' concluded:

> It is vital to protect the fruits of innovation. In economic terms, it has been clearly established that companies with specialized know-how which sell branded products and patented products or processes have a competitive advantage when it comes to maintaining or expanding their market share. (European Commission 1997: 1)

Most of the discussions of strong appropriability center on one particular form of IP, the patent, because it covers the fundamental idea rather than its expression, and also blocks independent invention by potential imitators. However, industrial innovation also makes use of copyright and trade secret protections, so these two are also potentially applicable to Open Innovation.[2]

6.3 IP Enables Open Innovation

For firms seeking to gain additional revenues through Open Innovation, Chesbrough (2003*a*: 155) identifies two key factors. First, licensing technology depends on the firm's IP strategy, which defines the role of the IP both for the innovator and any potential licensees. Second, the innovator must develop a business model consistent with both the value of the IP and the innovator's position in the value network (Chesbrough and Rosenbloom 2002).

Here I focus on innovations related to a core technology at the beginning of a complex system supported by complementary products.[3] For this model, the value network would consist of technological innovators, component suppliers, system integrators, suppliers of complementary products, and end customers (Figure 6.1). The core technology may be incorporated either into a component or directly into a product. If in a component, such components are then integrated with other components into a complete system (Hobday 2000; Prencipe et al. 2003), and firms succeed through the application of integrative competencies (See Chapter 3). The system, in turn, gains value through the

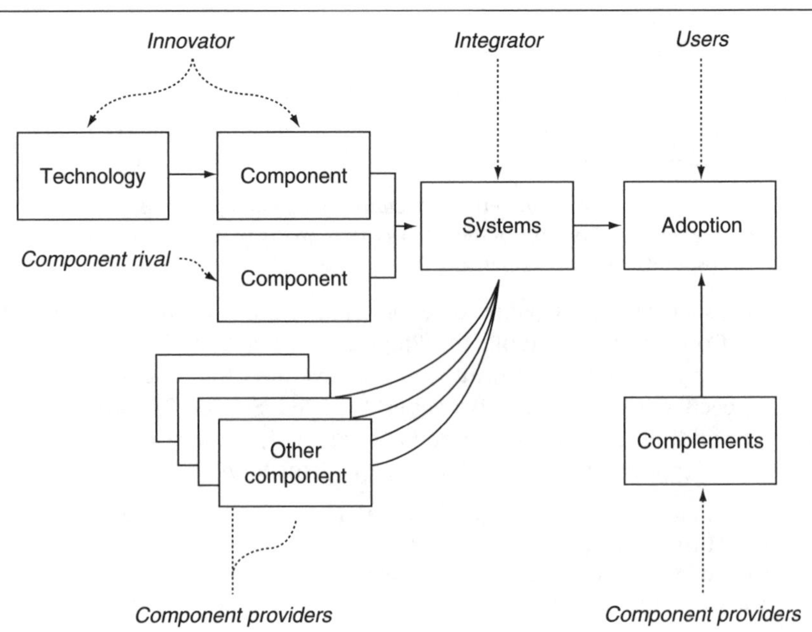

Figure 6.1. Incorporating technology innovations into complex systems

provision of complementary products customized to work with the system (Teece 1986; Shapiro and Varian 1999).[4]

This model corresponds to complex assembled systems, such as IT and machinery products. However, other forms of cumulative innovation exist with more or fewer intermediate levels in the value chain, as well as different forms of complementary assets. For example, in the pharmaceutical industry, essential complementary assets may include manufacturing, distribution, or service (Teece 1986).

6.3.1 *Vertical Integration Versus Open Innovation*

Innovators have multiple paths to gaining an economic return from their innovation:

- they can license the innovation to downstream suppliers to incorporate in products and components, as Qualcomm has done with CDMA-related patents and Rambus attempted to do with RDRAM memory technology (West 2002; Tansey et al. 2005).
- they can distribute them in components, which compete with other similar components, as Intel does with its microprocessors;
- they can incorporate them in complete solutions, as happened during the golden era of proprietary vertically integrated computer manufacturers such as IBM and DEC (Moschella 1997).

Large vertically integrated firms can create a systems innovation from beginning to end, which potentially limits the scope of imitation: if only vertically integrated firms can appropriate an invention, then there's a relatively small number of firms that have the necessary end to end capabilities. Once, the large computer makers all followed this model, but today only IBM, Sun, and Fujitsu can sustain it. Meanwhile, most of the computer industry is moving to the vertically dis-integrated, horizontally specialized model used by the PC industry since early 1990s (Grove 1996; Kraemer and Dedrick 1998), while using shared open networking standards that were developed through nonproprietary engineering committees (Simcoe 2006). More generally, modular decomposition of a technical problem has enabled separation of production between specialist firms (Langlois 2003*b*).

If an innovation is not protectable, firms may be able to combine such innovations with others that are protectable to gain indirect economic returns. For example platform vendors often bundle new applications with their systems or software to make them more attractive (as Microsoft bundled Internet Explorer with Windows 95 or Apple bundled iPhoto with OS X) and thus drive upgrade sales. However, this scenario encourages vertical integration and discourages Open Innovation because the innovation does not earn a direct return but only the indirect return through bundling; in this case, the cross-subsidies discourage other (otherwise profitable) innovation and experimentation in competing with the subsidized component.

6.3.2 *Scale Economies and Priming the Adoption Pump*

The production of many types of innovations are subject to scale economies, whether in amortizing the total costs of production (such as upfront R&D) or in demand-side economies of scale through positive network effects (Katz and Shapiro 1985; Arthur 1996). The latter are common in IT industries where two or more technologies compete for adopters and the provision of specialized complementary products (Teece 1986; Shapiro and Varian 1999).

Thus, for an industry with scale economies and a need for complementary innovation, firms need to expand the total value created by the value network rather than just maximizing their share of the existing value. In reviewing open standards strategies of computer producers, West (2003) concluded:

> These various strategies reflect the essential tension of *de facto* standards creation: that between appropriability and adoption. To recoup the costs of developing a platform, its sponsor must be able to appropriate for itself some portion of the economic benefits of that platform. But to obtain any returns at all, the sponsor must get the platform adopted, which requires sharing the economic returns with buyers and other members of the value chain. In fact, openness is often used to win adoption in competition with sponsors of more proprietary standards. (West 2003: 1259)

Of particular concern is making sure that investments are made to create assets specialized to work with the specific technology, as identified by Teece

(1986). If an innovator seeks to increase its returns through a strategy of openness—i.e. reducing its appropriability and thus value capture—one goal is to increase the investment by others in complementary assets. Such invesments may come from adopters, such as those encouraged by the 'software freedom' assured by free software licenses (Stallman 1999). Or they may come from suppliers of complementary assets, such as add-on software or services, who are more likely to invest in a technology where they have greater access and control.

This is not to imply that value capture and value creation are mutually exclusive. Grove (1996) identifies how Intel's strategy of horizontal specialization in microprocessors—selling processors to all systems integrators—was more efficient than the vertically integrated mainframe vendors that sold processors only for their respective computers. This strategy provided supply and demand-side economies of scale, reducing costs and maximizing adoption; it also maximized the availability of complementary assets. Yet, as Kraemer and Dedrick (1998) note, Intel and fellow component vendor Microsoft captured most of the profits from the PC industry value chain during the period Grove was advocating this model.

In other cases, component suppliers or system integrators will need to fund their innovations by finding a way to appropriate the value perceived for complements. As an example, videogame console makers sell their consoles at or below cost and recoup those costs through royalties on the key complement, i.e. videogame software (Takahashi 2002).

Thus, for an industry with scale economies and a need for complementary innovation, firms need to expand the total value created by the value network rather than maximizing their share of the value captured, that is, to worry about 'growing the pie' rather than 'slicing the pie'. Such a strategy reduces the risks of a self-reinforcing downward spiral of declining market share and scale economies associated with losing a technology contest associated with network effects, such as the Betamax or Macintosh standards (Cusumano et al. 1992; Arthur 1996; West 2005).

6.3.3 *Interdependence of Business Models in the Value Network*

A firm's business model depends not only its IP and value proposition—as explicitly identified by Chesbrough (2003*a*: ch. 1; Chesbrough and Rosenbloom 2002) but also implicitly on the corresponding business models of the suppliers, customers, competitors, and complementors throughout its value network.

Chesbrough and Rosenbloom (2002) identify six functions of a business model: articulate a value proposition, identify a market segment and its revenue potential, define the structure of the value chain, estimate the cost and profit potential, describe the position within the value network, and formulate the innovator's competitive strategy. Early stage Silicon Valley companies often refer to a 'revenue model', which correspond roughly to the first two functions of the Chesbrough and Rosenbloom formulation, without meeting the more stringent business model requirements of profitability and sustainability.

Because IP provides barriers to imitation, strong appropriability can be essential for firms to identify and establish the value capture (but not value creation) part of their business model. New technologies will tend to require new business models, when the technology changes the value proposition to customers, the value capture by the innovator firm, or the relationship of firms within the value network. A firm's competitive advantage thus is determined in part by its structural position relative to external organizations that play a role in its innovation (Teece et al. 1997; see also Chapter 11).

In fact, few innovators can determine their business model in isolation. The business model depends not only on the value perceived by customers, but also suppliers, competitors, customers, and complementors. A firm's ability to command its desired price (and thus extract value) depends on intrasegment rivalry and its negotiating power relative to buyers and sellers (Porter 1980), as when Microsoft and Intel used their quasi-monopolies to capture the profits in the PC value chain.

Firms that have influence over their business models thus will be concerned about entering into an Open Innovation value network where their exchange partner has strong enough IP to assure appropriation of rents. But such power will be rare: few partners—whether component suppliers or systems integrators—have the alternative of walking away from an unfavorable deal without enabling a potential competitor. One example of how IP claims shifted demand was the failed attempt by European telephone companies to compel royalty-free licensing of GSM mobile phone patents (Bakkers, 2001: 322). Another example comes with the monopsony buying power of US cable TV companies[5] who used that monopsony power to force commoditization by their suppliers (e.g. through cable modem standards). Given this, operators have been reluctant to procure a key component (settop boxes) from Microsoft, for fear that it would use its copyright (and trade secrets) to create supplier power in the cable TV industry comparable to that it holds in personal computers.

Conversely, Teece (1986) is concerned with the case when innovators have weak initial appropriability. In such cases, he posits that firms have a temporary window to improve their appropriability, vertically integrate, or otherwise build barriers to imitators. Teece et al. (1997) later concluded that in the absence of formal appropriability barriers, firms are best able to create advantage through superior 'dynamic capabilities' such as rapid learning, but such advantages would appear to be more rare and less sustainable than those provided by formal appropriability.

Another key issue is the use of cross-subsidies in business models. Such business models are increasingly common in complex systems (e.g. see Chapter 5) and such models can both create vulnerabilities for business models of other firms in the value network and, in turn are vulnerable to competition from such firms. For example, a firm's business model may be vulnerable to shifts in the business model of complementors. Netscape used a revenue model of licensing its market-leading web browser application to large corporate users that was consistent with other PC software.[6] However, this revenue model was decimated by Microsoft's decision to give away a directly competing product (Internet Explorer) as a free complement bundled with its Windows operating system (Cusumano and Yoffie 1998; Bresnahan and Yin 2004).

6.3.4 *IP and Information Search*

One key issue in inbound and outbound licensing of innovations is the information exchange necessary to evaluate the innovation. Open Innovation requires significant disclosure to match buyers and sellers for transacting the exchange, as O'Connor (Chapter 4) discusses in the context of Dupont's patented materials innovations.

The two parties to the potential exchange have conflicting interests:

- The potential in-licensor wants information to evaluate, judge its value, and compare the cost of buy versus build.
- The potential out-licensor wants to provide enough information to conclude the transaction; at the same time, it must be concerned about providing enough information to customers (or rivals) to invent around and bypass the seller.

This is consistent with Arrow's (1962) 'information paradox', a limiting case that—absent property rights—a seller disclosing information for evaluation by potential buyers allows the buyer to acquire that information at no cost.

IP potentially solves this problem, because it can protect a firm's ideas while they are disseminated in search of a market: IP is thus valuable in both shopping innovations, and also allowing them to be licensed. In particular, a fundamental trade-off in patent policy is that patenting requires disclosure of an innovation to enable subsequent cumulative innovation (Gallini 2002).

However, not all IP mechanisms are created equal. A patent provides the best protection for this sort of information disclosure; even so, such protection is incomplete in some industries. A copyright provides protection of the expression of an idea, but does not protect against independent invention that duplicates the functionality of that idea. Information disclosure is contrary to the basic principles of trade secret law, and thus provides very limited protection for innovators seeking to license their innovations.

Having rights to IP is not the same as asserting them. Some innovators may be more interested in winning adoption than minimize potential spillovers. This is certainly more likely for organizations that are 'innovation benefactors' (see Chapter 14). However, as Fabrizio (Chapter 7) notes, the desire to profit from basic science has caused universities to act less as benefactors (donating innovation) and more as information explorers (selling innovation).

Information search is easiest when there is no IP on the innovation: the US Federal government does not hold copyright while other governments (like the UK with Crown Copyright) do. But giving away innovation is not a business model. With adequate resources, an innovation benefactor can persuade its stakeholders that the social benefits of giving away innovations (such as unpatented public research) exceed the cost of doing so: a common argument is that the innovation creates spillovers that increase employment and economic development. Without such a rationale, innovation benefactors (public or private) become innovation investors, forced to justify a direct return on innovation spending.

6.3.5 *Limitations to IP-based Business Models*

In addition to the challenges of market competition, the IP-based business models of firms are vulnerable to potential conflicts with other public policy goals.

For example, if IP is strong enough to have anticompetitive effects regulators may weaken or waive IP protection to implement competition policy.[7] These may be part of a general pattern of reforms to balance innovation and competition policy goals (e.g. Farrell and Shapiro 2004). General exceptions include allowing reverse engineering exception to software copyright (West 1995) and policies speeding generic copies of patented pharmaceuticals (CBO 1998). IP may be weakened to address specific monopoly concerns, as with the 2004 European Commission decision to compel Microsoft to disclose server interfaces to competitors (Meller 2004).

The IP goals of firms may also come into conflict with a country's industrial policy goals. For example, developed country innovators have found their patents voided or subject to compulsory licensing in developing countries (such as Brazil, China, and India) seeking to use local imitation as a way to bootstrap the innovation capabilities of domestic producers.

Other regulatory goals may also override IP enforceability. For example, trade secret law in Silicon Valley provides only limited protection against the interfirm mobility of knowledge due to state labor laws that restrict noncompete covenants. Gilson (1999) argue that these key differences between Silicon Valley and Route 128 can be traced to historical differences in the respective state civil codes.

In addition to regulatory conflict, a second limitation to *de jure* IP protection is that it may not provide *de facto* protection. For example, strong copyrights have not protected against unauthorized copying of information goods such as music or software, particularly in developing countries (Burke 1996). Small firms holding patents infringed by larger firms may not be able to enforce them unless they can garner sufficient resources to credibly threaten litigation;

a rare example of this is Stac Electronics' 1994 landmark $120 million award against Microsoft (see Chapter 9).

If formal IP does not provide appropriability, then (as Teece 1986 predicted) firms may take other steps such as vertical integration to earn returns from their innovation, as when a firm incorporates its technology in a ready-to-use component. But even such component strategies have appropriability limits, particularly with information goods. Modular or component innovators run the risk of not getting paid by integrators, who are tempted to maximize their attractiveness to customers while minimizing their cost of inputs. This is particularly a problem with information goods, such as software utilities that are bundled with other hardware or software. Software vendors might seek to use technical means (tying their product only to use with the bundle), copy protection, or even a physical artifact such as the Windows 'Certificate of Authenticity'. The innovator can also modify its business model to provide components of limited utility and see profits primarily by selling an enhanced version of the innovation: the software industry refers to this as a 'teaseware' or 'crippleware' strategy.

Finally, a firm seeking to build a business model based on licensing IP-protected technologies and components may prove too successful, if its exit from vertical integration results in 'hollowing out'. In the 1960s and 1970s, RCA turned its consumer electronics emphasis from product innovation to licensing its patent portfolio to Japanese competitors—partly because such licensing produced high growth rates and profit margins, and partly because regulators compelled it to license IP free to domestic rivals (Chandler 2001). When the company attempted to extend this model to the next-generation high-definition television, the once dominant RCA was unable to produce technology competitive with rival HDTV systems (Brinkley 1997).

6.4 Does IP Conflict with Innovation Openness?

The Open Innovation paradigm assumes firms can extract income (whether through licensing or other forms) from their innovation, which provides both the revenues and incentives to produce that innovation. However, such payments are contrary to the expectations of what many consider to be an 'open' form of innovation, in which a shared (if not communal) external innovation is available without significant direct cost.[8] Conflicts between these two viewpoints have resulted in some of the most controversial IP issues of recent years related to innovation practice and policy.

The following chapters in Part II consider three key areas of conflict. In Chapter 7, Kira Fabrizio looks at the impact that universities have patenting their innovation has upon the cumulative production model embodied by 'open science'. In Chapter 8, Tim Simcoe considers the increasing conflict between patenting and 'open standards'. And in Chapter 9, Stu Graham and David Mowery look at the impact of patents on the software industry at large, including open source software.

In these three areas, there are two issues. One is the effect asserting IP has upon the (potentially zero sum) allocation of returns within the value network. The second is the net effect of this income transfer—whether the incentives to innovate by technology producers outweigh the innovative drag for technology consumers, either through increased search costs, transaction costs, or duplicative investment to 'invent around' innovation IP. Even under the same IP regime, there are more and less efficient solutions—as when Gallini (2002: 137) observes that increasing the appropriability of patents increases the transaction costs for producing cumulative innovation if firms are forced to separately negotiate licenses with each owner of potentially blocking patent.

6.4.1 *Evolving University Roles in Open Science*

One major source for firms seeking external innovations has been university research that is widely disseminated for firms to use as a building block in their innovation efforts. The exemplar of this policy was US federal funding of university research in the post-World War II era and the role of this research in enabling industrial innovation (e.g. Henderson et al. 1998*a*; Cohen et al. 2002; Colyvas et al. 2002).[9] However, many have lamented a decline in free innovation spillovers in recent years, tied to declining government support and an increasing emphasis by universities on licensing their innovations.

Explaining the declining importance of US government funded research is complex and controversial. During the heyday years of the 1960s, much of the federally funded research was tied to building complex systems for space exploration and military forces, and the relative importance of such systems declined during the 1970s–1990s (Jaffe 1996). Of course, the government decisions for funding R&D depend not only on those R&D funding requests, but also the availability of resources and other demands on those resources.[10] While government R&D and procurement helped fund the development of the US IT sector from the 1960s through the 1980s, the relative importance of federal R&D funding began a steep decline starting in 1988 (Fabrizio and Mowery, forthcoming B).

The issues extend beyond the US context. Even as US funding was declining, Pavitt (2000) called on the European Union to increase funding of university R&D to match the US success at creating new innovations, which he tied to spillovers from its university research.

Taxpayer money is not the only way to pay for university research. Particularly in areas such as biotechnology, university research is increasingly funded by the private sector. Of course universities and university faculty (at least in some countries) have always performed contract research for industrial firms; but overall the importance of such industry funding has been increasing over time (Jaffe 1996). In some cases, corporate funding offers publication and open disclosure of results comparable to government funding, as with Intel's research labs located next to leading US and UK universities (Chesbrough 2003*a*: 123; Tennenhouse 2003). But in other cases, the funding comes with restrictions or expectations of exclusivity.

This relates to a second trend, which is changing university attitudes towards their innovation IP. At one point, universities did not assert IP rights covering their research results, allowing that research to spillover to firms as the basis of subsequent industry innovations. For example, during the 1980s government-funded and industry-funded computer science research (at UC Berkeley and MIT, respectively) provided crucial technologies for Unix systems vendors of the 1990s (West and Dedrick 2005).

Today, universities are increasingly asserting IP rights (particularly patents) over their innovations and licensing them under a royalty-bearing license. Many have attributed the rise of patenting by US universities to the 1980 Bayh-Dole act, whose objectives and policies are summarized in Chapter 7. However, Mowery and Sampat (2001, 2004) argue that increased university rate of patenting was discernable during the 1970s, and thus the act reflected rather than initiated the trend towards increased university patenting.

In addition to demands made by corporate sponsors, universities have also shifted away from a model of free spillovers to that of technology transfer offices, in hopes that licensing revenues would replace declining revenue from other sources of income. In fact, the policy of universities with these offices (captured in research such as Siegel et al. 2003 and Bercovitz and Feldman 2003) is that any innovation not controlled by the technology transfer office constitutes a failure of the system or the individual researchers to capture the value of the innovation. This reflects a (largely unproven) assumption by these offices that university innovations will invariably lead to firm success and economic growth (cf. Miner et al. 2001).

Universities licensing technology to firms might raise the price of external innovations used by firms, but such a practice is entirely consistent with practices of Open Innovation: in the terminology of Chesbrough (2003*b*), universities would shift from innovation benefactors to innovation explorers. Exclusive licensing might reduce the number of firms that can benefit from a given innovation, but at the same time exclusivity provides greater incentives for licensees to invest in commercialization—a key justification cited in passing Bayh-Dole (see Chapter 7).

However, researchers have postulated another potential disadvantage of increased patenting. Incentives for academic research have encouraged the free flow of information through career incentives for publication. Assertion of IP rights on basic research output could restrict the flows of information between basic researchers, thus slowing or impeding the process of cumulative innovation that characterize 'open science' (David 2002; 2005).[11]

Is there any evidence of such deleterious effects? This is exactly the question Fabrizio (Chapter 7) attempts to answer. From prior research, she identifies two potential negative effects of greater university patenting.

First, restrictions on access to university IP slow attempts by other researchers to build upon university research. It is straightforward to predict the effect that exclusively licensing IP to a single firm has on other firms in the same industry. But in other cases, she notes that the transaction costs of dealing with technology transfer office inefficiencies means that even IP that is available for licensing becomes less accessible. Second, researchers involved in commercialization of university research are more secretive in sharing their results to protect the proprietary value of such data. As Fabrizio notes, this undercuts the fundamental basis of open science collaborative innovation, in which the output of one researcher becomes the input to another.

From her own analysis of citation patterns of firms' patents, she identifies two effects consistent with prior expectations. First, as public science (unpatented prior art) became more important in a technology class, firms separated into 'have' and 'have nots' in their access to public science, suggesting that some firms are doing a better job than others in their capability to access university research. Second, as university patenting increased, so did the lag of citations to cited prior research, suggesting a slowing down of firm exploitation of existing knowledge.

Finally, analyzing patents by biotechnology and pharmaceutical firms, she identifies two factors that explain the ability of firms to access public science. First, consistent with Cohen and Levinthal's (1990) concept of absorptive capacity, access increases with increased R&D. More interestingly, access also increases as the firm's scientists publish research coauthored with university scientists—but only up to a point. This implies that firms that always depend on university researchers do not fully develop their own internal scientific capabilities, while firms that rarely (or never) coauthor have access to less cutting edge basic science. This also implies a shift by industrial researchers away from relying solely on open publication towards using university colleagues to help identify and interpret the relevant output of open science—at least for these two industries.

6.4.2 *Open Standards*

Another venue where IP potentially impairs shared innovation is in the area of product compatibility standards, particularly in the IT sector. Of course, IP has long been a central part of the proprietary *de facto* standards strategies of firms from IBM to Microsoft. As with other firms, the exclusive rights provided by IP allows firms to gain economies of scale and earn a return on their R&D investments (Morris and Ferguson 1993; Shapiro and Varian 1999).

Open standards implemented by multiple vendors—whether created through formal standards development organizations (SDOs) or ad hoc research consortia—have similarities and differences to single-firm proprietary standards. By their nature, compatibility standards enable a modular subdivision of labor and thus a decentralized production of innovation (Garud et al. 2003; Langlois 2003*b*). Both types provide similar incentives to third parties producing complements, as well as for end users utilizing implementations of the standards (West 2006).

But in other key areas—the organization of innovation, ownership of intellectual property, and even cultural norms—'open' and 'proprietary' standards have been as distinct as 'open science' and its commercial counterpart. The open standards have historically focused less on individual firm competitive advantage and more on defining rules for interoperability for a common infrastructure. Adherents of this form of standardization have emphasized openness and transparency in the standardization process and outcomes (Krechmer 2006).

However, firms have increasingly sought commercial gain within open standards, not merely through superior implementations in commercial products, but also by negotiating to have their royalty-bearing IP incorporated into the required terms of the standard. These and other business models have blurred whatever bright line might have existed between open and proprietary standards (West 2006).

Such tactics are problematic for Open Innovation. The increasing use of licensing-based business models by specialist firms has fueled a three-way conflict between technology producers, system integrators, and the eventual technology users. On the one hand, a key example of Open Innovation identified by Chesbrough has been creating business models to gain returns on innovation through outbound licensing, and so standards committees (such as W3C) that mandate royalty-free IP licensing would help users and integrators (or vertically integrated producers) while potentially eliminating the Open Innovation business models of IP-only specialists. Even without such royalty-free mandates, the feasibility of such pure IP-based models can be limited by other SDO policies or weak appropriability (Tansey et al. 2005).

On the other hand, asserting IP on industry-wide standards has potentially anticompetitive impacts, such as when vertically integrated firms to increase the costs and reduce competition from potential rivals (cf. Bekkers et al. 2002). Even without such market power, the holdup of IP owners can disrupt standardization activities (see Chapter 8). And any royalty-bearing IP—whether from actual innovators or rent-seekers such as so-called 'patent trolls'—increases the cost of implementing a standard and thus the net cost to consumers.

What is the net effect of this increasing role of IP in standardization? In his various empirical studies on IP and formal standardization, Simcoe concludes that Open Innovation has delayed standardization, increased implementation, and coordination costs. For example, Simcoe (2006) shows that increasing assertion of patents in Internet standards has delayed standardization during the period 1993–2003.

Simcoe (Chapter 8) also considers the potential of an IP owner with blocking patents to 'hold up' a standardization effort by preventing implementation of a standard without payment of a licensing fee of the innovator's choosing. Such efforts are always controversial—due to vocal objections from licensees used to cross-licensing or otherwise avoiding patent liabilities. In some cases, such licensing is an essential way for a nonintegrated innovator to get compensated for its innovation, as Mock (2005) asserts is the case for Qualcomm's mobile telephone patents.

At the same time, when presented with the actual costs of innovation licensing, a standardization effort will often modify the standard to avoid infringing on a patent and thus the associated patent royalties. This is not possible if a firm tries to exploit the standardization process by not disclosing the existence or cost of IP, or if they modify patent claims after seeing the eventual standard. In Chapter 8, Simcoe interprets such tactics as failures of IPR policies, either by standardization groups or national patent examiners.

6.4.3 *Software and Open Source*

Patents are also playing an increasing role in the software industry, and open source software in particular. Like open standards, open source software reflects collaborative production between multiple organizations (and individuals). But unlike open standards, in open source the collaboration results in a shared implementation of a technology, rather than merely its technology.

West and Gallagher (Chapter 5) classify the different business models of firms used by firms sponsoring or leveraging open source development projects. Some firms use open source as a form of Open Innovation, while others use it as a way to win adoption of their technology and attract complementary products

But fundamentally open source software—and the related 'free software' movement—are about IP. The IP requirements of the latter are a superset of the former, such that anything classified as 'free' is also 'open source', but not vice versa. Both agree that source code should be publicly disclosed and that all recipients have a right to enhance and improve that code. However, there are important philosophical and cultural differences between the two groups that are embodied in their respective IP licenses (West and Dedrick 2001; 2005).

Open source licenses such as the Apache and BSD licenses impose few restrictions, and thus software licensed under these terms are attractive to firms to use as components in their own systems. 'Free software' licenses are more restrictive: licenses such as the General Public License require modifications to GPL-licensed technology to be publicly disclosed, to prevent firms from creating proprietary derivative works that eventually supplant the free alternative (West 2003). In fact, the restrictions of the GPL today are now used by firms to release innovations while making them less attractive for use by direct competitors (Välimäki 2003).

These two licenses thus represent two different approaches to shared innovation. The BSD-type licenses represent free spillovers that can easily serve as external innovations for firms in their own products; such commercial products may compete with the open source benchmark and could conceivably supplant it. On the other hand, the restrictions of the GPL assure that the shared innovation remains shared, while limiting the incentives for further commercial investment to develop and enhance the technology.

A second, emerging IP issue in open source software is that of software patents (cf. Nichols 1998). While patents have already had a demonstrable impact on open standards (Bekkers et al. 2002; Simcoe 2006), their impact on open source is far from resolved.

Many of the impacts of patents on software are not yet understood because patenting software dates only to 1981, and largely limited to the US. In analyzing US software patents during the period 1987–2003, Graham and Mowery (Chapter 9) subdivided the largest software innovators into two different business models: 100 software specialists whose primary business was selling software, and 12 manufacturers of electronics goods for whom software is but one component in an overall system. While the patent propensity of the former group has increased, they conclude that the latter group (dominated by IBM) has increased both its software patenting and overall share of those patents. In particular, in comparing the largest firm from each category, Graham and Mowery (2003; (see also Chapter 9) found that IBM not only has a higher patent propensity than Microsoft—as scaled by software R&D—but has been widening the gap during the past decade.

However, Graham and Mowery conclude that (despite specific examples such as the Stac case), we don't yet know what role software patents play in the overall IP strategies of the sampled firms. For example, are these patents intended for suing rivals, defending against lawsuits by rivals, or cross-licensing with various competitors and complementors? Each suggests a different business model based on the software innovation, as well as differing implication for other (typically smaller) organizations outside this patenting population. In particular, the offensive (suing rival) alternatives has potentially severe implications for open source projects that lack a direct firm sponsor or revenue stream. Some projects have attempted to preempt potential problems by adopting licenses that threaten retaliation against threats of patent litigation, but such licenses have yet to be tested in court. Also, as Graham and Mowery (Chapter 9) note, there is little such license can do to address patents held by other firms not a party to creating or using the software.

6.5 Case Study: Mobile Telephone Standards

An example of how shifts in Open Innovation both affects and is affected by IP policies can be seen in the increasing role of patents in mobile telephone standards across successive generations of mobile telephone standards.

Innovations in mobile telecommunications are constrained by the need for compatibility standards to provide interoperability. To be put into use, a technological innovation (such as digital encoding of radio signals) is incorporated into the formal specification of a standard; these specifications are then implemented in products (e.g. a mobile telephone or radio base station), which are then purchased and used (West 2002, 2006). By the same token, the industry has seen an increasing emphasis of royalty-bearing patents in *de jure* telecommunications standards, reflecting both shifts in industrial organization and the associated changes in business models.

6.5.1 *First Generation Closed Innovation*

In most countries through the first eight decades of the 20th century, a government-operated monopoly telephone company allocated equipment orders to one (or a few) domestic manufacturers; the one major exception was the US, where AT&T vertically integrated research, development, manufacturing, and telecommunications services. In a few cases, companies with small home markets (such as Sweden's Ericsson and Canada's Northern Telecom) exported their existing designs to other countries (Noam 1992; West 2000).

The rate of technological innovation in the wireline industry was slow, with capital investments in network equipment amortized over decades. With the lack of competition and monopsony buying conditions, most firms lacked incentives for patenting their innovations. And whether AT&T's vertical integration or the collaboration between government departments and their captive suppliers, few were examples of what we today call Open Innovation.

Limited scale mobile telephone systems had been deployed in major cities in the US and Europe during the three decades following World War II, but their capacity was limited to hundreds of users per city. In the late 1970s and early 1980s, microprocessors enabled subdividing a metropolitan area into cells, increasing the capacity of systems a thousandfold. Of the most widely adopted 1st generation analog cellular systems, those in Northern Europe and Japan were designed by operators and built by manufacturers. Vertically integrated AT&T designed and built its own system, while Motorola and other radio manufacturers built systems for competing operators. Some of these systems were exported to other countries, such as the UK, Middle East, and Latin America (West 2000; West and Fomin 2001).

6.5.2 *2nd Generation: Open Innovation*

To address unexpectedly large demand, during the 1980s cellular phone operators and manufacturers began investigating digital technologies to provide higher capacity and better security, among other features. These reflected a range of Open Innovation strategies: sourcing external innovation, shared innovation, licensing internal innovations, and a hybrid of vertical integration and licensing.

In Japan, the government-owned NTT DoCoMo designed its second generation PDC standard, but outsourced handset design and production to its four major suppliers. DoCoMo used buyer power and control of tacit information to both maintain control over these suppliers and gain competitive advantage over competing cellular operators (Funk 2003).

Two more Open Innovation models were used. For the European GSM and US D-AMPS (aka TDMA) standards, the technology was developed by multiple equipment manufacturers and operators through an industry standardization committee. The other major US standard, eventually branded cdmaOne, was largely developed by one firm, Qualcomm, that had its standard ratified by US and foreign standards committees; it earned royalties of approximately 4.5 percent on wholesale price of equipment using its patents, which amounted to nearly all CDMA equipment (West 2002).[12]

By far the most successful of the second generation standards was GSM (Table 6.1). It was the first to be deployed, and for most European countries marked the first significant deployment of cellular technologies (West and Fomin 2001). Much of the technology was designed by Ericsson and Nokia, who had the most home market experience of European manufacturers. However, to win approval in the by-country voting of the GSM committee, the design was modified at the last minute to incorporate technologies from French and German manufacturers (Bekkers 2001).

As ably documented by Iversen (1999) and Bekkers (2001; Bekkers et al. 2002), a second key goal was to reduce the threat of foreign (primarily Japanese) competition in the European market, and provide domestic manufacturers an advantage when exporting the technology worldwide. One key

Table 6.1. Market share of digital cellular technologies, June 2000

Standard	Origin	Subscribers (million)	Ratio (%)
GSM	Europe	331.5	57.4
CDMA	USA	67.1	11.6
PDC	Japan	48.2	8.3
D-AMPS	USA	47.8	8.3
Nondigital	various	82.8	14.3
Total		577.4	100.0

Source: Adapted from West (2002).

mechanism was to mandate the use of the GSM standard across the European Union (EU)—unlike Japan, US and China (among other countries), which allowed use of foreign technologies. The other was through GSM patent cross-licensing, which allow suppliers of key innovations incorporate in the standard to realize royalties on the equipment sales by competitors, both handsets (sold to consumers) and network infrastructure (sold to mobile phone operators). As the least integrated of the GSM IP licensees—as well as least established of selling products in Europe—Motorola in particular pursued an Open Innovation strategy that emphasized IP licensing over product sales (Bekkers 2001: 323).

The GSM standard is often held out as an exemplar of open standardization, particularly in competition with CDMA (West 2006). Both the GSM and CDMA standards reflected a form of Open Innovation in which innovators received licensing income from the standards they created. The Qualcomm case differed from GSM in two major ways. First, for CDMA only one firm (Qualcomm) paid the lowest royalty rate, versus at least five for GSM (Nokia, Ericsson, Motorola, Alcatel, Siemens). Second, for CDMA, all major makers had to pay patent royalties (and were rather vocal in their complaints) to Qualcomm at an undisclosed rate estimated at 4.5 percent of gross sales (West 2002).[13] By comparison, for GSM the major European makers (plus Motorola) were believed largely exempt from patent royalties through cross-licensing among fifteen key companies, while outsiders paid total royalties estimated at 10–13 percent (Loomis 2005).

As of this writing, what was the effect of these Open Innovation strategies?

- Nokia (and, to a lesser degree, Motorola) remained active vertically integrated manufacturers, developing both new technologies and continuing to be major global suppliers of cellular handsets.

Table 6.2. Performance of handset operations by key GSM patent holders

Firm	Country	Market Share† 1998	Market Share† 2004	Fate of Handset Division
Nokia	Finland	22.5%	30%	Still owned by original parent
Motorola	USA	19.5%	15%	Still owned by original parent
Ericsson	Sweden	15.1%	6%	In 2001, formed Sony Ericsson, a 50/50 joint venture with Sony of Japan, which pays patent royalties to Ericsson.
Siemens	Germany	n.r.	7%	In 2005, sold division to BenQ of Taiwan
Alcatel	France	4.3%	n.r.	In 2004, formed TCL Communication, a 45/55 joint venture with TCL Corp. (China); sold joint venture to TCL in 2005

† Global share for all standards; 1998 as reported by West and Fomin (2001), 2004 as reported by Testa (2005).

- Qualcomm exited equipment manufacturing to concentrate on a successful strategy licensing CDMA IP and selling chips to implement that IP.
- In the face of stiff price competition, Ericsson, Siemens, and Alcatel all exited the handset business to concentrate on selling network equipment.

For the equipment makers exiting the money-losing handset business, their cost advantage in patent royalties played a key role in selling their divisions to Asian competitors (Bekkers et al. 2002; Loomis 2005).

6.5.3 *3rd Generation: Learning the Wrong Lesson?*

Not surprisingly, the success of GSM IPR licensing created an increased interest by telecommunications firms in generating and patenting licensable innovations (Bekkers et al. 2002). More than fifty companies sought to get their patents established as 'essential' for implementing of WCDMA, the 3rd generation mobile phone standard created through the cooperation of the leading GSM vendors along with NTT DoCoMo. When the patents of the CDMA inventor (Qualcomm) were factored in, the high patent royalties put WCDMA at a cost disadvantage, with royalties estimated as being twice that of the leading competitor.

In response, the leading handset maker, Nokia, sought to cap total WCDMA patent royalties at 5 percent. But in the end, Nokia won only support for 'reasonable' licenses from DoCoMo and three European manufacturers. The remaining European and Asian manufacturers—as well as leading operators—formed the competing 3G Patent Platform Partnership (3G3P). North American participants in WCDMA standardization (Qualcomm, Lucent, Motorola, Nortel, TI) joined neither camp (Tulloch 2002; Lane 2003; Salz 2004). As of mid-2005, there is no reported solution to the problem, and additional patent claimants continue to be identified. Thus, the patent strategy used by the leading GSM manufacturers to profit licensing internal innovations in the 2G era is hindering their abilities to sell their main products in the 3G era.

Many of the same patents also apply to WCDMA's leading rivals, the Qualcomm-sponsored cdma2000 and China's competing TD-SCDMA, mitigating some of the competitive effects but overall likely to slow adoption of any 3G standard. As one component supplier said, 'The jury is out on whether 3G will be so compelling that consumers will pay the price for 3G handsets—and IPR is part of that equation' (Salz 2004).

The handset manufacturers face the same adoption versus appropriation trade-offs as West (2003) identified for computer systems. In this case, there are serious problems of collective action accommodating the heterogeneous royalty (i.e. business model) preferences of more than fifty actors due to a varied mix of equipment and IP revenues. This suggests that combining two Open Innovation strategies—shared innovation and licensing internal innovations—can dramatically raise coordination costs or, at worst, create an anticommons that fails due to the misalignment of individual and group incentives. Certainly in the trade-off space identified by Simcoe (Chapter 8), the WCDMA standardization effort has biased toward value capture over value creation.

6.6 Conclusions

Appropriability ties back to the fundamental question of who pays for innovation. Innovations can be directly subsidized by innovation benefactors, or cross-subsidized through vertical integration. Open innovation assumes the cooperation of two or more organizations—at least one generating an innovation and at least one utilizing it—with a viable business model for each.

Usually considered in the context of public policy, appropriability is what allows the innovator to capture a return from the value created by an innovation. For some classes of innovations, IP law plays a key role in providing appropriability, and thus allowing some open innovators to get returns on their internal innovations and others to have a supply of external innovation.

Open Innovation can thus be affected by changes to *de jure* IP protection, whether enacted directly by legislative statute, administrative policy, or judicial precedent. An example of the latter is patenting of software algorithms, as enabled by the US Supreme Court in the 1981 case *Diamond v. Diehr* (see Chapter 9).

But other institutional policies can also affect IP, appropriability and thus Open Innovation. For example, an innovation benefactor can change the rights allocated for contract research, as when the US government granted universities rights to contract research under the Bayh-dole Act (see Chapter 7). A cooperative technical organization may specify certain rules for how IP will be appropriated for the organization's joint product, as Simcoe (Chapter 8) considers for standards setting organizations. Finally, within a given appropriability regime, individual firms have broad discretion as to how much they choose to appropriate—tied to their overall value creation strategy—as illustrated by the preceding mobile phone case.

6.6.1 *Allocating the Returns of Innovation*

While discussions of appropriability focus on value capture, equally important in Open Innovation is value creation. For complex ecosystems such as those illustrated by Figure 6.1, this can require complex market (and nonmarket) coordination among multiple firms in the value network. A crucial part of the Open Innovation strategies of technology-component suppliers (such as Intel and Qualcomm) is proactively building ecosystems to attract systems integrators and complementors.

The appropriation decisions of the focal innovator can affect other firms in the ecosystem in two ways. First, as all four chapters of Part II note, the friction from the innovator appropriating the value of its innovation can hinder the process of Open Innovation if it discourages information search or cumulative innovation. Second, if suppliers, component producers, or complementors lack their own ability to capture value, then the value network may not create enough value to win customer adoption; such systemic innovation issues are the focus of Part III, particularly Chapter 12.

Thus, any Open Innovation business model must consider the relationship of value creation and value capture for all the participants in the value network (Chesbrough 2003*a*). This imperative is particularly important for technologies subject to network effects, where firms must trade off value appropriation against the demand-side economies of scale provided by widespread adoption (West 2003).

Nearly a decade ago, Brandenburger and Nalebuff (1996) suggested a game theoretic framework for trade-offs within what they call a 'value net'. However, research on the complex process of managing such value networks has either been highly simplified (as with Brandenburger and Nalebuff 1996) or highly particularistic (e.g. Kraemer and Dedrick 1998). More recent research—such as Staudenmeyer et al. (2000), Iansiti and Levien (2004*a*) and O'Mahony and West (2005)—has attempted to compare and generalize the processes of ecosystem management. But these have yet to provide a broader framework (comparable to Teece 1986) that explains the relationship of formal appropriability and voluntary appropriability waivers in the value creation and capture within an arbitrary value network.

6.6.2 *Unresolved Questions of Appropriability and Open Innovation*

At first glance, stronger IP regimes are directly associated with more Open Innovation. Gallini (2002: 141) summarizes the predicted relationship between appropriability and innovation: first, strong patents establish as willingness to out-license; second, that strong patents promote vertical specialization.

Consistent with this, based on a large-scale survey of UK industries, Laursen and Salter (2005) conclude that Open Innovation attitudes are strongest in industries with high appropriability (such as pharmaceuticals) and weakest in industries with low appropriability (such as textiles). One might be tempted to infer that there is a direct correlation (if not causal relationship) between high appropriability and high openness.[14]

Openness \ *Appropriablility*	*High*	*Low*
High	pharmaceuticals	open source
Low	†	textiles

† Does this case exist? Is is always suboptimal?

Figure 6.2. Does appropriability determine openness?

However, the case of open source software (see Chapter 5) raises questions about this relationship. Open source limits how much firms can appropriate and effectively forces openness (West and Dedrick 2005), and yet as West and Gallagher note, firms invest in open source-based Open Innovation strategies nonetheless. Does this undercut the correlation observed by Laursen and Salter, or are there problems with the generalizability (or even sustainability) of the open source business models? And if the combination of high openness, low appropriability is observable in practice, are there examples of the converse (high appropriability, low openness)? Or would we expect that in cases of high appropriability, the highly open strategy would always produce greater returns? (Figure 6.2).

The Laursen and Salter (2005) paper also raises a second issue, about the different forms of appropriability, both through government-granted IP and other means. They consider a combination of measures taken by firms to appropriate returns for their innovation, including strong formal appropriability (such as patents, trademarks, registered designs), weak formal appropriability (secrecy), and other means of securing competitive advantage (through design complexity and first mover advantage). They concluded that the six items (from an existing EU-designed survey) represented a single construct, but further research is needed to reconcile this with previous observations regarding the complementary relationship between formal and informal appropriability means.

This relates to a broader gap in the innovation (and Open Innovation) literature, which is the focus on patents as a means of appropriating value. Other researchers have observed that there are contexts where copyright and trade secrets are the most effective form of IP protection but the implications of these forms of IP upon Open Innovation have yet to be considered. If, as commonly assumed, such IP mechanisms provide less appropriability than do patents, then studying the use of these mechanisms in Open Innovation could also provide broader insights about how the nature and strength of appropriability mechanisms relate to effective Open Innovation strategies.

Notes

My thanks go to Rudi Bekkers, Henry Chesbrough, Kira Fabrizio, Tim Simcoe for engaging the work and providing many useful suggestions. The opinions (as well as all remaining errors) are mine alone.

1. There is a trend in US copyright law towards quasi-permanent monopolies for key entertainment content, epitomized by the 1996 copyright term extension act dubbed the 'Mickey Mouse Copyright Law' (Slaton 1999). The effect of such term extensions has on incentives has not been established, but one analysis concluded that 'in the case of term extension for existing works, the sizable increase in cost is not balanced to any significant degree by an improvement in incentives for creating new works' (Akerlof et al. 2002: 3).
2. In addition to patent, copyright, and trade secret, Besen and Raskind (1991) list three additional mechanisms in US IP law: trademark law, the Semiconductor Chip Act of 1984 (a specialized form of copyright), and misappropriation (a rarely used common law doctrine regarding unfair competition). For this discussion of Open Innovation, I concentrate on the three IP mechanisms most often used to protect innovations.
3. External innovations can be incorporated not just at the beginning of the development funnel, but at every stage from invention to final sale (see Chapter 1). The IP issues faced by innovators are similar for all these stages, but for simplicity's sake this discussion focuses on innovations at the beginning of the funnel.
4. For Teece, almost any remaining portion of the value equation is a complementary asset. For the large body of standards research building upon the Katz and Shapiro 'hardware-software paradigm', a complementary product has a specific meaning of a separate product that adds value to the base innovation (e.g. Bresnahan and Greenstein 1999; Shapiro and Varian 1999). Unless specifically noted, I use 'complement' in the latter sense here.
5. Multiple companies exist in the US cable TV industry, and thus this concentration corresponds more to an oligopsony than monopsony situation. However, each firm effectively has enjoyed a monopsony in its respective geographic territory; intermodal competition is reducing but not eliminating this oligopsony power.
6. As with other Internet start-ups of the era, Netscape's business model relied on unproven assumptions of customer value, profitability, and sustainability. Netscape was forced to exit from web browsers (see Chapter 5) before those assumptions could be (dis)proven in the marketplace.
7. In the US, competition policy is called 'antitrust' policy for historical reasons dating back to the targets of the first major competition law, the 1890 Sherman Antitrust Act.
8. Even if they do not directly pay for innovations, firms may pay other costs to use external innovations—including the costs of developing absorptive capacity, search costs, technology transfer, and investments in technologies that do not yield commercial returns. For example, even when a university or government lab has a strong bureaucratic mandate to get innovations 'out the door', this is not sufficient to establish that the technology is actually being used, let alone has a significant market impact (Bozeman 2000).
9. In a dissenting view, Goolsbee (1998) argued that federal funding increased the wages of R&D workers and not the amount of research being done.

10. As with most other requests for government spending, both industry and academic pleas for additional R&D expenditures are usually made in isolation, without identifying additional sources of revenue or opportunities to reduce expenditures in other areas.
11. In addition to surrendering control of university IP to private firms and impairing the process of cumulative innovation, researchers looking at increased patenting also have at least implicit concern that increased private funding will be used as an excuse for reducing less restricted government research expenditures.
12. Started without venture capital, Qualcomm used an innovative business model to fund creation of its patent portfolio. It presold licenses to its research, which was valuable to telecommunications carriers because of the (eventually realized) promise of higher capacity utilization of scarce regulated radio spectrum (Jacobs 2005).
13. To gain government approval for CDMA usage in China, Qualcomm cut the royalty rate dramatically for Chinese manufacturers selling to the domestic market (West and Tan 2002).
14. I am grateful to Henry Chesbrough for originally sharing this interpretation of the Laursen and Salter paper.

7

The Use of University Research in Firm Innovation

Kira R. Fabrizio

7.1 Introduction

The Open Innovation paradigm focuses attention on the importance of firms' identification and use of ideas and knowledge from outside the boundaries of the firm. As documented by Chesbrough (2003*a*), firms in many industries have recognized the value of looking outside of their borders for ideas, knowledge, and sources of innovation. This value depends on the existence and depth of the knowledge landscape in which the firms operate. The characteristics of the knowledge landscape are determined by the knowledge flowing out of other firms and organizations and the intellectual property (IP) environment. The open innovation literature has focused primarily on the knowledge and ideas flowing from one firm to another. In this chapter I focus on a second important source of knowledge and ideas useful to the open innovation processes of firms: universities.

Many industries owe their technological foundation to federally-funded research performed in university laboratories (Fabrizio and Mowery forthcoming A). The body of science represented by university-based research is an important and growing contributor to industrial innovation. University research is not automatically transferred to industry researchers. The transfer of this knowledge from universities to firms is affected by the appropriability regime, the nature of the knowledge, and the competencies developed by firms to identify and exploit this external knowledge (Teece 1986; Chesbrough 2003*a*).

Recent federal policy changes have altered the interface between US universities and companies that make use of the research results generated at these universities. Specifically, policies embodied in the Bayh-Dole Act of 1980 allowed and encouraged universities to pursue formal IP right (i.e. patent) protection to research results developed using federal funds, which at the time accounted for 70 percent of US university research funding. As a result of this and related policies, patenting of US university research results has exploded during the intervening decades.

During the same period, firms in many industries increased their reliance on research and technology developed outside of the firm (Chesbrough 2003*a*). Increased licensing activity, collaborative alliances, and outsourcing of research activities have highlighted the importance of effective knowledge transfer across the boundary of the firm. In this sense, study of the university–firm interface is one example of the more general knowledge transfer activity undertaken in several contexts. The recent substantial increase in formal property rights associated with university research in the US provides a quasi-experiment that allows empirical investigation of the effect of increasing formal IP rights on the use of university research in industrial innovation.

The goal of the Bayh-Dole Act was to promote commercialization of university research results that were seen as going to waste sitting on the shelves of university laboratories. The increased patenting and commercialization activity, however, brought with it concerns over increased secrecy, restrictions on follow-on research, and destruction of the open-science norms on which the institution of academic science relies. How has the increase in formal property rights altered the transfer of knowledge from universities to industry? How can firm managers position their firm to take advantage of the important contributions that reside in university research? Answering these questions contributes to the Open Innovation paradigm by shedding further light on how the use of external knowledge occurs, how it is affected by the prevailing IP regime, and what managers can do to improve performance in this regard.

In this chapter, I summarize relevant empirical literature and present new empirical evidence relating to the impact of recent changes in the IP regime associated with university research results. Results demonstrate that the exploitation of openly published scientific research became more unequal across firms in technology areas experiencing the greatest increases in average reliance on university research by industry. This highlights the variation across firms' abilities to identify and exploit external (especially university-based) research knowledge. However, the increase in university patenting does not itself appear to increase the disparity across firms.

Further evidence suggests that exploitation of openly published scientific research is in fact beneficial for at least one dimension of firm innovative performance. By examining the lag time between existing patented knowledge and the firm's new inventions building on that knowledge, I find evidence that as university patenting increased in a technology area, the length of this lag time increased. This suggests that the pace of knowledge exploitation by firms is slowing as universities are increasing their formal IP claims to their research results. This may be due to increased formalization and required lengthy negotiations at the university-industry boundary. Finally, I report evidence suggesting that firms that are exploiting more openly published scientific research experience a shorter lag time, consistent with innovative performance benefits associated with use of this external knowledge.

To successfully embrace the Open Innovation paradigm, firms must develop the ability to identify, assimilate, and make use of external knowledge and ideas. In the case of university-based research knowledge, publications and dissemination of research results have traditionally contributed to the knowledge landscape surrounding firms. However, seeking out and making effective use of this knowledge requires firm investments in building internal research expertise and collaborative networks with external scientists. This chapter considers evidence that points to some of the research activities that enhance the ability of a firm to take advantage of public science.

The remainder of this chapter proceeds as follows. Sections 7.2 and 7.3 describe several aspects of the university–industry interface, including the importance of university research in industry innovation, recent policy changes that have influenced the intellectual property rights environment at this interface, and the expected implications of these changes for firms' use of university-based research result. Section 7.4 empirically test some of these implications and reports the results of the analyses with respect to both changes in the patterns of knowledge exploitation and how firms can enhance their ability to take advantage of university research. Section 7.5 concludes and discusses how this research relates to the Open Innovation paradigm.

7.2 University Knowledge and Industrial Innovation

7.2.1 *An Important Source of External Knowledge*

Public science supports the productivity of private science in multiple ways. Industry researchers across many industries rely on universities for research findings, instruments, experimental materials, highly trained human capital, and research techniques (Cohen et al. 2002). Industry researchers report that linkages with university researchers provide benefits in terms of keeping abreast of university research, gaining access to the university researchers' expertise, and receiving general assistance with problem-solving (Rappert et al. 1999). The successes and failures from basic research at universities provide information useful for guiding applied research in the direction of most promising opportunities, avoiding unfruitful areas, thereby increasing the productivity of applied research (David et al. 1992). Access to a stronger knowledge base facilitates more efficient and effective search for new innovations by firm researchers (Nelson 1982; Cockburn and Henderson 2000).

Existing studies have documented the reliance of industrial innovation on university-based research. Industrial patents heavily cite university-generated published basic research, and the citation linkage between universities and industries has been growing over time (Narin and Olivastro 1991, Narin et al. 1997). Universities were reported to be the most important sources of external technologies by British and Japanese firms (Tidd and Trewhella 1997). In a study of US industry researchers, respondents report that approximately 10 percent of their product and process innovations could not have been developed without substantial delay in the absence of academic research inputs (Mansfield 1991, 1998). Although all industries report some reliance on university research results, the importance of this research is particularly strong in some high-tech areas, including drugs, computers, semiconductors, and medical equipment (Cohen et al. 2002; Mansfield 1998).

In terms of the channels through which university research reaches researchers in industry, open publication of research results in the scientific literature dominates. Consistent with the expected importance of complementary, uncodified research results, more interactive channels of knowledge transfer (such as conferences, consulting, and informal interactions) are also important for effectively transferring university research results to industry (Cohen et al. 2002). Collectively, these studies highlight not only the contribution of university science to industrial innovation, but also the critical importance of informal, open, non-IP-related knowledge transfer mechanisms at this interface.

7.2.2 *Traditional University Research Environment*

University research has traditionally been held apart from private science research. The research performed at universities is generally taken to be more basic in nature (as opposed to applied, development-focus research), more important, and of larger impact than research performed by private companies (Trajtenberg et al. 1997). The norms and practices associated with the 'open science' nature of the academic research environment provide incentives for researchers that are consistent with the cumulative development of scientific knowledge (David 1998). The reputation-based reward system, associated priority claim, and review by peers support a system of rapid disclosure and broad dissemination of new research results by scientists. Rewarding a scientist for being first to discover encourages both inventive drive and disclosure. Disclosure and peer review allow validation of the research results. A reputation-based reward system encourages dissemination of research and the production of meaningful, contributory science on which others can build.

This system avoids excessive duplication of research efforts, promotes information sharing, and allows the development of a strong public knowledge base from which following researchers can draw. Importantly, the open science system encourages both the dissemination of codified research results (through publication and the like) and the transfer of the complementary know-how that remains uncodified, through collaboration, interaction, and discussions between researchers. This system has clear benefits for open innovation, as it encourages contribution of research knowledge to the knowledge landscape from which firms can draw and interactions which facilitate knowledge transfer to industry.

The open science environment can be contrasted with a system of private science, characterized by restricting access to knowledge in order to appropriate rents from research (Dasgupta and David 1994; David 1998). The norms and rewards mechanisms of the two systems differ considerably, and result in different behaviors and outcomes. As I return to below, the increasing focus on property rights, appropriation of rents, and commercialization on the part of university faculty and administrators has (perhaps unavoidably) brought some of the private science incentives into the traditionally open science research community of academia.

7.2.3 *Markets and Transfer of External Knowledge*

The prevailing theory regarding how formal property rights influence the market for technology and knowledge assets is primarily based on transferring the technology between parties both seeking to profit from it. Creating value from innovations and new technologies requires complementary assets to bring the innovation through development, commercialization, marketing, and distribution. The firm that generates the innovation often does not hold all of these pieces of the value chain in-house, and therefore some interfirm transactions are necessary and desirable. As the open innovation framework makes clear, the best way for a firm to gain value from innovations that do not fit the firm's own set of complementary assets is to look outside of the firm for a licensee or spin-off to develop the innovation (Chesbrough 2003*a*).

In general, the markets for technology and knowledge assets between organizations are assisted by the ability to protect the value of the knowledge asset from expropriation while also being able to effectively transfer the technology and related knowledge. Patent rights and the associated disclosure and exclusion rights allow parties to negotiate over a clearly defined and specified piece of technology without the worry that potential buyers will walk away from the deal once they internalize the knowledge contained in the patent (Teece 1986).[1] Formal IP associated with the codified portion of a technology also is expected to aid in the transfer of complementary tacit knowledge (Arora 2002).

In addition, strong patent rights encourage specialization (Lamoreaux and Sokoloff 1999). Researchers can specialize in creation of intellectual assets, which they can then be compensated for through the licensing process. Licensees can specialize in the development, marketing, and delivery of the technology or associated product. By reducing the transaction costs associated with identifying and negotiating for technologies created outside of the firm, formal property rights may encourage firms to license and utilize technology from outside of the firm boundaries (Gallini 2002). Therefore, strong IP right protection encourages disclosure and promotes efficient trade in the market for technology.

Viewing technology transfer from universities in this light provides some interesting insights. University researchers do not generally possess the complementary assets necessary to bring the often early-stage research results through development into a commercialized product that is marketed and distributed to consumers. The researchers tend to specialize in the creation of knowledge assets, the commercialization of which is typically left to other organizations. In most cases, the technology must be transferred to a holder of complementary assets if development and commercialization is to occur. Given the necessity of technology transfer, are strong IP rights necessary or beneficial to such trades?

The traditional open science environment of university research makes this transfer of knowledge assets different than transfer between two profit seeking firms, since university researchers have different incentives than a firm that has generated an innovation or technology. University scientists seeking recognition or reputation rewards are not concerned with protecting their intellectual contributions—in fact, they openly publish and distribute their contribution in the hope that others recognize the value of their work and build upon it. Therefore, under the traditional stance of university researchers, a lack of property rights does not create a desire for secrecy and inhibit knowledge transfer, as one might expect in the case of firms generating knowledge assets.

The IP concerns come instead from the firms to which the innovations are flowing. Transferring and developing the university-based innovations often requires significant investment on the part of the firms, and firms may be hesitant to make these investments if the innovations which they are receiving are not protected by IP rights due to a fear of imitation or expropriation (Thursby et al. 2001). The traditional model of openly publishing and disseminating university research results was therefore considered an impediment to effectively transferring university discoveries into commercial products. The hazard in this transaction falls on the acquirer of the technology, who does not want to be copied by competitors, rather than the originator of the technology. How does an increase in formal property rights affect transfer across this type of interface?

7.2.4 *Changing Intellectual Property Regime*

The Bayh-Dole Act of 1980 (The Patent and Trademark Amendments of 1980, Public Law 96–517) standardized the process by which universities could acquire patent protection for research conducted with federal funding. Prior to this policy, universities could obtain patent protection for research results only by applying to the federal government for permission to do so, and the university's activities were constrained by the case-by-case allowances of the government. The Act provided blanket permission and standardized procedures for universities to apply for patent rights covering the results of federally funded research, license the patents to interested firms, and collect royalty payments. In addition, the Act supported the negotiation of exclusive licenses to university patents resulting from federally funded research (Mowery and Ziedonis 2001). This policy change standardized the procedure for university patenting and encouraged university patenting as a means to achieving technology transfer between university and industry (Mowery et al. 2001).

The motivation for the Bayh-Dole Act was to increase the commercialization of publicly funded research that occurred at universities and government laboratories. At the time, 70 percent of US university research was federally funded, so policies affecting this portion of university research had a significant impact. The justification for increasing formal property rights to the outcomes federally funded research was based on the belief that many of these (potentially commercializable) research outcomes were going undeveloped due to a lack of property rights. By granting formal IP protection to federally funded university research results, allowing universities to license these results and collect royalty payments from the licensee firms, and providing an appropriability mechanism to the investing firms, the Bayh-Dole Act aimed to provide the necessary incentive structure to get more of the federally funded university research off the shelf (or out of the laboratory) and into industry (Henderson et al. 1998*a*, 1998*b*).

The response, in terms of the amount of university research protected by patents, was dramatic. In 1965, there were 96 US patents granted to 28 US universities. By 1992, nearly 1500 US patents were granted to more than 150 US universities (Henderson et al. 1998*a*, 1998*b*). Since 1975, the growth in the number of university-assigned patents granted by the US Patent and Trademark Office has far outpaced the increase in the general population of US patents (see Figure 7.1). Not surprisingly, the increase in university patenting has been concentrated in fields where licensing is a relatively effective mechanism for acquiring new knowledge (Shane 2004), such as drugs and medical, electronics, and chemical fields.

It is important to note that other changes also contributed to the overall increase in university patenting during this time. US patent protection generally was increased by the Federal Court Improvements Act of 1982, which created the Court of Appeals for the Federal Circuit to hear all patent case appeal. This court was broadly seen as favoring the patent holder in cases of infringement (Jaffe 2000; Gallini 2002). The Bayh-Dole Act was followed by a

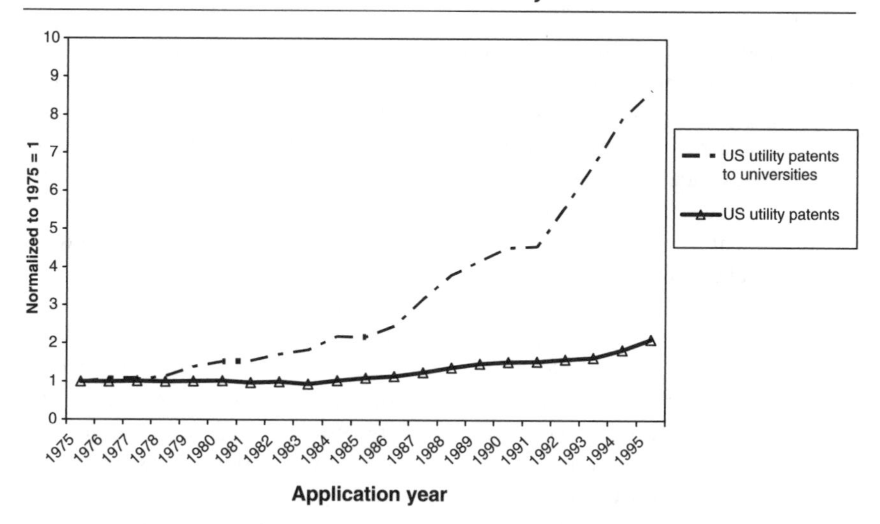

Figure 7.1. The increase in university patents has outpaced growth in patenting

law in 1984 that expanded the patent rights of universities further and removed some restrictions contained in the Bayh-Dole Act. In part because of the Bayh-Dole Act, the number of universities with formal offices dedicated to technology transfer from universities grew following 1980. In addition, due to changes in technology, decreasing federal funding, and an increased focus on technology transfer, industry funding of university research increased as well (Henderson et al. 1998*b*). In fact, the increase in university patenting and license appears to predate the Bayh-Dole Act in some fields, such as the biomedical area, where university patenting grew significantly between 1975 and 1979 (Mowery et al. 2001). The Act, however, did encourage universities that had not been involved in patenting previously to begin patenting their research outcomes (Mowery et al. 2001).[2]

It is also important to recognize that universities differ considerably in their implementation of internal policies regarding patenting of research results. Federal policy left universities considerable leeway in forming IP policies, and resulting university policies vary with respect to the resources dedicated to technology transfer, the percent of revenue generated that is allocated to the inventing faculty member, rules regarding faculty members starting companies based on university research, and the goals of the technology transfer office (Thursby et al. 2001; Siegel et al. 1999; Siegel et al. 2003; Kenney and Goe 2003; Debackere and Veugelers 2005). Some universities focus on patenting as many research results as possible and extracting maximum licensing revenues, while others have used patenting selectively for inventions that would be less likely to be transferred to industry without a license. Similarly, some universities grant exclusive licenses to many of their patenting inventions, while others promote a more open licensing approach, granting nonexclusive licenses to many licensees. For example, when Stanford University patented the Cohen-Boyer recombinant DNA research toll in 1979, it offered nonexclusive licenses to all interested parties for a modest fee, in order to disseminate the critical research method widely. The majority of university patents and the great majority of revenue earned through technology transfer are concentrated at relatively few universities (Graff et al. 2002).

The policy changes relating to university patenting in the US are more dramatic, but similar to, changes taking place in other developed countries. Cesaroni and Piccaluga (2002) and Geuna and Nesta (2003) describe the recent changes in European countries, including increased patenting by universities and more interactions between industrial and university researchers, but also point out that European universities in general are not as active as US universities in either activity. As in the US, university patenting activity is concentrated in the biotechnology and pharmaceutical related fields. Collins and Wakoh (2000) document the recent regime changes in Japan, aimed at creating a system similar to the post-Bayh Dole environment in the US. Although the existing empirical work and the results I report here primarily focus on the US case, similarities among the policies in other developed countries suggest that the discussion and qualitative results presented here may be relevant to other regions.

7.3 Implications for Industry Exploitation of University Research

The increase in the number of patents on university research reflects an increase in the percentage of university research that is patented, while the underlying generation of university research has remained relatively stable (Henderson et al. 1998*b*). Dasgupta and David (1994) warn of the potentially detrimental consequences associated with altering the norms established in the open science environment in favor of property rights and commercialization of university science. By altering the system that so effectively produces and disseminates the body of public scientific knowledge, follow-on innovation based on this knowledge may be inhibited. As Nelson (2001: 16) reflects after interviewing industry researchers:

> [M]y strong suspicion is that a good share of the technology transfer that has occurred would have proceeded as widely and rapidly as in fact it did, even if there had been no claiming of intellectual property rights by the university. And in some cases, it would appear that such claiming probably has made technology transfer more costly and time consuming for the firms involved.

These concerns, evidence regarding their validity, and implications for open innovation are reviewed in the following sections.

7.3.1 *Increasing IP and Transfer of University Research to Industry*

The historically differing norms between the academic environment and the private science carried out in industry have collided in the technology transfer process. In their survey of various stakeholders in the university-industry transfer process, Siegel et al. (2003) found that the dominant complaint of industry managers was a lack of understanding of corporate culture and norms by university technology licensing officials. Similarly, university technology licensing officials complained that their corporate partners failed to understand and appreciate the goals and norms of the university. The technology transfer process forces the inherent conflicts between open science and private science norms to the surface. Although the Bayh-Dole Act was intended to facilitate commercialization of university-generated technologies, there are several concerns that have been voiced related to the potential for limited availability of upstream university research and the destruction of norms that have supported the cumulative, open, and basic nature of scientific discovery associated with university research.

7.3.2 *Fencing Off Upstream Research*

One concern that has been raised with respect to increasing university patenting is that downstream research will be stifled due to the unavailability of upstream research inputs, especially in complex industries that require many, potentially overlapping, IP-protected inputs (Heller and Eisenberg 1998). This 'anti-commons' problem at worst leaves industry researchers unable to access the needed inputs to their own innovation process, and at best requires time consuming negotiations plagued by hold-up hazards.

Because of the early-stage nature of many university inventions, pressures from industry, and revenue-seeking by universities, technology licensing offices often grant exclusive licenses to patented university research, limiting follow-on development to a sole licensee. In many cases, a patent and an exclusive license are necessary to provide the incentives for industry development. However, in other cases, industry researchers have indicated that they would have utilized or developed the university research even in the absence of patent protection or an exclusive license (Thursby et al. 2001). Patenting and licensing in these cases increases the costs and time resources required to make use of university research results and may unnecessarily limit the set of firms utilizing university research in their own innovation. This is especially true for research tools, which make up a significant (if not majority) share of the inventions patented at universities (Gelijns and Their 2002).

Even university patents that are licensed on a nonexclusive basis still require the negotiation of licenses and potentially the payment of upfront and royalty fees, both which may restrict the set of follow-on innovators relative to an open science environment. Murray and Stern (2005) find that following the grant of a patent covering university research that is also contained in a publication, citations to the publication are lower than would otherwise be predicted, suggesting diminished follow-on research associated with a patent grant. In contrast, Walsh et al. (2003) find that industry researchers in the biomedical sector do not report that the increase in patenting has decreased the accessibility of research tools, which are often generated by university research. However, over a third of the respondents in that survey reported that the increase in research tool patents caused delays and increases in costs associated with their own research.

Increasing IP concerns in an arena previously characterized by open knowledge sharing may create barriers and administrative burdens that can be a drag on innovation. Firms are forced to wade through the increasingly crowded and complicated IP rights surrounding their own research and identify and negotiate access to relevant technologies. This process is time consuming and costly, and can slow down the research activity of the firm (Walsh et al. 2003).

Industry researchers report difficulty negotiating for licenses or access to IP-protected university-based research. In a significant number of cases, IP concerns presented an insurmountable barrier to firms joining with a university in a research partnership (Hall et al. 2001). Industry researchers experience the increasing formalization of university technology transfer as detrimental to the (more effective) knowledge transfer through informal, collaborative channels (Rappert et al. 1999). One researcher in a biotechnology firm interviewed by Walsh et al. (2003) reported that university patenting of research tools causes them to work around the university IP, often slowing down their research progress. More generally, industry researchers in that study reported that high licensing fees or exclusive licenses to research tools could limit access to upstream university research and that wading through the increasingly complex IP landscape to identify the relevant property rights added time to the research process.

7.3.3 *Restricted Dissemination of University Research*

Aside from property rights protection, increased patenting and commercialization activity by faculty members may be associated with less willingness to openly discuss and share research results and data within the scientific community. Louis et al. (2001) find that life sciences faculty members that are more involved in the commercialization of university research are more secretive about their research than other faculty members, all else equal. That is, they are more likely to deny requests for information about their research from other researchers. Faculty members in the biotechnology field with industry funding are more likely to keep research results secrecy to protect their proprietary value, more likely to take commercial applicability into account when choosing research projects, and more likely to produce research results that could not be published without the sponsors permission (Blumenthal et al. 1986).

This lack of willingness to share results, materials, and findings suggests a shift in the norms of the scientific community on which the progress of the academic system has been built. Recall that industry researchers report that publications are the most important source of university research that they rely on in their own research. In the absence of peer review and publication, dissemination of the important component of research knowledge than is contained in publications may be restricted or slowed. In addition, less willingness of researchers to share results and collaborate may mean that the informal interactions critical to transfer of uncodified or unpublished research may be inhibited.

7.3.4 *Implications for the University-Industry Interface*

The expected benefits of formal IP protection and these potential negative consequences of increased university patenting have many far-reaching effects. The innovation system in the US, as well as other regions that draw on research generated at US universities, may be affected by a change in the pattern of dissemination or any facilitation or delay of knowledge transfer from US universities. The possibility that some industry innovations that would have occurred under the traditional (pre-Bayh-Dole) system will now not occur is impossible to test. However, it is possible to investigate changes in the pattern of use of university research and the pace of knowledge exploitation in industrial innovation as university patenting has increased.

In particular, if university research is becoming increasingly 'fenced off' such that only those firms with license to the patented university research results may use and build upon university research, we would expect the use of university research to become more restricted. That is, the use of university research will be more concentrated at some firms, leaving other firms without this input. In addition, if the increase in university patenting is inhibiting the transfer of university research result to firms, the Open Innovation processes of firms may suffer. Given the importance of university research to the innovation processes in industry, limited or delayed access to this important input may result in slower exploitation of existing knowledge for new inventions by the firm. Empirical evidence relating to the consequences of the increase in university patenting on the transfer of research results from universities to industry is scant. In Section 7.4, I draw on a large-scale panel data set to offer a preliminary evaluation of two concerns associated with university patenting: fencing off university research and slowing industrial innovation.

7.4 Empirical Evidence

7.4.1 *University Patenting and Patterns of Knowledge Exploitation*

In an effort to investigate the relationship between increasing university patenting and these two concerns associated with knowledge transfer and industry innovation, I conduct an empirical investigation of patent activity during the 1975–1995 period. I use the information contained on the front page of each patent application to the US Patent Office to examine the relationship between the growth of university patenting in a technology area and both the exploitation of openly published scientific research in firms' inventions in that technology area and the pace of knowledge exploitation evidence in a firms' patents.

To explore exploitation of scientific research, I examine the citations to public science, which I define as citations to the 'non-patent' prior art listed on firms' patents. These citations list the prior art related to the invention covered in the patent, and they typically contain references to scientific journal articles, textbooks, and other codified, nonpatent research reports.[3] The number of these citations to openly published scientific work in each

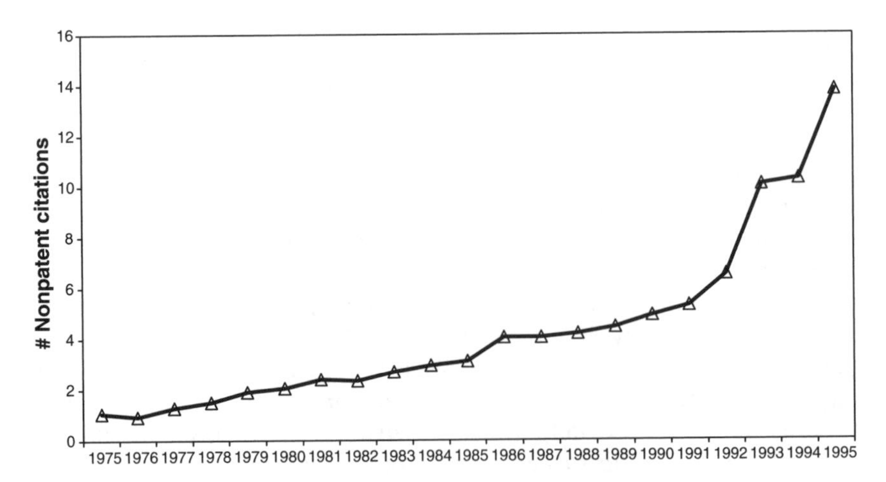

Figure 7.2. Citations to nonpatent prior art increased; pharmaceutical patents example

patent increased substantially in many fields during the same period as university patenting increased. For example, Figure 7.2 plots the average number of such citations per patent for patents in the main pharmaceutical patent technology class.[4] The upward trend in the number of these citations in industry patents is clear.

During the same period, patenting by universities increased substantially. For example, between 1975 and 1995, the percent of all US patents in the main pharmaceuticals class that were university-assigned increased from about 2.6 percent to about 8.6 percent. The parallel increases in citation to public science and patenting by universities likely reflect an increase in the amount of university research in this area and an increased applicability of the university research results to the Open Innovation processes in the pharmaceutical sector, as well as an increase in the patenting propensity of universities.

The concerns outlined above suggest that increased formal IP rights may be associated with less open dissemination of research results and increased limitations on the use of university research results in industry. If these concerns are true, we would expect that some firms (those with an advantage in terms of accessing university research or those that are able to gain access through licensing patented university technologies) will become increasingly advantaged relative to other firms. That is, if open dissemination, through publication, informal interaction, and other means, declines as university patenting increases, university research results may become increasingly channeled to some firms relative to others. Again using pharmaceutical patents as an example, Figure 7.3 displays the upward trend of the variance across firms of the average number of nonpatent prior citations per patent. This suggests

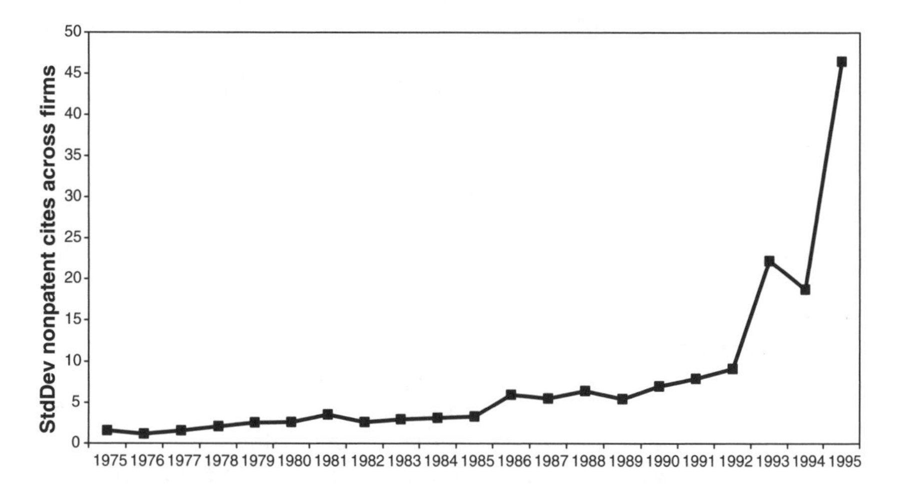

Figure 7.3. Inequality of nonpatent citations across firms increased; pharmaceutical patents example

that during a period in which reliance on university science and university patenting were increasing, some firms were exploiting this important knowledge source more than others, and the difference across firms increased over time. The following empirical analysis explores this relationship further by looking across all technology areas over time.

I use a panel data set containing information of all US utility patenting assigned to firms over the 1975–1995 period in 620 World International Property Organization (WIPO) assigned international patent classes to evaluate the relationship between the amount of university patenting, the citation of public science, and the variance of citation to public science by firms in each technology class over time. For each technology class and year, I proxy for the amount of *university patenting* with the percentage of patents assigned a university, with a lag of one year. For each firm that patented during this period, I calculate the average number of citations to nonpatent prior art in the firm's patents in each technology class.[5] Then, for each technology class-year observations, I calculate the standard deviation of the number of nonpatent citations across firms to arrive at a measure of the *variance of citations across firms* in the technology class-year (see Fabrizio 2005*a* for detailed description of the data construction).

I use a technology-class fixed effects model to estimate the relationship between the percent of patents that are assigned to universities in the technology class and the variance of citation of public science across firms in the class. By controlling for technology-class averages, the fixed effect analysis effectively compares each technology class to itself over time. In the analysis, I also control for the *number of firms* in the technology class-year observations, because the variance could be affected by firms exiting or entering.[6]

Results of the fixed effect analysis of the variance of nonpatent citations across firms are reported in column (1) of Table 7.1. The significant and positive coefficient on *university patenting* suggests that as university patenting increased in a technology class, citation to public science became more unequally distributed across the firms in that class. In other words, the exploitation of public science in firms' inventions became increasingly concentrated at some firms relative to other firms in the technology areas where university patenting increased. However, this does not control for changes in the reliance on public science over time.

The variance of citation across firms to public science across firms may increase as reliance on public science increases in a technology area if some firms are better able to identify and exploit university research results than are other firms. As university research results gain importance in industry innovation, firms with an advantage with respect to exploiting university science will increase their use of this resource while other firms are left behind. This suggests that a control for the applicability of university science to industrial innovation should be included.

Table 7.1. Effects of increased university patenting 1976–95

Dependent variable	Variance of citations across firms (1)	Variance of citations across firms (2)	Backward citation lag (3)	Backward citation lag (4)
University patenting$_{k,t-1}$	27.67*(11.76)	−9.27* (4.32)	1.25** (0.36)	1.30** (0.36)
Avg. Cites to public science (Class-year)$_{k,t}$	—	1.69** (0.17)	0.05** (0.02)	0.08** (0.02)
Avg. Firm cites to public science (Firm-class-year)$_{j,k,t}$	—	—	—	−0.05** (0.01)
Firms$_{k,t}$	0.05* (0.02)	0.01 (0.01)	—	—
Avg. patent cites (per patent)$_{j,k,t}$	—	—	0.33** (0.02)	0.34** (0.02)
Patents in class$_{k,t}$	—	—	−0.05** (0.02)	−0.05** (0.02)
Constant	0.36 (0.47)	−0.37 (0.23)	1.78** (0.12)	1.76** (0.13)
Tech. Class FE	Yes	Yes	Yes	Yes
Observations	6,090	6,090	107,893	107,893

*Significant at the 5 percent level ** significant at the 1 percent level *Subscripts: j*: firm, *k*: class, *t*: year
Robust standard errors (clustered by technology class) in parentheses.
All equations include year fixed effects and technology class fixed effects.
Eq (1) and (2) are at the technology class-year level, with 6,090 observations. *Variance of citations across firms* is the standard deviation of the average number of nonpatent citations per patent across firms in the technology class-year.
Eq (2) and (3) are at the firm-technology class-year level, with 107,893 observations. *Backward citation lag* is the natural log of the average number of years between the application year of the patent and the grant year of the cited patents for patents of a given firm in a given year in a given technology class.
University patenting: percentage of patents assigned to universities in the class, lagged one year.
#Firms: number of firms in the technology class-year observation.
Avg #cites to public science: Average number of nonpatent citations per patent for either the class-year or firm-class-year observation. Natural log is used in eqs (3) and (4).
Avg #patent cites: Natural log of the average number of patent prior art citations for patents in class-year observation.
 # patents in class: Natural log of the number of patents in class-year observation.

One way to control for changes in the applicability of university science to industry over time is to include the *average number of citations* to public science across firms in each technology class-year observation. Results are reported in equation (2) of Table 7.1. The coefficient on *university patenting* is now negative and significant, while the positive relationship is attributed instead to a change in *average number of citations* to public science in the technology class. Both the magnitude and significance are greater for the coefficient on the *average number of citations* to public science. These two measures are highly positively correlated with university patenting, both in the cross section (across technology areas) and in terms of changes over time within a technology area. The results here suggest that the increase in variance of citations across firms is more closely related to a change in the importance of public science to industry than to a change in patenting of university research results. The negative coefficient on *university patenting* suggests that controlling for the average reliance on public science, more patenting by university researchers is associated with citation of public science more equally across firms.

Consistent with the survey results reported by Walsh et al. (2003), the preliminary results reported here suggest that patenting by university researchers does not restrict the use of public science by firms that continue innovating in the technology area. Instead, as university research results increase in importance to firm innovation, some firms increasingly take advantage of university science more than other firms. In this case, a rising wave does not float all boats equally. From an Open Innovation perspective, a firm that enhances its ability to exploit this research will have an advantage over other firms. Although universities may increase patenting as the importance of their research to industry grows, it appears from this analysis that the increase in patenting itself does not restrict the use of university research to only some firms.

This result may not, however, reflect a direct causal relationship, but instead a common antecedent or unmeasured confound. First, we can only observe citation of public science in innovations by firms that successfully innovate. If firms that are 'cut-off' from public science are not able to generate innovation that they would have otherwise, we can not capture that effect with this data. Second, there is an omitted variable that may affect both the variance of citations to public science and patenting by universities, namely the usefulness of university research to industry. The average citation to public science may not fully capture this. Without a better control for this underlying factor, it is possible that the coefficient estimates will be biased if the amount of patenting at universities is endogenous to the usefulness of university research to industry. Patenting of university research may respond to the applicability of university research to industry because universities will patent where industry is more interested in licensing the inventions. The applicability of university research to industry may also respond to university patenting if university researchers focus in more applied areas in order to pursue patents and licensing opportunities. Further research exploring the determinants of the level of and increase in university patenting would be interesting in its own right and could also serve as the 'first-stage' of an instrumental variables analysis to explore this possibility.[7] The results here should be treated as suggestive, pointing to areas for important future research.

7.4.2 *Delays in Accessing Innovations*

Turning to the second concern, I am interested in exploring the relationship between an increase in university patenting and the ability of firms to innovate. If dissemination and use of university research is delayed by increasing formal property rights, we might expect the industrial innovation would be slowed as well. On the other hand, if university patenting facilitates knowledge transfer, industrial innovation may be sped due to enhanced access to an important resource.

In order to investigate this concern, I estimate the relationship between the increase in university patenting and the amount of time that passes between existing knowledge and the new firm innovations that build on that knowledge. I construct a *backward citation lag* measure similar to the 'technology cycle time' measure described by Narin (1994) by calculating the average number of years between the application year of a given patent and the years in which the patented 'prior art' listed in that patent were granted. If a patent cites significantly older relevant prior art, it took longer for the inventors to build on the prior art in their new invention.

As an example, the distributions of this *backward citation lag* for patents applied for in 1985 in four technology classes are displayed in Figure 7.4. The distributions that peak quickly and drop off represent classes in which patents cite relatively recent prior art heavily, and do not cite older (now obsolete) prior art. These two classes, Medical Preparations and Semiconductors, are classes that we would expect to be progressing at a relatively rapid pace in term of technological advance. The distributions of backward citation lags for patents in the other two classes, Stone Working and Hinges, are much flatter, suggesting a slower pace of advance and a longer period until technological obsolesces. In the following analysis, I examine changes in the *backward citation lags* of patents in each of the 620 technology classes over time to evaluate changes in the pace of knowledge exploitation in each class.

By looking at this *backward citation lag* for all industrial patents across all technology classes for the 1975–1995 period, I evaluate whether the pace with which researchers developed new patented inventions slowed or sped up with

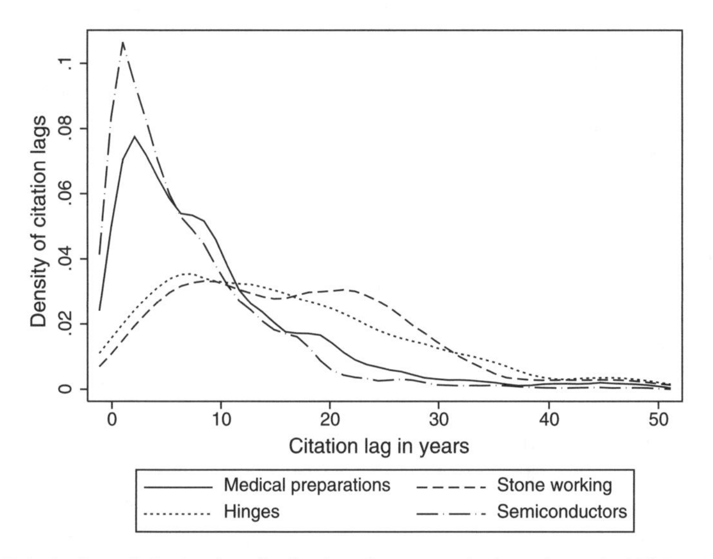

Figure 7.4. Backward citation lag distributions for patents in four classes in 1985

increased university patenting. Analogous to the analysis above, I use a technology class fixed effects analysis to examine the relationship between changes in the university patenting in a given technology class and changes in the backward citation lags of patents in that class. Between-technology class differences are controlled for with the technology class fixed effect. In this analysis, the dependent variable is the average *backward citation lag* for patents by a given firm in a given technology class-year observation.[8] I control for various characteristics of the patents expected to affect the lag, including the *average number of citations made* and the *number of patents* in each technology class-year. As above, I control for the average reliance on university science with the average number of nonpatent citations per patent in each class-year observation.

Results of this analysis, presented in column (3) of Table 7.1, demonstrate that the *backward citation lag* of firm patents increased as *university patenting* in the technology area increased, suggesting a slow down in the pace with which firm exploited existing knowledge in new inventions (see Fabrizio 2005*a* for more detail and further analysis). As university patenting increased, the time between the relevant existing knowledge and the firm's new patented inventions lengthened, even controlling for the *average number of citations* to public science in the technology class-year. This may reflect slow-downs due to negotiation for rights to patented technologies or materials, time spent inventing around patented upstream technologies that a firm did not license, and slow down due to a lack of research knowledge inputs.

If this apparent slow down is related to reduced or delayed access to university research, then we would expect that firms with an advantage in terms of accessing and exploiting university research would demonstrate some advantage with respect to the pace of knowledge exploitation. To evaluate this prediction, I reestimate the backward citation lag model including the *average number of citations to nonpatent prior art in the patents by the firm* in each technology class-year observation. This variable reflects the firm's exploitation of public science in their inventions in that technology area and year.[9] Results, reported in column (4) of Table 7.1, demonstrate that firms whose patents contain more citations to nonpatent prior art, suggesting more exploitation of public science, have patents with significantly shorter backward citation lags. In other words, exploitation of public science is associated with an advantage to firms while overall the pace of knowledge exploitation slows with an increase in university patenting.

These results suggest that the pattern of knowledge transfer from universities to industry has changed as industry relies more heavily on university research results and these research results are increasingly associated with formal IP claims. In addition, the pace of knowledge exploitation in industry inventions on average is slowing as university patenting increases. Firms that are better able to exploit university science do not experience as much of a slow down. This is consistent with possible detrimental effects of limiting access to important upstream research, restricted dissemination of such research, and slower transfer due to lengthy negotiations over increasingly complex university IP concerns.

7.4.3 *Firm Research Strategies and Knowledge Transfer*

If the exploitation of university research results is becoming increasingly unequal across firms, and use of university science is beneficial to firm innovation processes, what can firm managers do to improve their access to and use of university research results?

Even without restrictions imposed by patent protection, use of external knowledge requires some investment by the firm. When university science is published in the open literature, the fact that research results are theoretically available for use by other researchers does not imply that all potential users are equally able to identify and make use of the research. Firms must develop and maintain the ability to 'plug in' to these research communities.

Geographic proximity to high quality university scientists enhances the firm's ability to capture the 'spillovers' of knowledge from the university (Zucker et al. 1998), but firms can also take an active role to develop their ability to exploit public science. The capability of a firm to identify, assimilate, and exploit external knowledge was termed 'absorptive capacity' by Cohen and Levinthal (1989), who recognized that this ability requires some in-house research capability and expertise on the part of the firm researchers. Simply being aware of relevant basic research discoveries generated at a university necessitates the ability to understand cutting edge science, even if the research results are promptly published (Rosenberg 1990). The ability to assimilate and exploit such research results requires even more expertise, and may also require interaction with university scientists in order to fully understand and implement the research. In addition, active collaboration between university and firm researchers may facilitate more complete and faster transfer of tacit research knowledge.

The President of Centocor (a biotechnology firm now a subsidiary of Johnson & Johnson) recognized this need and the expected benefits when he stated that 'Centocor should know about major research at least a year before it's published' by stressing collaboration and giving their scientists the freedom to make contacts and stay on top of relevant scientific research. In addition to finding out about new research results, contacts with university researchers may give a firm access to results that would not be published (such as lessons learned from failed experiments) and knowledge related to published research that can not or is not written down.[10] How a firm can develop its ability to identify and exploit public science is the subject of research that evaluates the benefits of geographic proximity, founder experience, firm research, and collaboration for performance outcomes such as production of patents.

Several empirical studies demonstrate the significant positive effect of internal basic research on firm productivity (see e.g. Griliches 1986).[11] Gambardella (1992) finds that pharmaceutical firms whose researchers produce more scientific publications also generate more patented inventions, controlling for the scale of firm R&D. Henderson and Cockburn (1994) find that for a small sample of large pharmaceutical firms, firms that promote researchers based on scientific publications and standing within the scientific community generate more important patents. In addition, pharmaceutical firms that collaborate with university scientists generate more important patents (Cockburn and Henderson 1998). Firms in the biotechnology sector that coauthor scientific publications with top university scientists produce more patents and more important patents (Zucker et al. 2002). These findings suggest some inventive performance benefits associated with a firm's investment in basic science research, the scientific focus of its research culture, and the collaborative connections developed through researcher interactions.

These studies interpret the superior innovative performance as a benefit of superior absorptive capacity stemming from the internal basic research and university collaborations of the firm. Liebeskind et al. (1996) provide some support for this with a case study of two biotechnology firms. The authors find that the firms rely on social network connections as the dominant channel for transfer of scientific knowledge from universities. This suggests that there are open innovation advantages to being well connected to the scientific community.

In order to evaluate the relationship between firm research activities and the firm's exploitation of public science, I again explore the citation to public science in firm's patents. By focusing on firms in the pharmaceutical and biotechnology sectors, I am able to collect more detailed firm level data to investigate whether there is direct evidence that firms investing in particular research strategies possess superior absorptive capacity with respect to public science.

Following the existing empirical literature, I proxy for firm basic research expertise with a count of the *number of scientific publications* generated by each firm in each year. I proxy for the collaborative focus of the firm with the *percentage of these publications that were coauthored with a university researcher.*[12] Controlling for the firm's *size, research expenditures, research intensity,* and *age*, I estimate the relationship between the *number of nonpatent citations in a firm's patents* and that firm's basic research focus and collaborative research efforts (see Fabrizio 2005*b* for more detail regarding the sample, method, and results). Results of a negative binomial estimate, reported in column (1) of Table 7.2, indicate that pharmaceutical and biotechnology firms investing more in basic research and collaborating more with university scientists do cite more public

Table 7.2. Firm basic research and collaborations are associated with more citations to public science

Dependent variable	Citations (1)	Citations (2)	Citations (3)
% Pubs w/Univcoauthor$_{j,t}$	0.96** (0.33)	2.26** (0.80)	
% Pubs w/Univcoauthor$_{j,t}^{2}$		−1.70* (0.72)	
Pubs/100$_{j,t}$	0.07* (0.03)	0.07* (0.03)	
% Pubs w/Univco-Author$_{j,t}$*Pre1985			0.72 (0.47)
% Pubs w/Univco-Author$_{j,t}$*Post1985			1.09** (0.32)
Pubs/100$_{j,t}$*Pre1985			0.25** (0.07)
Pubs/100$_{j,t}$*Post1985			0.06* (0.03)
ln(claims)$_{i}$	0.04 (0.03)	0.04 (0.03)	0.04 (0.03)
% self-citations$_{i}$	−0.04 (0.20)	−0.05 (0.20)	−0.05 (0.20)
ln(Min. distance to univ.)$_{j}$	−0.10^ (0.06)	−0.10^ (0.05)	−0.12* (0.06)
Foreign firm dummy$_{j}$	−0.32* (0.15)	−0.31* (0.15)	−0.37* (0.15)
ln(R&D/Employee)$_{j,t-1}$	0.18 (0.11)	0.16 (0.11)	0.14 (0.11)
ln(employ)$_{j,t-1}$	0.02 (0.07)	0.00 (0.07)	0.00 (0.07)
Firm age$_{j,t}$	−0.02 (0.01)	−0.02 (0.01)	−0.01 (0.01)
Biotech dummy$_{j}$	1.15** (0.38)	1.17** (0.38)	1.13** (0.39)
Observations	24,610	24,610	24,610

^ Significant at 10 percent level *significant at 5 percent level ** significant at 1 percent level *Subscripts*: *i*: patent, *j*: firm, *k*: class, *t*: year
All equations are estimated at the patent level, including all patents for the 82 pharmaceutical and biotechnology firms in the sample. Robust standard errors, clustered by firm, are reported in parentheses.
All equations include year dummy variables.

science in their patented innovations, consistent with the prediction that these activities enhance the firm's absorptive capacity. Also note that being farther away from a research university is associated with less exploitation of public science in the firm's innovations, consistent with the diffusion benefits associated with geographic proximity noted in other studies.

In order to explore the potential diminishing returns to additional collaborations, I also estimate this model including the *square of the percentage of publications with a university coauthor.* Results of that estimation are reported in column (2) of Table 7.2. The relationship does in fact demonstrate diminishing returns. Holding all other variables constant at their mean, the maximum number of citations to nonpatent prior art is reached by collaborating with university scientists on about 75 percent of a firm's published research projects. A firm that generates publications only when the researchers are collaborating with a university does not exploit public science as much as a firm that generates some publications without university coauthors. This suggests that a firm can not rely solely on external collaborations as a means to incorporate public science. The firm researchers also need to develop a level of internal scientific ability and expertise in order to most effectively assimilate and exploit public science research results.[13]

In light of the evidence described above relating to the changes associated with increased reliance on university science and the increase in university patenting, it is of interest to explore how the apparent benefits of these firm research activities change over time. Because all firms in this sample are (potentially) drawing from the same fields of university research, cross-class analysis of changes in university patenting is not possible here. Instead, I examine whether there is any difference in the relationship between firm research activities and the citation of nonpatent prior art before and after 1985.[14]

The final column of Table 7.2 reports the results of the analysis including the pre- and post-1985 indicator variables interacted with the publications and collaborations variables. The basic science research of the firm, as represented by the number of publications generated by firm researchers, provides significantly more benefits, in terms of the number of nonpatent citations, during the pre-1985 period.[15] The coefficient on collaborations with university scientists, as reflected in the coauthored publications, is only significant in the post-1985 period. These results suggest that, relative to the 1975–1984 period, the basic scientific expertise of the firm is less important and the collaborations with university scientists are more important during the 1985–1995 period. This is consistent with increasing importance of connections with university scientists, in order to gain access to university research and participate in licensing, as university research became more important and access to that research was increasingly limited by university patents.

These results demonstrate that by investing in certain organizational and research practices, firm managers in the biotechnology and pharmaceuticals sectors can enhance the ability of the firm to identify, assimilate, and exploit the public science research available in the open literature. These research activities appear to be associated with superior knowledge transfer from a knowledge source that is particularly important in this sector. However, the usefulness of the research activities for sourcing external knowledge may depend on the IP regime in which the firm operates.

7.5 Conclusions

As the Open Innovation framework highlights, firms in many industries are increasingly reliant on external knowledge and ideas in their innovation processes. One important source of these knowledge assets is university-based research. The Open Innovation literature suggests that the market for technologies and know-how across firm boundaries is assisted by, and even depends upon, protection of IP rights. This conceptualization assumes that the owners of the knowledge assets seek to protect the appropriable rents from their innovation. Therefore, they will not disclose the innovation to potential buyers if the risk of imitation is too high.

In the case of university research, the increase in formal IP protection is entering a system traditionally characterized by open disclosure and rapid dissemination of research results. When considered relative to an open science environment where the incentive structure promotes dissemination, collaboration, and knowledge transfer, stronger IP rights may come with substantial costs. Although US federal policy encouraging university patenting was motivated by a desire to generate more commercial applications of university research results, the increase in patent protection to university research also presents cause for concern. Formal IP protection may have the benefit of encouraging firm investment in university technologies, but also has the potential downside of fencing off upstream university research, slowing down technology transfer, and creating incentives for secrecy among university researchers.

This chapter describes some of these concerns and their implications for Open Innovation by firms. Evidence presented here looks for the first time at changes in the pattern of exploitation of public science as university patenting increased. Although preliminary, results suggest that as reliance on university science increased in a technology area, exploitation of public science became increasingly concentrated at some firms relative to others. More detailed analysis of the pharmaceutical and biotechnology sectors suggests that firms that are more collaboratively connected to university scientists may be better positioned to exploit university science. The analysis here does not suggest, however, that the increase in IP rights to university research results itself restricts the use of university research in industry innovation, although more research is needed to further investigate this relationship.

Results here also demonstrate that, on average, the pace of knowledge exploitation evidence in industrial patents has slowed as university patenting increased, even controlling for the average reliance on university science in a technology area. University patenting may be slowing the transfer of university research and therefore slowing open innovation in industry. Firms that are exploiting more public science in their patents appear to be at an advantage in terms of exploiting existing patented knowledge more quickly in their innovations. Overall, this evidence is consistent with a slow down in knowledge transfer and restricted or slowed dissemination of public science as university patenting increases. Further research investigating the possible causes of this apparent slow down will help differentiate property rights related concerns, such as exclusive use of patented research results, from institution related concerns, such as time-consuming negotiations for licenses.

If federal funding of basic research in universities continues to be replaced by industry funding (see Chapter 6 and Fabrizio and Mowery forthcoming), and universities increasingly pursue patents for research results, spillovers from university research to industry will be increasingly associated with formal IP rights. The increased focus on commercialization and licensing revenue at some universities may also encourage faculty to focus on more applied and patentable areas of research, potentially at the expense of the basic research that has served as a building block for industry innovation. There remains concern that the spillovers from university research generated under the open science tradition will become less available, or available more slowly, to industry as universities focus more on applied research and generating an IP portfolio. Further research to separate the effects of an increase in IP rights from the effects of potential shifts in research focus would be useful both for policy-making and assessing the impact of the changing university environment on open innovation in industry.

Increased patenting of university research may also increase the university research knowledge potentially available for purchase or license. For example, universities increasingly engaging in and marketing more applied research may make substantial technologies available for firms to acquire. University researchers seeking royalty revenue may be more motivated to work with the licensee firms to effectively transfer the technology (assuming an appropriate contract is put in place), and as a result the firms may gain better access to a growing body of innovation. The Open Innovation processes of firms would benefit from the opportunity to find out about and license university research, although the firms would have to spend time in negotiation and pay royalties for the use of the patented research results. Firms with better connections to university scientists will likely benefit the most from an increasing body of university research applicable to industry.

The net outcome of these multidimensional factors depends at least partially on the patenting and licensing practices of universities. If universities pursue patents on technologies that are of low or highly uncertain value to industry, it serves only to clutter the IP landscape and drain the resources of the university. If all patented university inventions are licensed exclusively, the university may gain more income but the societal costs due to limited access are much larger than if the university pursues nonexclusive licenses at reasonable royalty rates. A blanket approach that treats each potentially patentable university research result the same is unlikely to provide an optimal solution to these trade-offs, because each technology and market situation is different. However, a case-by-case approach may become overly time consuming and prevent efficient negotiations. Further research is needed to assess the impact of various technology transfer strategies on the use of university research results by industry.

The results reported here highlight the fact that increased patenting does not by itself solve the knowledge transfer problem. Even in the biotechnology and pharmaceutical sectors, where patenting is seen as an important mechanism to appropriate rents from an innovation and an important source of information about public research (Cohen et al. 2002), patenting university research results does not by itself assure technology transfer to industry. Interaction between university researchers and industry researchers, as well as continued investment in basic science by firms, is important to the knowledge and technology transfer process.

The goals of increased transfer and commercialization of university research might be best accomplished with selective patenting, nonexclusive licensing when possible, and (perhaps most importantly) effective technology transfer management that is flexible enough to allow interaction between industry and firm researchers, promote speedy license negotiation for access to university research results, and protect the incentives of university researchers to engage in basic science. This would continue to provide a rich knowledge landscape for Open Innovation, create formal IP protection when it is needed to facilitate knowledge transfer, and encourage the interactions with university scientists that assist firms drawing from this knowledge source.

Notes

1. The ability of others to imitate or duplicate the technology also depends on the characteristics of the technology, such as the complexity or tacitness of the related knowledge. For example, if understanding the technology requires scientists to work with the innovator and learn aspects of the related knowledge that would be difficult to capture in written description, it is considerably more difficult to copy the technology without this interaction. However, the same characteristics that make the knowledge easier to protect may also make it more difficult to transfer.
2. Several studies have examined the effect of the increase in patenting on the quality of university inventions. Quality of patents is typically proxied for using a count of the future patents that refer to the original patent as cited prior art. Patents relied upon more by follow-on innovation, the argument goes, are more important and of a higher 'quality.' Universities new to patenting received patents for inventions that were less important and less general, as compared to the patents of universities that were involved in patenting prior to the Bayh-Dole Act, but the patents of these 'entrant' universities improved over time (Mowery and Ziedonis 2001; Mowery et al. 2002). Research has demonstrated that citations to university patents are coming with increasing lags, relative to other patents, but the overall quality (as measured with a count of citations to the patent) of the university patents is not decreasing over time (Sampat et al. 2003).
3. Citations contained on the front page of patent applications have been used in existing literature to evaluate the importance of a patent (Hall et al. 2000, Trajtenberg 1990), trace knowledge transfer and diffusion (Jaffe and Trajtenberg 1996), proxy for characteristics of the patented technology (Trajtenberg et al. 1997), and compare the pace of innovation and obsolescence in an industry (Narin 1994), and a firm's closeness to science (Narin et al. 1989; Deng et al. 1999).
4. This Figure and Figure 7.3 rely on patents in international patent technology class A61K.
5. In order to exclude only occasional patenters, I restrict this analysis to firms in the technology class with at least twenty-one patents over the twenty-one-year period in each class.

6. The relationship of the variance with the number of firms depends on where the entering or exiting firms fall in the distribution of the number of nonpatent citations. I don't make any prediction about the sign of the relationship here, and simply include the number of firms as a control variable.
7. Shane (2004) provides an interesting approach to modeling the increase in university patenting at the line of business level and finds no relationship between the closeness to science and the amount of university patenting. However, he finds that the annual proportion of university research that is devoted to applied projects is significantly and strongly related to the amount of university patenting. A valid instrument in this context would be correlated with changes in university patenting over time but uncorrelated with the applicability of university research to industry.
8. I use the same sample of patents here as in the variance regression for consistency.
9. Note that the inclusion of the technology class fixed effect controls for the average citations to nonpatent prior art in each class. Therefore the firm-year level measure reflects differences across firms within each class.
10. The codified information resulting from research, such as publications, patents, or blueprints, may not be sufficient for a researcher to recreate or implement the research results described. In many cases, additional tacit knowledge, held by the original researcher, is required (Dasgupta and David 1994).
11. Most of the studies of knowledge transfer from university to industry either take an aggregate look across many industries or explore in detail the firm activities in the pharmaceutical and/or biotechnology sectors. This industry selection is (in part) because it is a fertile context for such study: Reliance on published university-based research has been shown to be the highest in the pharmaceutical and biotechnology sectors (Cohen et al. 2002; McMillan et al. 2000).
12. It is important to remember that these measures, publications, and collaborations, are indicator variables that represent the underlying organization routines and strategy of the firm. Firms that generate more publications are doing more basic science, but they also are likely to promote individual scientific inquiry, value scientific contributions, and build organizational practices that support the sharing of knowledge both with and across firm boundaries. Firms that collaborate with university researchers are also likely to build informal networks both within and across the boundaries of the firm. All of these unobserved characteristics likely contribute to the effects attributed to the indicator variables here.
13. As described in Fabrizio (2005*b*), results utilizing a model with firm level fixed effects suggest that an increase in publication activity or an increase in collaborations with university scientists by firm researchers is associated with an increase in citation to public science, as would be expected if these activities enhance the absorptive capacity of the firm. As a firm increases its internal basic research activities or builds its network of collaborations with university scientists, its exploitation of public science increases as well.
14. Although 1985 is an arbitrary break point, the increase in university patenting and number of citations to public science is most dramatic following 1985, and this is when the greatest increase in the variance in citations to nonpatent prior art across firms occurs.
15. The equality of the coefficients on the pre-1985 and post-1985 publications variable is rejected at the 1 percent level.

8

Open Standards and Intellectual Property Rights[1]

Timothy S. Simcoe

8.1 Introduction

Compatibility standards are used to govern the interaction of products and components in a technological system. In other words, they are the shared language that technologies use to communicate with one another. Standards are particularly important in the information and communications technology industries, where there are large numbers of interdependent suppliers and a very rapid pace of technological change. This chapter explores the inherent tension between cooperation and competition in the standards creation process, with a special emphasis on the role of intellectual property (IP) rights. These issues are closely linked to several key themes of Open Innovation, including the growing significance of IP-based business models, and the trend towards vertical dis-integration between technology development and commercialization.

While new standards can emerge from a market-based technology adoption process, this chapter focuses on the role of voluntary Standard Setting Organizations (SSOs). These organizations provide a forum where firms can collaborate in the design and promotion of new compatibility standards. Most SSOs promote the adoption of open standards—where the term 'open' implies that technical specifications are widely, perhaps even freely, available to potential implementers.[2] However, openness can pose a dilemma for individual firms hoping to benefit from SSO participation. While openness increases the probability of coordination on a particular standard (and hence its total expected value), it can also increase the intensity of competition, making it harder to capture that value once the new standard is introduced. As a result, SSO participants are often tempted to take actions that 'close off' a standard when those actions also give them a competitive edge in the product market. To put it crudely, SSO participants usually want all of the technology needed to implement a standard to be open—except for their own.

This tension between value creation and value capture—a key concern of open innovation—is also an inherent feature of standards creation, and is particularly evident in the ongoing debate over intellectual property rights (IPR) in the standard setting process. On one side, proponents of the open source model are working to create a set of legal institutions that make it impossible for firms to capture value through IP licensing. On the other side, some firms are actively 'gaming' SSOs in an effort to ensure that industry standards will eventually infringe on their own patents. Meanwhile, SSOs and policymakers are stuck in the middle trying to devise a framework that balances the legitimate interests of the various interested parties.

This chapter's central argument is that changes in the nature of the innovation process—particularly an increase in the number of specialized technology developers whose business models rely heavily on IP—have led to an increasingly contentious standard setting process. While there is nothing inherently harmful about the fact that the trade-off between value creation and value capture has become more severe, SSOs and policymakers need to be aware of this change in the economic and technological landscape when formulating IP policies and enforcing regulations.

The chapter begins by reviewing the literature on nonmarket standard setting and developing a framework for thinking about the relationship between standards, technologies, and implementations (i.e. products). It goes on to consider a number of strategies that technology developers may use to capture the value created by new compatibility standards. Firms that do not rely heavily on IP rights to capture value are often praised by standards practitioners for 'cooperating on standards and competing on implementation'.[3] However, the chapter uses a number of examples to illustrate how many firms appear to be moving away from cooperation on standards and towards business models that emphasize IP ownership as a primary source of revenues. The leading example of this phenomenon is the well-known Rambus case, where a new entrant successfully manipulated the standard setting process by exploiting loopholes in the patent system (Graham and Mowery 2004; Tansey et al. 2005).

The evidence of increasing conflict over IPR in the standard setting process raises the question: What has changed to make 'cooperating on standards and competing on implementation' less effective? The emergence of an innovation system characterized by Open Innovation provides a potential answer. In particular, the broad trend towards increased specialization in technology development and commercialization has created a more active technology input market, which many firms now rely on to procure standards-based inputs and/or monetize their inventions. However, many of the entrepreneurial but undiversified firms that supply the technology input market do not 'compete on implementation' (because they specialize at supplying technology) and therefore have few incentives to 'cooperate on standards'.

How should SSOs respond to a less cooperative standards creation environment? In the past, most SSOs stayed away from questions related to the licensing of IPR, for fear of alienating members or coming under the scrutiny of antitrust authorities.[4] However, over the last few years, a number of SSOs have experimented with changes to their IPR policies in an attempt to maintain a balance between encouraging cooperation and ensuring participation. The creation of an explicit antitrust 'safe harbor' for *ex ante* (i.e. prestandards) licensing negotiations should also be considered as a way of encouraging SSOs to govern the trade-off between the collective benefits of high-quality standards and the legitimate interests of IPR holders more effectively.

8.2 An Overview of Standards Creation

Compatibility standards are a set of rules for the design of new products. These rules facilitate coordination between independently designed products or components by establishing a common interface to govern their interactions. Much of the existing literature on compatibility standards has focused on network effects, and their ability to create positive feedback in the technology adoption process. This often leads to intense competition between technologies and the emergence of a single dominant technology or design as the industry standard. This process is often referred to as a 'standards war', and the list of well-known examples includes VHS versus Betamax in video recording, Apple versus Windows in operating systems, and Explorer versus Netscape in Internet browsers. The competitive dynamics of standards wars have been studied extensively, and the interested reader should see Varian and Shapiro (1999) for a thorough and easily accessible survey of this literature.

This chapter emphasizes the role of voluntary nonmarket Standard Setting Organizations as an alternative to standards wars. In their survey of the economic literature on standardization, David and Greenstein (1990) used the term *de jure* standard setting to describe the work of SSOs. Although this term suggests that SSOs have legal authority, in reality this is rarely the case. Most SSOs are voluntary associations with little or no power to enforce the technical rules they produce. However, because these groups operate in industries where the demand for coordination is large, SSOs can have a considerable impact on the rate and direction of technological change—primarily through their influence on the bandwagon process that leads to the adoption of a particular technology as the industry standard.

The term SSO can be applied to a broad range of institutions. At one end of this spectrum, there are a number of large well-established Standards Developing Organizations (SDOs) like the International Telecommunications

Union (ITU) or the Institute for Electrical and Electronic Engineering (IEEE). Many of these groups have been practicing collaborative innovation (i.e. technology sharing) for hundreds of years. In the middle, there are a number of smaller industry- or technology-specific groups—often labeled consortia. At the other end of the spectrum are the relatively informal standards developing communities that comprise the open-source software movement.[5] While these groups approach the problem of standardization in very different ways, their common goal is to create new technologies that will be widely implemented and adopted.

There is a relatively small body of formal theory related to nonmarket standard setting in SSOs. Much of this work is focused on issues of bargaining and delay, and emphasizes the fact that removing the standard setting process from the marketplace does not eliminate self-interested or strategic behavior by the sponsors of competing technologies. Farrell and Saloner (1988) use a simple model of standard setting based on the war of attrition to compare standard setting in markets and committees. They conclude that while markets are faster, committees are more likely to produce coordination on a single compatibility standard. Farrell (1996) and Bulow and Klemperer (1999) generalize and extend this model. Simcoe (2005) develops a slightly different model that emphasizes the role of collaborative design as well as competition in the committee standard setting process. His basic conclusion is that the process of design-by-committee will produce long delays and 'over design' when there are significant distributional conflicts over competing proposals. There are also a number of papers that theorize about other aspects of SSOs. For example, Lerner and Tirole (2004*a*) study the process of 'forum shopping' in which firms seek an SSO that will endorse their own technology; Foray (1994) considers the importance of free-rider problems in collaborative design; and Axelrod et al. (1995) examine alliance formation among the sponsors of competing technologies in a hybrid (market and committee) setting.

Most of the empirical evidence on the committee standard setting process is based on case studies. Examples include Weiss and Sirbu (1990), Farrell and Shapiro (1992), and Bekkers et al. (2002). There is also a large literature outside of strategy and economics—primarily written by standards practitioners—which sheds some light on committee standard setting. The leading authors in this literature include Cargill (1989, 1997), Krechmer (2005), and Updegrove (www.consortiuminfo.org). Recently, however, a number of large-sample of empirical studies of SSOs have started to appear. These include papers by Simcoe (2005) on the distributional conflicts and delay; Rysman and Simcoe (2005) on the economic and technological impact of SSOs; Toivanen (2004) on committee choices in cellular standardization; and Dokko and Rosenkopf (2003) as well as Fleming and Waguespack (2005) on technological communities and standards committee participation. The empirical work most closely related to this chapter are the empirical case studies by Bekkers et al. (2002)

and West (2003), that examine the IP strategies of SSO participants, and the question of how 'open' to make a standards-based product.

This chapter focuses on the trade-off between openness and control in standards creation. While this is a central theme in the literature on standards, it has not received a great deal of attention from empirical researchers.[6] This partly reflects the fact that openness is hard to define (e.g. West 2006). For some, openness means that anyone has a right to participate in the standards developing process. This 'open process' view is particularly common among large and well-established Standards Developing Organizations. For others, openness means that anyone who wants to implement a standard can do so on reasonably equal terms. This is the pragmatic 'open outcomes' view taken by many consortia, and some larger SSOs (such as the IETF). Finally, there are those who believe that a standard is not truly open unless it can be *freely* adopted, implemented, and extended by anyone who wishes to do so. The strongest advocates for this viewpoint are found within the open-source software community, which has developed a number of innovative legal institutions to safeguard the widespread availability of its work (Lerner and Tirole 2005).

For this chapter, it is important to note the somewhat subtle distinction between SSOs' use of the term 'open' and that of Open Innovation. In particular, Chesbrough (2003: xxiv) describes open innovation as, 'a paradigm that assumes firms can and should use external ideas... and external paths to market'. Open standards and Open Innovation both refer to a process that involves sharing or exchanging technology across firm boundaries. The difference is that the objective of open standard setting is to promote the adoption of a common standard, while the objective of Open Innovation is to profit from the commercialization of a new technology. In other words, Open Innovation might take place in a regime of either open or closed standards.

Why, then, do firms participate in 'open' standards development? The short answer is that open standards usually produce more value than closed standards. For consumers, open standards create value by promoting competition between implementations. This leads to a combination of lower prices and improved product quality. For the firms selling products that implement a standard, openness increases demand by resolving the uncertainties associated with potential coordination failures. Openness can also reduce implementers' costs through explicit restrictions on the 'tax' that can be imposed by technology licensors or through *ex ante* (i.e. prestandardization) competition between the sponsors of rival technologies.

The situation is somewhat more complicated for firms that produce the technologies used to implement a standard. It is reasonable to assume that these firms also participate in SSOs because they hope to capture some of the value associated with the creation of a new compatibility standard. Moreover, these companies will benefit from the additional value created by adopting an open specification. However, these firms might be willing to adopt a closed specification that produces less total value when it allows them to capture a larger share of the pie. In other words, they might settle for being the 'tax collector' in a world of closed standards—particularly when the alternative looks something like perfect competition.

Firms that develop standardized technologies must confront the trade-off between openness and control in developing a business model for commercializing their innovations (Chesbrough 2003: 64). In particular, firms that choose to specialize in developing input technologies and licensing them to implementers will bear the costs associated with a closed standard—including the possibility that firms will search for open substitutes to their proprietary technology. However, these costs may be tolerable for some firms, particularly small companies that cannot easily access the complementary assets needed to 'compete on implementation' (Teece 1986; Gans and Stern 2003).

To clarify this idea, Figure 8.1 depicts a world in which there is a continuous trade-off between openness and control. This trade-off is represented by a curve that indicates the share of value that a firm sponsoring a particular standard could capture as a function of the total value produced by that standard. In the limiting case of a completely open standard, there is a great deal of value created but the firm does not capture any. Conversely, when a standard is completely closed, it produces little or no value but the firm captures all of it.

The objective of a profit-maximizing firm is to choose a spot on this curve that maximizes the total amount of value it captures (i.e. the rectangle underneath any spot on this curve). The objective of an open SSO is to maximize the total value produced by the standard.[7] However, while an SSO sets the rules under which a standard is chosen, it cannot simply mandate the socially

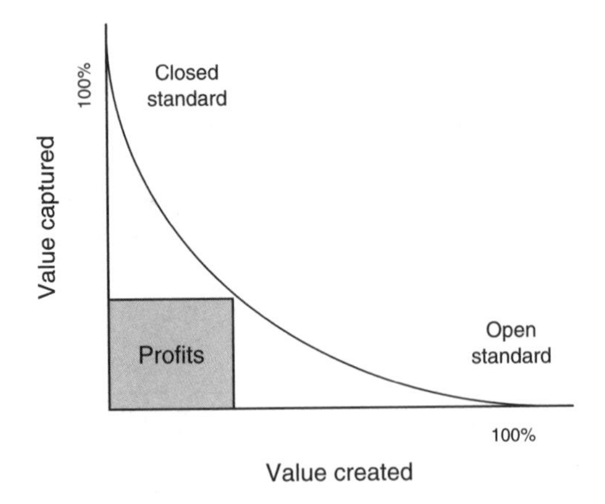

Figure 8.1. Openness versus control

8.4 Intellectual Property Strategies in Standards Creation

While the term 'intellectual property' encompasses patent, trademark, and copyright protection, this section will focus on patents, which are the vast majority of standards-related IPR. Patents give an inventor the right to exclude others from using their invention for a specified period of time (see Chapter 9). From a policy perspective, the role of a patent system is to create incentives for innovation by providing a legal solution to inventors' appropriability problems. This incentive will clearly be especially important for firms that cannot easily access or acquire the complementary assets required to profitably commercialize their inventions. As a result, patents play an important role in promoting vertical specialization in research and development by limiting the hazards faced by specialized technology developers with business models that call for selling inputs rather than implementations.

On the other hand, any administrative process granting potentially valuable property rights will almost certainly create some rent-seeking behavior. Over the last two decades, there has been a notable increase in the number of US patent applications. The majority of these applications have been granted, which has led to an increase in the scope of patentable subject matter and arguably a decline in average patent quality. A number of authors have considered various explanations for this surge in patenting and explored a number of its effects (e.g. Jaffe and Lerner 2004, and works cited therein).

Standards developers face a fundamental challenge with respect to IPRs. While patent proliferation means that more parties now have the right to impose a 'tax' on implementation, the shift towards Open Innovation has created an environment where 'taxation' appears to be a more attractive strategy. Increasingly, SSOs and their participants are facing difficult questions about how and when to reveal information about patents; the rights and obligations associated with SSO participation; the precise meaning of SSO policies; and whether the government will play an active role in enforcing them. It is not clear whether the existing framework of self-governance will be adequate to handle these changes.

Between 1995 and 2005, there were a number of legal disputes over the appropriate use of IPRs in the standard setting process. The two most significant examples, *Dell* and *Rambus*, both involved allegations that a firm failed to disclose essential IPRs—in violation of SSO policy—and then sought to license the undisclosed technology to potential implementers.[11] Both led to actions by the FTC.[12] These cases and several others have led to a growing interest among legal scholars in the antitrust and IP issues associated with standards creation. For an overview of these legal and antitrust issues, interested readers should refer to the standard setting bibliography prepared by the American Bar Association (ABA 2003), and the online transcripts from a series of hearings held in February 2002 (FTC 2002).

While a number of economists and strategic management scholars have also taken an interest in standard setting and IPRs, this literature remains small and somewhat fragmented.[13] For example, while Bekkers et al. (2002) and Rysman and Simcoe (2005) present some evidence of increasing IP disclosures at specific SSOs, no one has collected the data to illustrate any systematic increase in the number of standard-related patents or IPR disputes. Table 8.1 offers a brief overview of several IPR strategies that seem to be emerging at various SSOs, or have been discussed in the legal, practitioner, economics, or strategic management literatures.

The strategies listed in Table 8.1 can be separated along two dimensions. The first dimension corresponds to whether the strategy's objective should be

Table 8.1. Intellectual property strategies in standards creation

Strategy	Description	Examples	Open/Closed	Transparency
IPR contribution	Give away IPRs (royalty-free license) to promote implementation of a standard	Ethernet	Open	Yes
Defensive patent pools	Aggregate essential IPRs in the public domain to lower implementation costs	Cable Labs	Open	Yes
Open-source licensing	Require implementers to freely license any follow-on innovations	Linux, Apache, etc.	Open	Yes
Anticipatory standard setting	Create standards early to establish prior art and avoid commercial pressures	Early IETF	Open	Yes
Participatory licensing	Disclose patents in standard setting process and license to implementers	RSA cryptography patents	Closed	Yes
Ex post licensing	Conduct a search for standards-related IPR and approach implementers about licensing	Eolas vs. Microsoft BT hyperlink suit	Closed	No
Active hold-up	Participate in SSO without disclosing IPR and then pursue *ex-post* licensing opportunities	Rambus	Closed	No
Royalty-generating patent pools	Pool IPRs within a centrally administered licensing authority	MPEG-LA, Via Licensing	Closed	Sometimes
Cross-licensing alliances	A series of bilateral cross-licenses that has the effect of patent pool	GSM Semi-conductors	Usually closed	Sometimes
Disclosure strategies	Using information about IPRs to influence the pace and direction of SSO deliberations	Cisco MPLS?	Open	Sometimes

characterized as open or closed. Open strategies, such as IPR contributions, anticipatory standard setting, and defensive patent pools encourage value creation by enhancing the availability of the underlying technology. Closed strategies, such as licensing or hold-up, use IPRs as a mechanism to capture a share of the value created by a standard. The second dimension corresponds to the *transparency* of the strategy (i.e. whether other SSO participants are meant to know what the firm is doing). While all of the open strategies are transparent, this is not true for closed strategies. Some closed strategies—such as disclosure and licensing, or the formation of a royalty-generating patent pool—are consistent with a reasonably transparent standard setting process. Other closed strategies—such as secretly amending patents to cover a standard contemplated by an SSO, or conducting after-the-fact patent searches focused on exploiting industry standards—rely on secrecy and the informational advantages associated with holding a patent or pending application.

The simplest example of an open IPR strategy is the decision to disclose, but not assert, essential patents. For all of the attention paid to more aggressive IPR tactics, there are still a large number of firms who disclose the existence of their IPR to SSOs in a timely manner and make it available for free.[14] The decision to give away IPRs is usually based on an explicit recognition that doing so will improve the odds of a standard's success in either a committee or the marketplace. For example, the original sponsors of the Ethernet protocol (Digital, Intel and Xerox, sometimes called the DIX alliance) made a conscious decision not to pursue patent royalties before submitting the technology for standardization through the IEEE (von Burg 2001). Each of the companies in the DIX alliance was clearly in a position to benefit from the rapid dissemination of a free networking standard, given their large stake in complementary lines of business like computers and printers. Moreover, each of these firms had reason to fear the emergence of a proprietary protocol as the de facto local area networking standard.

One of the weaknesses of the traditional 'disclose but don't assert' strategy in a world of rapidly proliferating IPRs is that it requires a great deal of coordination. This is because patents apply to technologies rather than standards (see Figure 8.2). When a number of different technologies are needed to implement a single standard, it only takes a single firm asserting their IPR to create considerable uncertainty about potential costs. Royalty free patent pools are an open strategy that attempts to address this coordination problem by aggregating the IPRs needed to implement a standard. For example, the Cable Labs consortium maintains a royalty-free patent pool containing a number of patents needed to implement standard cable modem protocols (Lo 2002). In addition to ensuring access, these pools can lower potential implementers' IPR search and transactions costs.

SSOs with a royalty-free IPR policy, such as the World Wide Web Consortium, can be thought of as a de facto patent pool. There are even reports that

IBM has contemplated the creation of a 'public patent pool' in order to provide a formal mechanism for placing IPR in the public domain (Lohr 2005). The open-source licensing model (see Chapter 5) is a logical extension of royalty free patent pooling. The innovative feature contained in most open-source licenses is a 'grant-forward' provision which tries to make openness a self-sustaining feature of the technology by limiting implementers' ability to develop proprietary extensions.

While open-source licensing and variations on the royalty-free patent pool are innovative open strategies, it remains to be seen whether any of these approaches will actually solve the problem of a lone patent-holder's ability to hold a standard hostage. Some practitioners advocate 'anticipatory' standard setting (i.e. developing standards well ahead of the market) as a simpler approach to this problem (Baskin et al. 1998).[15] The advantages of anticipatory standards are twofold. First, they help to establish a body of prior art that can prevent companies from pursuing opportunistic patents designed to cover standards-related technology. In this sense, the anticipatory strategy closely resembles the practice of preemptive patenting or publication. Second, the anticipatory standard setting process may actually run smoother because it is further from the pressures created by imminent commercialization. The weakness of anticipatory standard setting is that it requires a great deal of foresight (and probably some good luck). Ongoing changes to the patent system, the process of university technology transfer, and the pace of commercialization also threaten to limit the scope of this strategy.

The most straightforward 'closed' IPR strategy is to license one or more patents for an essential technology to standards implementers. However, it is important to draw a distinction between firms who disclose their patents during the standards creation process, and those who wait until the process is over. One of the best known examples of the disclosure and licensing strategy comes from public key cryptography. In the late 1970s, a firm called RSA obtained a number of extremely strong patents covering the basic methods of public key cryptography. RSA regularly disclosed these patents—which were fairly well known in any case—to SSOs working on computer security or cryptography standards. Even though most SSOs have a preference for standards that do not require the use of IP unencumbered technology, the significance of RSA's invention and the scope of its patents led to the adoption of a number of specifications that required implementers to seek a license from RSA.[16]

While RSA's patent licensing strategy was carried out within the open standard setting process, some firms do not disclose their IPRs prior to the adoption or implementation of a standard. By waiting for a standard to be implemented and perhaps widely adopted before demanding royalties, these firms can take advantage of switching costs that naturally arise in many settings. These costs include product designs, specialized investments in manufacturing or distribution, and the accumulated experience with a particular technology. Section

8.2 described how these endogenous switching costs can *change* the value of an essential technology. This is an example of the 'hold up' problem, which has a long history in economics (e.g. Farrell et al. 2004).

Table 8.1 distinguishes between two slightly different variations on the 'hold-up' strategy. The first of these strategies, labeled 'active hold-up' is exemplified by Rambus' actions in an SSO that developed standards for computer memory. Rambus participated in the SSO but failed to disclose that it had a number of pending patent applications related to technology under consideration.[17] The firm then demanded that implementers license the patents after the standard was established and its pending applications were granted. The Rambus case generated a great deal of controversy—much of it centering on the company's efforts to subvert the transparency of the standards creation process.

There is another variation on the hold-up strategy that is labeled 'ex-post licensing' in Table 8.1. In this strategy, firms that do not necessarily participate in an SSO use the creation of new standards as an opportunity to extract rents from their existing patent portfolio. For example, in 1999 a small firm called Eolas sued Microsoft for including so-called 'applet' and 'plug-in' technologies in its Internet Explorer web browser, and was initially awarded over $500 million. In response, the W3C appealed for a USPTO review of the patent in question, suggesting that, 'the impact will be felt ... by all whose web pages and applications rely on the stable, standards-based operation of browsers threatened by this patent'. Another example of this rent-seeking strategy was British Telecom's attempt to assert a patent on the method of hyper-linking that is the basic method of creating links between pages on the World Wide Web. Recently, firms that specialize in acquiring patents purely for litigation—often derided as 'patent trolls' by the targets of their lawsuits—have emerged as significant players in some technology markets.

The apparent increase in *ex post* IPR licensing strategies may also be related to the reemergence of patent pools. In 1995, the Department of Justice (DOJ) issued guidelines relaxing its prior restrictions on the formation of patent pools. This ruling appears to have opened the door for patent pools to serve as a coordinating mechanism for firms who see standards as a tool for boosting licensing revenues from an existing patent portfolio. This takes place through firms like Via Licensing, which has issued a 'call for patents' to solicit potential licensors of technology related to a variety of established standards, such as the IEEE's 802.11b standard for wireless networking. In some cases, the goal of creating a royalty-generating patent pool is an explicit part of the initial standards creation effort. This practice is common for media-format standards, such as MPEG, CD, and DVD.

It is often difficult to evaluate the competitive implications of patent pools or cross-licensing agreements. These arrangements can encourage competition—particularly when they solve the 'patent thicket' problem by reducing the transaction costs associated with repeated bilateral licensing for complementary technologies (Lerner and Tirole 2004*b*) On the other hand, it seems clear that they can also be used by incumbent firms to create entry barriers or raise rivals costs. These issues are addressed by a number of authors who have written about the widespread use of cross-licensing agreements in high-tech industries (Grindley and Teece 1997; Hall and Ziedonis 2001). However, Bekker's study of GSM alliance formation is one of the only papers to explore how these arrangements both influence and respond to the creation of new compatibility standards.

The final category of Table 8.1, 'disclosure strategies' describes a variety of tactics that firms may use in the standards creation process. This chapter briefly discussed the distinction between transparent IPR strategies and 'active hold-up'. However, it is clear that there are a number of more subtle approaches to IPR disclosure. For example, some practitioners claim that Cisco used IPR disclosures at the IETF to discourage the adoption of a new routing protocol that might emerge as a competitor to its preferred technology (Brim 2004). The issue of disclosure strategy raises a host of questions related to the costs and benefits of delay, forum shopping, and competition between standards. Many of these issues call for additional research.

In describing a number of different standards-related IPR strategies, this section has suggested several possible reasons for the apparent proliferation of IPR issues at many SSOs. This chapter has focused primarily on a single explanation—the trend towards an innovation system of Open Innovation that involves a greater reliance on IPR-based business models. However, there have also been changes in the quantity and average quality of issued patents as well as the increase in standards-related patent pools. Moreover, the success of a few firms like Qualcomm and IBM at licensing their standards-related IPR may have raised firms' awareness of the strategic possibilities in this area.

It seems likely that each of these explanations for the increasing awareness of the strategic possibilities of IPR in compatibility standards is at least partly correct, and they may be working to reinforce one another. The actual size of the increase in IPR controversy and precisely how much can be attributed to each of these explanations is a subject for future research. What is clear is that the increasing controversy surrounding IPR strategies in standard setting presents a clear challenge for SSOs. Section 8.5 examines how these organizations are responding.

8.5 Intellectual Property Rules at SSOs

The simple framework developed in Section 8.3 described how the creation of new compatibility standards can influence the value of technologies used to implement them—and by extension any IPRs that 'read on' those technologies. Section 8.4 presented some evidence that firms are becoming increasingly sophisticated in their efforts to gain a competitive advantage through the interaction of IPR strategy and participation in the standards creation process. This raises the question of how SSOs deal with IPR concerns during the standard setting process.

By joining an SSO, individual members incur a set of obligations that are outlined in the charter and bylaws of the organization.[18] The goal of these rules is to ensure that participants can make an informed decision between alternative technologies. To a large extent, the role of IPRs in the open standards creation process is governed by SSO-specific rules and procedures that can be divided into three types: search, disclosure, and licensing. Broadly speaking, these rules are designed to provide a set of procedural safeguards that will prevent SSO participants from adopting a standard that exposes them to *ex post* hold-up by patent holders offering a license that would not have been accepted in an *ex ante* negotiation.

Much of our knowledge about SSO practice comes from recent work by Lemley (2002), who surveyed the IPR policies of roughly forty SSOs. Lemley's survey found that while most SSOs with a formal IP policy have some kind of disclosure rules, relatively few require their members to conduct a search of their own files or the broader literature in order to identify relevant IPRs. The survey also revealed considerable heterogeneity in the substance of disclosure rules. The most general rule requires SSO participants to disclose any patents that they could 'reasonably' be expected to know about—particularly those owned by their own employers. While this raises significant questions about what constitutes reasonable knowledge of a firm's IP portfolio (consider the different situations faced by a sole proprietor and an employee of IBM) SSOs do not typically address this issue. Most of the SSOs surveyed by Lemley required the disclosure of granted patents but not pending patent applications, in spite of the growing lag between patent applications and grant dates.

There are a number of explanations for the apparently limited use of search and disclosure rules by many SSOs. The most straightforward explanation is that these rules can impose a significant burden on SSO members and participants. This is particularly true of search rules, which may require legal skills and expertise that most of the engineers who participate in SSOs do not have. Moreover, search costs will be heavily skewed towards firms with large patent portfolios. These are often firms that the SSOs are anxious to have participants, since they can play an important role in promoting a completed standard. In addition to the concern that larger firms would respond to strict search and disclosure rules by refusing to participate, there is the possibility that they would simply provide 'blanket' disclosures containing so much information that they are essentially useless. In some cases, search and disclosure rules may be weak simply because it is easy for the SSO or its participants to learn about potential IP—in which case it is easy to make an informed decision without the burden associated with formal rules. Finally, the lack of strong search and disclosure rules may reflect a combination of historical bias and organizational inertia, since many SSOs adopted their rules and bylaws at a time when the economic and technological landscape was quite different. There is some evidence that a number of SSOs are responding to the various examples discussed above by updating their IPR rules.

SSOs have also sought to ensure the openness of their standards through licensing rules, which restrict the terms sought by SSO participants for IPR that is included in (or essential to) a compatibility standard. Licensing rules can be motivated by a number of different goals. First, they encourage adoption of the standard by offering a guarantee to potential implementers. Second, they can reduce inefficiencies and incentives to engage in rent-seeking behavior (such as the manipulation of information) in the standard setting process. Finally, they reduce the level of uncertainty inherent in the standards creation process by removing worries about pending patent applications, infringement, or the scope of granted claims.

There are essentially three types of SSO licensing rules. The most popular by far is the 'reasonable and non-discriminatory' (RAND) licensing requirement. In practice, this requirement is fairly vague. While it is clear that a RAND rule implies that IPR holders cannot refuse to grant a license, it leaves them with fairly wide latitude to set prices that can even vary by licensee. Moreover, most SSOs do not actually make any determination about the 'reasonableness' of a license, but rather presume that this criteria has been met as long as a license has been granted. A few SSOs, such as the W3C or some IETF Working Groups, go beyond RAND and require participants to grant a royalty-free license for any technology incorporated into a standard. Finally, there are a handful of SSOs with rules requiring patent holders to assign their IPRs to the SSO.

The fact that SSO licensing policies appear to be clustered at the 'corner solutions' of RAND and royalty-free is somewhat puzzling. While RAND places a very limited set of restrictions on the SSO participants, royalty-free licensing requirements are plainly quite severe. Why haven't SSOs adopted a range of intermediate solutions, such as an *ex ante* 'single-price' rule that would require a firm to commit to a single set of verifiable licensing terms before their IPR is included in a standard?[19] The attractiveness of RAND could come from its minimal impact on SSO participation. Stricter rules might drive organizations with large IPR portfolios out of the SSO, or even worse, lead to a standards war—although the adoption of a royalty-free policy does not appear to have had a major impact on participation in the W3C.

Another possibility is that most SSOs have adopted RAND licensing policies because they worked reasonably well in the past. Lemley found that the rules governing IP were specified in much greater detail for JEDEC and VESA—the two SSOs involved in the Dell and Rambus disputes. Moreover, there is some evidence that these recent controversies have started to have an impact. W3C and OASIS adopted royalty-free licensing policies in 2003 and 2004, respectively. While it was nearly impossible to obtain information on IPR disclosures a few years ago without going directly to individual participants, a number of SSOs have recently made their disclosure data available on the website.

A final explanation for the popularity of RAND and royalty-free licensing policies is that SSOs worry about the antitrust implications of adopting alternatives. In particular they may fear that policies leading to explicit negotiation over royalties can be construed as facilitating collusion. RAND requirements are too vague to be construed as collusive and royalty-free licenses are not an issue (since firms rarely collude to set prices at zero). The problem with this outcome from a policy perspective is that RAND leaves open the door to hold-up while royalty-free licensing rules may damage innovation incentives by preventing IPR holders from capturing the value associated with their inventions.

The strongest threat available to an SSO is to withhold or withdraw its endorsement of any standard sponsored by a firm that fails to comply with its rules. However, this will have little impact if the breach is not revealed until a specification is well on its way to becoming a de facto standard.[20] Much stronger compliance incentives are created by SSOs bylaws that contain explicit language to the effect that participants who violate search and disclosure rules forfeit their future rights to assert IP in a given standard. However, it is up to government agencies and the courts to enforce this type of rule. The legal outcomes in the Dell and Rambus cases suggest that antitrust authorities are inclined to intervene in support of SSOs when there appears to be a violation of disclosure rules. However, these cases also suggest that SSOs must be far more explicit in the construction of their own charters and bylaws if they hope to see them upheld in court.

Although the data available to answer this question are rather limited, explaining the variation in SSO IPR policies is an important and interesting topic for research. Several authors (e.g. Lerner and Tirole 2005) have speculated that much of this variation can be explained by competition between SSOs to offer an attractive standards creation environment. However, these authors seem to reach different normative conclusions. While Teece and Sherry (2003) argue that SSO competition should lead to an efficient distribution of IPR rules, Lemley (2002) concludes that, 'diversity [in IPR policy] is largely accidental, and does not reflect conscious competition between different policies'.

Perhaps the most interesting explanation for the exiting variation in SSO policy is offered by Cargill (2001), who suggests that SSOs have undergone a type of 'organizational evolution' in response to a broader imperative for faster standards development. He suggests that the different types of SSO described at the beginning of Section 8.2 correspond to different phases in the history of standard setting. From this perspective, the current controversy over IPR policies is at least partly the result of the convergence of the information and communications technology (ICT) industries. The culture of open source software development is very different from that of telecommunications engineering. Both of these cultures have established a set of norms and routines that reflect the logic of the respective industries. However, they have very different ideas about the appropriate use of IPR or what constitutes a legitimately 'open' standard. Cargill's thesis suggests that as the computing, telecommunications, entertainment, and information businesses continue to combine in new and unexpected ways, we will continue to see strong differences of opinion over the issue of open standards and IP rights.

8.6 Conclusions

While there are a wide variety of different SSOs, all of them face a basic trade-off between collaboration and competition, or between openness and control. To analyze this trade-off, it is crucial to understand the distinction between standards, technology, and implementations. This chapter has argued that the increasing controversy surrounding IPR strategy and policy is an indication that the trade-off between collaboration and competition has become more severe. While there are several potential explanations for this increasing severity, I focused on the importance of the broad shift towards a system of Open Innovation. Open Innovation is characterized by increasing vertical specialization in technology and development and commercialization and has led to a proliferation of firms whose business models rely heavily on IPR because they lack access to the manufacturing and distribution capabilities required to 'cooperate on standards and compete on implementation'.

As the US innovation system continues to evolve towards the Open Innovation model, it is important for firms in industries where standards are important to recognize the potential costs associated with their IPR strategies. In particular, aggressive IPR strategies can reduce the expected value of a standard and slow down the standards creation process. While these aggressive strategies may be perfectly legitimate from a legal perspective and strengthen (at least in theory) the incentives for small-firms to commercialize their innovations, it is important to recognize that these strategies increase the severity of the trade-off between value creation and value capture.

SSOs also need to understand how the trend towards Open Innovation potentially complicates their job of providing a forum for informed decision-making in the creation of new compatibility standards. IPR policies based on little search, vague licensing rules and lax enforcement are likely to lead to time-consuming and expensive controversies along the lines of the Rambus case. While SSOs are clearly constrained by the need to encourage participation—and potentially by competition with other SSOs—they should strive to update their IPR policies in a way that promotes transparency in the standard setting process while respecting the legitimate rights of IPR holders.

SSO IPR polices must balance the goals of providing incentives to select the best available technology (which includes encouraging participation), ensuring that the standard setting process is reasonably efficient (which includes not placing too large a burden on participants), respecting the legitimate rights of IPR holders, and encouraging widespread diffusion and implementation of the standard. In order to achieve these outcomes, SSO must ensure that participants in the standards creation process are well informed. However, they have often been reluctant to allow firms to negotiate *ex ante* commitments to licensing terms, partly because it wasn't necessary when most of the participants already had existing cross-licenses and partly because of antitrust fears. The first of these conditions has changed, and it would probably be useful for antitrust authorities to offer some type of SSO 'safe harbor' guarantee to eliminate the latter concern. This would encourage SSOs to take a more active role in resolving conflicts over IPR.

Finally, there is a clear need for more research into a number of the open questions raised in this chapter. For example, is there evidence for a connection between Open Innovation and the prevalence of IPRs in standard setting (e.g. has the growth in IPR disclosures been driven by small and/or vertically dis-integrated firms)? What is the link between SSO or firm characteristics and the choice of IP rules or strategies? What is the role of competition between SSOs in shaping the standard setting environment? At a broader level, there is a great opportunity to develop a research agenda that examines the links between features of the innovation and standard setting environment, the strategies and behaviors of SSO participants, and the performance of the standards creation system. These issues have a broad significance that extends beyond the creation of compatibility standards and will potentially deepen our understanding of nonmarket strategy and the institutions of self-governance.

Notes

1. I would like to acknowledge the generous financial support of Sun Microsystems, and the University of Toronto Connaught new faculty grant program. Many thanks to Catherine McArthy and Andrew Updegrove for taking the time to discuss these ideas with me. My editors, Joel West and Henry Chesbrough, also provided a host of useful comments and suggestions. All errors and omissions are, of course, the responsibility of the author.
2. The precise meaning of 'open' in the context of compatibility standards is highly contested. Moreover, the meaning of 'open' in this context is different from that employed by Chesbrough in Open Innovation—a point which will be elaborated below.

3. See, for example, IBM's patent licensing statement at http://www.ibm.com/ibm/licensing/standards/.
4. Indeed, the majority of SSOs continue to have so-called RAND ('reasonable and nondiscriminatory') IPR policies that encourage them to take a fairly passive stance on these issues (Lemley 2002).
5. Although one might exclude open source developers from the definition of an SSO on the grounds that they are focused on implementations rather than standards development per se. However, a reasonably broad definition of SSO should make room for open source projects that are truly multilateral, consensus-based efforts to develop new technology platforms.
6. One notable exception is Chiao et al. (2005).
7. The SSOs problem strongly resembles social dilemma created by issuing patents. Patents exist to provide an incentive for innovation, but once an innovation is in place the presence of patents leads to distortions of the market. In a similar fashion, 'closed-ness' can provide an incentive for firms to participate in standards development. However, once a standard is developed, society—and hopefully the SSO—would be better off if it were open.
8. Chesbrough and Rosenbloom (2002) make a similar point about technology more generally. They argue that the value of a technology is not realized until it is commercialized through a business model.
9. The formal models of Farrell, Farrell and Saloner, and Simcoe (discussed above) build on this idea.
10. Of course, 'competing on implementation' does not completely remove the incentive for firms to try and assert control over a standard—in Figure 8.3, a completely open standard still generates zero profits.
11. Dell—a large firm that was well positioned to compete on implementation—eventually agreed to license its patents freely, while Rambus—a small firm specializing in technology development—fought a long and bitter court battle.
12. *In the Matter of Dell Computer Corporation*, 121 F.T.C. 616 (1996) was settled by consent order, *while In the Matter of Rambus Incorporated*, 121 F.T.C. Docket # 9302 remained an active adjudicative proceeding before the FTC as of mid-2005.
13. There is, however, a large related literature in economics on technology licensing and collaborative R&D, much of it theoretical.
14. In practice, a firm that 'gives away' its IPRs usually agrees to grant any implementer a royalty-free license, which may contain a number of clauses related to 'reciprocity' and 'grant-back' (i.e. a promise to offer the original patent holder a royalty free cross-license for any improvement to the underlying technology). The overall impact of a royalty-free license is to place the underlying technology in the public domain.
15. The development of the original Internet protocols is a good example of anticipatory standard setting (Mowery and Simcoe 2002*a*).
16. The willingness of some SSOs to adopt standards based on RSA's technology is likely due to the fact that these patents were set to expire just as public-key encryption was set to become a critical part of the Internet.
17. In fact, Rambus utilized a loophole in the patent system known as a continuation filing to actively amend its pending applications, ensuring that it would own IPRs in the eventual standard (Mowery and Graham 2004).

18. For a full treatment of the contractual issues related to organization membership, the reader is referred to Lemley (2002).
19. In practice, some firms do commit to license their IP on specific terms (usually free) as part of the standards creation process. However, this is impractical in many cases—particularly when the IPR is contained in a pending patent application whose scope is highly uncertain.
20. Indeed, the fact pattern in the Rambus case involves an alleged failure to disclose IP followed by the creation of a JEDEC standard that probably infringed on Rambus' IP. Rather then withdraw a standard into which its members had sunk significant resources, JEDEC contended that Rambus' actions had led to a forfeit of that IP.

9

The Use of Intellectual Property in Software: Implications for Open Innovation[1]

Stuart J. H. Graham
David C. Mowery

9.1 Introduction

The sine qua non of the 'open innovation' paradigm is the use by firms of both internal and external sources of knowledge in their quest for competitive advantage in the marketplace (Chesbrough 2003*a*). The movement from the 'closed' to the 'open' innovation paradigm has spanned decades.[2] Chesbrough catalogues changes in firms' strategies during the 1980s and 1990s for sourcing knowledge and managing their innovation processes, with firms moving increasingly from a 'closed' to an 'open' paradigm. This shift to an 'open' paradigm encompasses change in industry structure, firm strategy, and managers' perception and uses of intellectual property (IP).

In this chapter we analyze management of IP in the software industry during the 1980s and 1990s. Our analysis of the development of IP protection in software highlights the tension between Open Innovation and changes in US IP policy that have strengthened the rights of patentholders and have raised the economic value of patents in many sectors, including software.

Software technology has been one of the main drivers of dynamism and controversy in the 'new' IP regime of the post-1980 period. Software is a good candidate for investigating how intellectual property rights (IPRs) have created an environment for 'open innovation.' First, software is an important element of the information technology (IT) 'general purpose technology' that is used across many diverse economic sectors. Second, the introduction and adoption of software technologies and the growth of an industry to support and supply software products have occurred during a period of growing emphasis by corporations on Open Innovation. In fact, many of the firms that Chesbrough highlights as practitioners of 'open innovation' are among the most significant users and producers of software technologies.[3]

IP protection creates a platform for the transfer of knowledge assets that spring from inventive activity. The external sourcing of knowledge is crucial to the operation of Chesbrough's system of 'open innovation' and frequently relies on the 'markets for intellectual capital' that may benefit from stronger formal IPRs. Transactions in knowledge are complicated by the characteristics of knowledge, an intangible asset characterized by substantial uncertainty. Uncertainty over the boundaries and uses of knowledge assets limits the recognition by buyers and sellers of opportunities and increases transaction costs in the transfer of these assets. Transaction costs rise because heterogeneity in these assets is relatively high, their boundaries are often fuzzy, and the disclosure of their attributes is relatively difficult (Teece 2000). IPRs often support the development of markets for the sale and licensing of knowledge.

Computer software is a technological field in which patenting, particularly in the US, has grown rapidly in recent years (Graham and Mowery 2003) despite the fact that the underlying 'technology' is several decades old. As we note below, much of this increased patenting reflects changes in industry structure and in US policy toward software patents that have increased the economic value of patents. This chapter builds on our earlier studies of software patenting (Graham and Mowery 2003; Graham et al. 2003; Graham and Mowery 2005) that analyzed trends in such patenting during the 1980s and 1990s. With our analysis in this chapter, we hope to shed additional light on the causes of the growth of US software patenting while investigating the environment in which 'open innovation' occurs in this and other sectors.

9.2 Development of the US Software Industry: Vertical Dis-integration

The growth of the global computer software industry has been marked by at least four distinct eras spanning the post-1945 period (Graham and Mowery 2003). During the early years of the first era (1945–65), covering the development and early commercialization of the computer, software as it is currently known did not exist. Even after the development of the concept of a stored program, software was largely custom-developed for individual computers. During the 1950s, however, the commercialization and widespread adoption of 'standard' computer architectures supported the emergence of software that could operate on more than one type of computer or in more than one computer installation. In the US, the development of the IBM 650, followed by the even more dominant IBM 360, provided a large market for standard operating systems and application programs. The emergence of a large installed base of a single mainframe architecture occurred first and to the greatest extent in the US. During this period, however, most of the software for early mainframe computers was produced by their manufacturers and users.

During the second era (1965–78), independent software vendors (ISVs) began to appear. During the late 1960s, producers of mainframe computers 'unbundled' their software product offerings from their hardware products, separating the pricing and distribution of hardware and software. This development provided opportunities for entry by independent producers of standard and custom operating systems, as well as independent suppliers of applications software for mainframes. The changing structure of the industry created a supply of innovation from newer, smaller entrants. Unbundling occurred first in the US and has progressed further in the US and Western Europe than in the Japanese software industry, and is consistent with Chesbrough's description of an increasingly dis-integrated innovation environment.

Although independent suppliers of software began to enter in significant numbers in the early 1970s, computer manufacturers and users remained important sources of both custom and standard software in Japan, Western Europe, and the US during this period. Some computer 'service bureaus' that had provided users with operating services and programming solutions began to unbundle their services from their software, providing yet another cohort of entrants into the independent development and supply of traded software. Sophisticated users of computer systems, especially users of mainframe computers, also created solutions for their applications and operating system needs. A number of leading suppliers of traded software in Japan, Western Europe, and the US were founded by computer specialists formerly employed by major mainframe users.

During the third era (1978–93), the development and diffusion of the desktop computer produced explosive growth in the traded software industry. Once again, the US was the 'first mover' in this transformation, and the US market quickly emerged as the largest single one for such packaged software. Rapid adoption of the desktop computer in the US supported the early emergence of a few 'dominant designs' in desktop computer architecture, creating the first mass market for packaged software. The independent vendors that entered the desktop software industry in the US were largely new to the industry. Few of the major suppliers of desktop software came from the ranks of the leading independent producers of mainframe and minicomputer software, and mainframe and minicomputer ISVs are still minor factors in desktop software.

Rapid diffusion of low-cost desktop computer hardware, combined with the emergence of a few 'dominant designs' for this architecture, eroded vertical integration between hardware and software producers and opened up great opportunities for ISVs in the marketplace, and supported a distributed supply base of innovation in software technologies. Declines in the costs of computing technology have continually expanded the array of potential applications for computers and supported a vibrant environment for innovation; many of these applications rely on software solutions for their realization. A growing installed base of ever-cheaper computers has been an important source of dynamism and entry into the traded software industry, because the expansion of market niches in applications has outrun the ability of established computer manufacturers and major producers of packaged software to supply them.[4]

The packaged computer software industry now has a cost structure that resembles that of the publishing and entertainment industries much more than that of custom software—the returns to a product that is a 'hit' are enormous and production costs are low. And like these other industries, the growth of a mass market for software elevated the importance of formal IPRs, especially copyright and patent protection.

An important contrast between software and the publishing and entertainment industries, however, is the importance of product standards and consumption externalities in the software market. Users in the mass software market often resist switching among operating systems or well-established applications because of the high costs of learning new skills, as well as their concern over the availability of an abundant library of applications software that complements an operating system. These switching costs typically are higher for the users who dominate mass markets for software and support the development of 'bandwagons' and the creation through market forces of product standards. As the widespread adoption of desktop computers created a mass market for software during the 1980s, these de facto product standards in hardware and software became even more important to the commercial fortunes of software producers than was true during the 1960s and 1970s. In some instances strong IP protection aided firms in establishing control of a proprietary standard (see Chapter 8), further raising the value of formal instruments for IP protection.

The fourth era in the development of the software industry (1994–present) has been dominated by the growth of networking among desktop computers, both within enterprises through local area networks linked to a server and among millions of users through the Internet. Networking has opened opportunities for the emergence of new software market segments (e.g. the operating system software that is currently installed in desktop computers may reside on the network or the server), the emergence of new 'dominant designs,' and potentially the erosion of currently dominant software firms' positions. Some network applications that are growing rapidly, such as the World Wide Web, use code (html) that operates equally effectively on all platforms, rather than being 'locked into' a single architecture. Like the previous eras of this industry's development, the growth of network users and applications has been more rapid in the US than in other industrial economies, and US firms have maintained dominant positions in these markets (See Mowery and Simcoe 2002*b*).

The Internet has provided a new impetus to the diffusion and rapid growth of open source software. Although so-called 'shareware' has been an important form of software in all of the eras of the software industry, the Internet's ability to support rapid, low-cost distribution of new software and the centralized collection and incorporation into that software of improvements from users has made possible such widely used software as Linux and Apache (see Kuan 2001; Lerner and Tirole 2002; and Lee and Cole 2003).

This brief overview of the development of the US software industry's structure illustrates the ways in which the emergence of independent software vendors and the growth of 'mass markets' for standard (packaged) software have elevated the importance of formal IPRs for firms in the industry. At the same time, however, changes in the policy regime governing overall US IP protection, as well as shifts in the coverage of software by such formal instruments for IP protection as patents and copyright, have important implications for the future evolution of industry structure. Section 9.3 describes the evolution of this formal IPR regime for software.

9.3 The Evolution of US Patent Protection for Software

This section and the next present a descriptive summary of the growth of software patenting during the 1987–2003 period. Although this discussion focuses on patents, other forms of IP protection, including copyright,[5] trade secret,[6] and trademark[7] remain important in software.

Federal court decisions during the past decades have consistently broadened and strengthened the economic value of software patents.[8] Although early cases during the 1970s supported the stance of the US Patent and Trademark Office (USPTO) in stating that software algorithms were not patentable,[9] the landscape has significantly shifted since then (Samuelson 1990).[10] In the cases of *Diamond v. Diehr*[11] and *Diamond v. Bradley*,[12] both decided in 1981, the Supreme Court announced a liberal rule that permitted the patenting of software algorithms. Both the courts and the USPTO have supported this policy, strengthening patent protection for software (Merges 1996). A vivid example of the effects of this stronger patent regime is the 1994 court decision that found Microsoft guilty of patent infringement and awarded $120 million in damages to Stac Electronics, the plaintiff. The damages award was hardly a crippling blow to Microsoft, but the firm's infringing product had to be withdrawn from the market temporarily, compounding the financial and commercial consequences of the decision (Merges 1996). Sony Corporation faced a similar ruling in 2005 affecting its Playstation user-control joysticks, after a US federal court determined that Sony had infringed two patents held by

Immersion Corporation. Sony was ordered to pay $91 million and cease US sales of its Playstation consoles and forty-seven of its software titles (Williams 2005).

9.4 Patenting Trends in the US Software Industry, 1987–2003

In order to better illuminate the IP environment in which 'open innovation' in software occurs, we examine trends in US software patenting during 1987–2003, focusing on the product areas we believe have been most affected by changes in the legal treatment of software patents—packaged software. As our previous work emphasized (Graham and Mowery 2003), no widely accepted definition of 'software patent' exists, and the problem of identifying software patents is not made easier by USPTO changes in patent classification schemes. Other researchers have chosen to define a 'software patent' by reference to certain key words in the patent disclosure (Bessen and Hunt 2004), but we rely instead on the classification decisions of USPTO patent examiners. Empirical support for the comparative accuracy of our 'classification' method is forthcoming (Layne-Farrar 2005).

One difficulty that arises when using patent office classifications is the rapid growth in the number of software-related USPTO patents during the period of this analysis. Because we are interested in analyzing changes over time in the number of software patents, we seek to insulate our sample from any 'reclassifications' of patents from 'all other' to a 'software-related' category. Such an analysis is made more difficult by the USPTO's unannounced reclassifications. In order to eliminate the impact of reclassifications, we use a 'snapshot' in time of the US classifications for US-issued patents. All patents in our analysis reflect their classifications on the same date.

9.4.1 *Method for Defining a 'Software' Patent*

Using the USPTO's 'Cassis' database, we collected data on all patents issued in the US from 1987 to 2003, relying on the USPTO classification scheme as of December 31, 2003, with patents issued on previous dates updated to reflect current USPTO classifications, thus allowing us to compare trends in patenting over time. These classes, disclosed in Table 9.1, were identified by examining US patenting in the years 1987–2003 by the six largest US producers of personal computer software, based on their calendar 2000 revenues as reported by *Softletter* (2001).[13] These twelve USPTO classes account for 67.9 percent of the patenting of the 'top 6' firms, while these same six firms account for 88.4 percent of the patenting of the *Softletter* 'top 100.' Table 9.1 contains data on the distribution of the patents in this sample among these twelve USPTO classes.[14]

Our definitional scheme does not cover all software patents, but it does provide longitudinal coverage of a particularly dynamic and important

Table 9.1. Patenting by the Softletter 100 (2001), by USPTO patent class, 1987–2003 (total patents = 3,891)

US patent class	Patent count	Share of all firm patents (%)	Cumulative total (%)
345	730	18.8	18.8
707	624	16.0	34.8
709	363	9.3	44.1
382	157	4.0	48.2
713	141	3.6	51.8
704	125	3.2	55.0
717	118	3.0	58.0
714	85	2.2	60.2
711	80	2.1	62.3
710	59	1.5	63.8
358	52	1.3	65.1
715	25	0.6	65.8

segment of the overall software industry, inasmuch as IDC estimated that global packaged software revenues amounted to $179 billion in 2004.[15] The data in Figure 9.1 indicate that the share of all US patents accounted for by software patents grew from 2.1 percent to 7.4 percent of all issued US patents between 1987 and 1998, and the share of patents in these twelve US classes has remained between 6.9 and 7.5 percent of overall patenting during 1999–2003. This slowing in the rate of growth in 'software' patenting as a share of total US patenting occurs in virtually every USPTO class included in our definition of software patents, and may reflect the effects of the post-2001 downturn in the IT industry.[16]

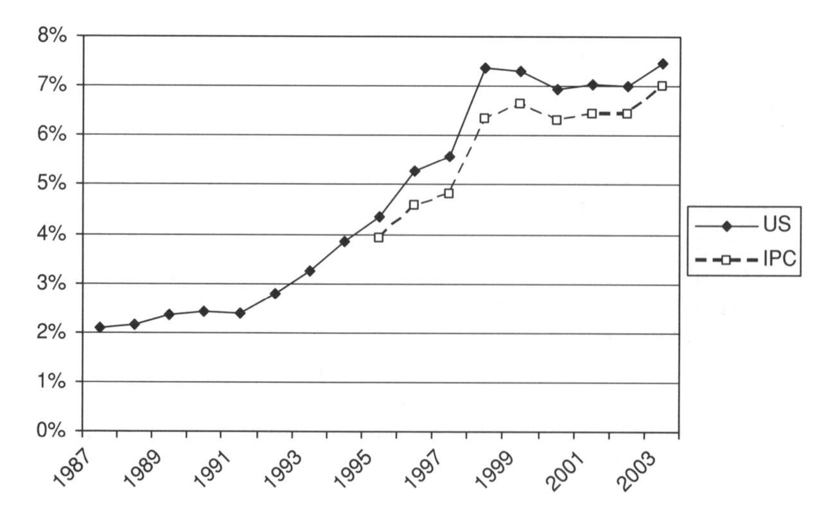

Figure 9.1. Software patents as a share of all issued US patents, 1987–2003 (Comparing two definitions: US patent classifications and international patent classification)

There are several potential explanations for the slowdown in software patent growth, relative to overall US patenting, after 1999. In other work (Graham and Mowery 2004), we noted that 1995 changes in the legal patent term of protection (changing the term of protection from seventeen years from date of application to twenty years from date of issue) created strong incentives for patent applicants to pursue 'continuations' in their applications, which (among other advantages) enabled applicants to extend the length of application secrecy. Because we showed that software-patent applicants made extensive use of continuations, it is possible that the 1995 changes in patent term may have reduced incentives for software inventors to seek patents. Moreover, the negative effects on applicants' incentives to pursue continuations have been intensified by more recent requirements to publish many patents within eighteen months of application.

It is also possible that the accumulation of experience by USPTO examiners in dealing with software-patent applications, as well as the expanding body of patent-based prior art on which examiners rely in part, have led to lower rates of issue for software patent applications. The fluctuations in growth in software patents, however, do not appear to be associated with fluctuations in the 'pendency' of patent applications (the length of time required to review and grant or deny patent applications), since the average pendency of applications for issued software patents, which is greater than the average for all patents, has increased steadily through the 1995–2003 period.[17]

9.4.2 *Software-related Patenting by Packaged Software and Electronic Systems Firms, 1987–2003*

In this section, we analyze patenting by US software firms during 1987–2003, focusing on leading US packaged software firms identified by Softletter in their 2001 tabulation of the 100 largest US packaged software firms (based on revenues). We focus on these firms because, unlike the electronics systems firms, inventive output is more likely to be purely software-related. Figure 9.2 displays trends during 1987–2003 in the share of all US software patents held by the 100 largest US packaged software firms, comparing trends that both include and exclude the largest player in the industry, Microsoft. Figure 9.2 demonstrates that these firms increased their share of overall software patenting during the 1987–2003 period, from less than 0.06 percent in of all software patents in 1988 to nearly 4.75 percent in 2002, declining to 4.13 percent of software patents in 2003. Eliminating Microsoft from the figure reveals more modest growth, with shares growing from less than 0.06 percent in 1987 to 1.35 percent in 2000 and declining to 1.0 percent in 2003. Similar to Figure 9.1, the data in Figure 9.2 suggest rapid growth in software patenting through the late 1990s, followed by no growth or declines after 2000.

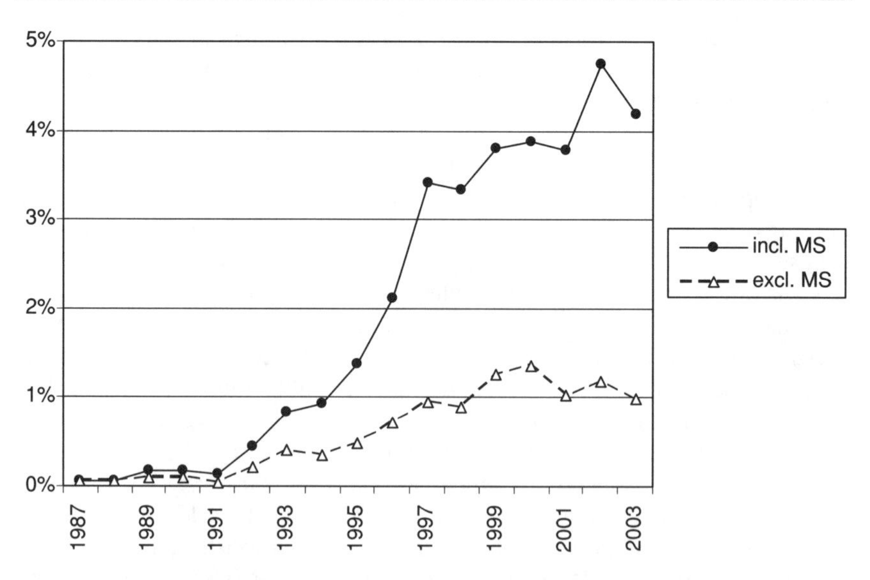

Figure 9.2. Large packaged-software firms' software patents, as a share of all US software patents, 1987–2003 (Comparison: US-class defined 'software' patenting by the 100 largest packaged-software firms [Softletter 100], including and excluding Microsoft)

Although patenting by large packaged-software firms has grown since the late 1980s, it is interesting and surprising to note that electronic systems firms account for a larger share of software patenting as we define it. Both our USPTO and IPC classification methods show that the share of overall 'software' patents accounted for by large electronic systems firms (IBM, Intel, Hewlett-Packard, Motorola, National Semiconductor, NEC, Digital Equipment Corporation, Compaq, Hitachi, Fujitsu, Texas Instruments, and Toshiba) considerably exceeds the share of 'software' patents assigned to specialist packaged-software firms. Our data analysis demonstrates that the share of 'software' patents assigned to our sample of twelve 'electronics systems' firms fluctuates between a low of 21 percent in 1990 and a high of 28 percent in 1994 before falling to 21–23 percent of all software patenting for 1998–2003.[18]

We calculate the share of all patents issued to these firms that we classify as 'software' patents during 1987–2003. Figure 9.3 displays the time trend for the share of these patents within these twelve systems firms' patent portfolios during 1987–2003. Software patents' share of overall firm patents increases during the 1987–2003 period for all of these firms, from roughly 14 percent in 1987 to 25 percent of their overall patent portfolios by 2003. Even more striking, however, is the level and growth of software patenting by IBM, the largest US patenter,[19] which increases its software patenting from 27 percent of its overall patenting in 1987 to 42 percent in 2003. In contrast to the software

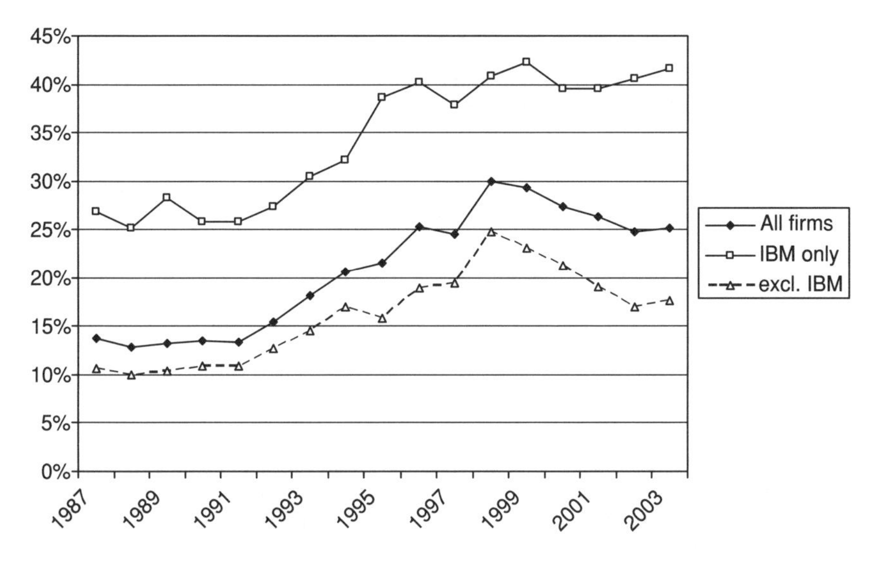

Figure 9.3. Large systems firms' software patents, as a share of firm patents, 1995–2003 (Comparing US-class defined 'software' patenting by IBM, Intel, Hewlett-Packard, Motorola, National Semiconductor, NEC, Digital Equipment, Compaq Computer, Hitachi, Fujitsu, Texas Instruments, and Toshiba)

patenting of the other eleven systems firms, IBM's share of 'software patents' in its annual patenting increases through 2003.

Inasmuch as electronic systems firms appear to account for a larger share of patenting during the 1987–2003 period than do packaged-software specialists, a comparison of patenting propensities between systems and software-specialist firms would be very interesting. Unfortunately, the absence of detailed line-of-business reporting of their research and development (R&D) investments means that we have data on software-related R&D spending for only one of the twelve systems firms included in Figure 9.3, IBM.

Figure 9.4 compares the patent propensities of IBM and Microsoft for the 1992–2002 period. The figure is presented on a log scale, and shows that IBM's software patenting per software R&D dollar spent is substantially greater than Microsoft's, dominating Microsoft's propensity by a factor approaching or exceeding an order of magnitude (a factor of 10) in every three-year interval. Furthermore, Microsoft's patent propensity has 'plateaued' at 0.10–0.12 patents per $100 million during the 1996–2003 period, but IBM's has continued to grow, climbing from 0.7 patents per $100 million R&D during 1997–1999 to nearly 1.0 patents per $100 million R&D during 2001–2003.[20] Some of the reported growth in IBM's patent/R&D ratio reflects shrinkage in the firm's reported software R&D budget during 1997–2002, a period of growth for Microsoft R&D investments. Nevertheless, the figure suggests considerable

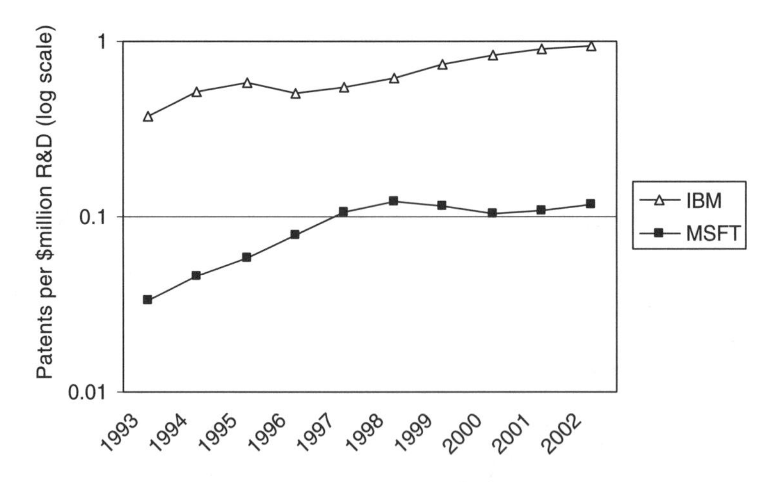

Figure 9.4. Comparison of IBM and Microsoft's software patent propensity, firms' 'software' patents per 'software' R&D expenditures, 1987–2002 (3-year moving averages; patenting limited to each firms' defined software patents)

contrast between the patenting behavior of the largest packaged-software specialist and the largest software producer among US electronic systems firms.

In earlier work (Graham and Mowery 2003), we showed that the growth in software patenting by both packaged-software 'specialists' and electronics system firms in the US was associated with a decline in the use of copyright protection for software. Is this growth in software patenting that we document consistent with a shift toward 'open innovation' in this technology? IBM's release of over 500 patents to the 'open source' community (discussed below) suggests that patents can support the creation of an IP 'bazaar' that itself advances open-source software development, although the ultimate significance of IBM's recent initiative remains to be seen.

There are several potential explanations for the rapid growth in software patenting during this period. First, software patenting may have grown along with overall patenting in the US simply because the returns to investment in innovation had increased or because of the broader strengthening of patentee rights that resulted from congressional actions and judicial decisions during the 1980s and 1990s (Kortum and Lerner 1999). But this explanation cannot account for the fact that software patenting grew as a share of overall US patenting during this period, more than doubling its share from 1.7 percent in 1987 to 3.9 percent in 1997 (Graham and Mowery 2003).

Another explanation for the growth in software patenting argues that increased patenting, especially by large firms such as Microsoft and IBM, reflects the growing importance of 'defensive patenting' in software. Competing firms may seek patents less to support the commercial development of specific invention than as a means of avoiding costly litigation (See Hall and Ziedonis 2001 for a discussion of 'defensive patenting' in semiconductors). 'Defensive patenters' apply for a large number of patents for exchange in cross-licensing agreements, thus preserving their freedom to innovate.

Growth in software patents alternatively might reflect a decline in the rigor of USPTO review of the increased number of software-related patent applications that followed the changes in the legal treatment of software patents during the 1980s. Lacking patent-based prior art to guide their evaluation of a much larger flow of applications, USPTO examiners may issue low-quality patents. Such explanations suggest that the 'quality' of software patents should have declined during the 1980s and 1990s, reflected in declining rates of citation to these patents in subsequent patents. But the 'defensive patenting' explanation predicts that the patents assigned to large software firms should exhibit particularly significant declines in quality, whereas the 'weakened review' explanation predicts an across-the-board decline in the quality of all issued software patents. In fact, evidence presented by us elsewhere (Graham and Mowery 2005) cast doubt on any such fall in patent software quality, at least in the patents issuing to software-specialist firms and electronics systems firms.

9.4.3 *Open Source Software and 'Open Innovation'*

Because in this chapter we focus on software, we are able to examine another mechanism by which firms are accessing external knowledge: open source. While other chapters in this book examine external pathways exploited by firms' Open Innovation strategies (see Chapter 7, and Chapter 8), the 'open source' mechanism has not been covered in detail. The 'open source' development model—essentially one in which developers are liberated to access and build upon the efforts of others—is being increasingly exploited by leading firms in their quest to 'open' corporate innovation.

As an exemplar, consider IBM. The giant computer systems, software, and services firm has taken a proactive stance toward the 'open source' model, both creating chief officers with responsibility over the firm's open source strategies, and investing in 'open source' research centers around the globe. In April, 2005, IBM's Vice President of Worldwide LINUX Business Strategy summed up the company's use of 'open source' in its larger Open Innovation strategy: 'Linux is not really about being free, it's about freedom—freedom to collaborate and innovate' (Kerner 2005). IBM has responded with investments, creating in Bangalore, India, a Linux Solution Centre (one of seven worldwide) and an IBM Linux Competency Centre (one of four in Asia). Moreover, in January, 2005 IBM released 500 of its patents to the open source community, allowing software engineers to freely use the ideas embodied in the patents without paying royalties to the company.

The finding that IBM is patenting more heavily within software relative to its R&D investment, while simultaneously 'opening up' a portion of its software-related patent portfolio, raises some questions about the applicability to IBM of the Open Innovation framework. Alternatively, the role of patent strategy within an 'open innovation' strategy remains to be developed. It is likely that these IBM patents are not among the firm's most valuable software patents, although that hypothesis remains to be explored. Nevertheless, IBM's recent action suggests that patents are not incompatible with open source software development, although the recent litigation between IBM and Santa Cruz Operation (SCO) centered on claims that IBM's use of 'open source' software had run afoul of preexisting IPRs held by SCO.

As open source software has sparked increased interest by developers, firms, and academics alike, a recurrent theme is the need to limit the strength and improve the clarity of proprietary IPRs in order to ensure open access and design freedom. The open source development model is undermined when the development community is blocked from using 'fountainhead' innovations, and when developers are uncertain about the extent or strength of protection of such key innovations. Indeed, the software innovation process may be unusually sensitive to IP roadblocks due to software's character as a 'cumulative innovation' technology, meaning that innovation is closely linked to and builds upon prior generations of the technology. Moreover, the attractiveness for would-be adopters of open-source software could be severely reduced if the open source code, once adopted, were subject to threats of litigation from third-party owners of IPRs implicated in the open source software. Moreover, the lack of procedures within the US patent system for administrative (as opposed to litigation-based) procedures for challenging the validity of patents once issued means that the quality of many software patents is uncertain, which can have a chilling effect on both development and deployment of open source software (OSS) innovations.

Open source licenses create a legal relationship between the creator of the software and its voluntary users, but the open source license cannot preclude the existence of various IPRs in the software. The successful General Public License (GPL or 'copy left') employed by the Free Software Foundation rewarded developers' collaboration while limiting the disincentives created by commercial expropriation (Lerner and Tirole 2002). Nevertheless, the GPL binds only those parties to the agreement—it does not apply to innovators who are not party to the GPL or have developed the same technologies independently. GPL-compliant users may relinquish certain IP rights in their derivative works under the terms of the license, but property rights, insofar as they define a relationship between the property holder and the world, cannot be eliminated by a bilateral license. In fact, the restrictive character of IPRs create the foundation for the operation of the open source license: copyright and patent rights, to the extent that the latter have been sought, are held by the inventor of the open source software who then 'passes' these rights on to other developers, allowing these voluntary adopters to use the rights under the terms of the open source license.

The innovation environment in software is complicated by significant variation among open-source licenses in the treatment of IPRs to works derived from the original open-source software. Licenses run the gamut from restrictive to permissive in their accommodation of creation by adopters of IP rights in derivative innovations. The popular GPL is relatively restrictive, reflecting the license document's expressions of hostility to software patents in its preamble.[21] The terms of the GPL largely ignore software patents, however, with the exception of restrictions in Section 7 that prohibit the distribution of any OSS subject to a patent infringement action initiated by a third party or a court order.[22] By contrast, the license offered under the Berkeley Software Distribution (BSD) is a bare-bones OSS license, including terms that require a notice of copyright and disclaimer of warranties (UC Regents 1999), but otherwise allowing the commercialization of derivative works with no restriction on patent rights per se (Simon 2003). An 'intermediate' variant is the Mozilla Public License (MPL), more permissive than the GPL but more restrictive than the BSD license in its treatment of IP rights. The MPL treats the patent rights of the originators explicitly in the terms of the license,[23] recognizing that the licensing of patent rights to complementary or build-on propriety applications may be necessary.[24] The MPL has been used by at least one software company to support 'taking the code private' into its own proprietary software.[25] While 'taking open-source code private' strikes at the heart of the bargain that OSS adopters make with the open source community, the texts of the GPL, MPL, and BSD license demonstrate that proprietary innovations arising from or built upon the core open-source software are nevertheless anticipated.

This variation in the legal treatment of 'open source' innovations demonstrates the uncertain environment in which development under this model occurs. For software firms engaged in the type of 'open innovation' that Chesbrough describes, stronger patent rights may have offsetting effects. On the one hand, strong IPRs may create more efficient markets for IP, thereby facilitating the purchase by firms of innovations from external sources. At the same time, however, strong IPRs may impede firms' access to 'open-source' software innovations from software developers outside the firm.

9.5 Conclusion

Spurred by favorable judicial decisions, software patenting has grown significantly in the US since the 1980s, although the available data suggest that growth in software patents' share of overall US patenting has slowed since approximately 2000. Scholars have produced little evidence to suggest that increased patenting has been associated with higher levels of innovation in the US software industry, although virtually no evidence has likewise been raised to suggest that increased patenting has proven harmful to innovation in this important sector of the 'post-industrial' economy. The vertically specialized structure of the US software industry, populated by firms specializing in software only, is a dramatic shift from the vertically integrated structure that characterized the US and global computer industries in the 1960s. But stronger patent protection for software emerged in the 1980s, well after the transformation of this industry structure that began in the late 1960s. The links between stronger formal protection for IP in this industry and the development of its vertically specialized structure thus are weak. In this sector, the connections between the increasing proliferation of innovators suggested by Chesbrough and the role of patents as transactional mechanisms requires further study.

Electronic systems firms appear to account for a larger share of overall software patenting, in our definition, than do the packaged-software specialist firms during the 1987–2003 period. It is possible, although we have no direct evidence to support this argument, that systems firms are patenting their software-related IP for strategic reasons, for example, to support complex cross-licensing agreements similar to those in the semiconductor industry that are discussed in Hall and Ziedonis (2001). There is less evidence of such cross-licensing agreements among software specialists, although the recent agreement between Microsoft and Sun Microsystems (Guth and Clark 2004) provides one such example. As Hall and Ziedonis note, much of the cross-licensing that provides incentives for extensive patenting by firms is motivated by the prospect or the reality of litigation. Evidence from software patent litigation cited in Graham (2004) indicates that packaged-software specialist firms account for a small fraction of software patent litigation, by comparison with computer hardware firms and firms from a diverse array of other industries. Thus, the strategic motives of firms' patenting, the function of defensive cross-licensing and litigation, and the role these play in Chesbrough's observations of changing innovation remain open questions.

Chesbrough's 'open innovation' paradigm raises many questions for researchers, including the manner and mechanisms of structural change in industries, and the role played by the transactional environment for knowledge. This chapter offered both a case study in the development of the software industry, and an analysis of patenting in an important source of innovation in the new economy, software. With it, we intended to offer a view into the system of innovation in one important sector as a means of raising questions about the generalizability of Chesbrough's 'open innovation' paradigm. We leave the development of these to further research, and researchers.

Notes

1. We acknowledge the helpful comments made on an earlier version by conferees at the American Enterprise Institute/Brookings Joint Center for Regulatory Studies 'Intellectual Property Rights in Frontier Industries' Conference, held April 30, 2004, with special thanks to Starling Hunter. Research for this chapter was supported by the National Research Council, the Andrew W. Mellon Foundation, the Alfred P. Sloan Foundation, the Kauffman Foundation, and the Tokyo Foundation.
2. Chesbrough suggests that the 1970s were characterized by a closed paradigm (p. 45) while the 1980s and 1990s were decades of change in which the Open Innovation paradigm was evolving (pp. 45–9).
3. Chesbrough highlights Xerox, Intel, Lucent, and IBM as adopters of the 'open' innovation paradigm.
4. Bresnahan and Greenstein (1995) point out that a similar erosion of multiproduct economies of scope appears to have occurred among computer hardware manufacturers with the introduction of the microcomputer.
5. Copyright offers protection from the moment of authorship, and remains an important protection for software 'writings.' *Computer Associates Int'l v. Altai, Inc.*, 982 F.2d 693 (2d Cir. 1992); *Whelan Associates v. Jaslow Dental Laboratory*, 797 F.2d 1222 (3rd Cir. 1986).
6. A trade secret is formally some information used in a business which, when secret, gives one an advantage over competitors. The secret must be both novel and valuable. *Metallurgical Industries, Inc. v. Fourtek, Inc.*, 790 F.2d 1195 (1986).
7. A trademark protects names, words, and symbols used to identify or distinguish goods and to identify the producer. *Zatrains, Inc. v. Oak Grove Smokehouse, Inc.*, 698 F.2d 786 (5th Cir. 1983).
8. The European Union currently is debating the extent to which 'software' will be afforded patent protection, an issue that has been reasonably well settled in the US since the mid-1980s.
9. *Gottschalk v. Benson*, 409 US 63 (1972).
10. Samuelson (1990) argues that the Patent and Trademark Office was at odds with the Court of Customs and Patent Appeals (CCPA) throughout the 1970s over the patentability of software and concludes that the CCPA's views in favor of patentability ultimately triumphed.
11. 450 US 175 (1981).
12. 450 US 381 (1981).
13. Reported in *Softletter* (2001) this group of 100 companies includes Microsoft, Novell, Adobe Systems, Autodesk, Intuit, and Symantec. We are grateful to *Softletter* for permission to use these data. The 2001 tabulation was the last year to date in which *Softletter* produced this report.
14. We also constructed a sample of software patents that was based on the International Patent Classes associated with patenting by large packaged software specialists during this period of time. Overall, the IPC-based definition yields similar results and trends, and we omit discussion of it because of space limitations.
15. Estimated by International Data Corp. and reported in *Software & Information Industry Alliance* (2005). In the US, 2003 software-only industry revenues were surveyed by

the Census Bureau to be $70 billion, with an average 12 percent growth during the 1990s, and employment growth showing 5 percent. Ibid.

16. The only class that did not show a reasonably flat growth trajectory after 1998 was the IPC group 'G06F 17' which is the newest 'software' class, and showed steady growth as a share of all patenting 1995–2003.
17. Pendency for software applications at issue for all software patents compared to all nonsoftware patents from 1995–2003 are:

Year	1995	1996	1997	1998	1999	2000	2001	2002	2003
Pendency (%)	130	127	122	120	122	128	136	145	146

18. Data analysis using the IPC-definition method found substantially the same trend in these firms' patenting shares.
19. The US Patent and Trademark Office (2002) shows that IBM (International Business Machine) is the largest patenting organization in the US, having been assigned 3,411 and 2,886 patents in 2001 and 2000, respectively. These numbers were 75 percent and 43 percent larger, respectively, than the next largest patentholder in these years (NEC).
20. Graham and Mowery (2003) point out two problems with these data in comparing the 'software patent propensities' of IBM and Microsoft: The R&D data reported by these two firms may not be strictly comparable, since a portion of Microsoft's total reported R&D investment may cover some fixed costs of maintaining an R&D facility that are not included in IBM's reported software-related R&D investment (Although IBM does maintain 'software only' research facilities around the globe; Its Bangalore, India LINUX facilities are but one example). In addition, an unknown portion of Microsoft's reported R&D spending includes development programs for hardware, and these data therefore may understate the Microsoft software-related patent propensity and overstate that for IBM.
21. Free Software Foundation (1991), Preamble: '[A]ny free program is threatened constantly by software patents. We wish to avoid the danger that redistributors of a free program will individually obtain patent licenses, in effect making the program proprietary. To prevent this, we have made it clear that any patent must be licensed for everyone's free use or not licensed at all.'
22. Free Software Foundation (1991), Section 7: 'If, as a consequence of a court judgment or allegation of patent infringement or for any other reason (not limited to patent issues), conditions are imposed on you (whether by court order, agreement, or otherwise) that contradict the conditions of this License, they do not excuse you from the conditions of this License. If you cannot distribute so as to satisfy simultaneously your obligations under this License and any other pertinent obligations, then as a consequence you may not distribute the Program at all. For example, if a patent license would not permit royalty-free redistribution of the Program by all those who receive copies directly or indirectly through you, then the only way you could satisfy both it and this License would be to refrain entirely from distribution of the Program.'
23. Mozilla.org (1999), Section 2.1: 'The Initial Developer hereby grants You a world-wide, royalty-free, non-exclusive license, subject to third party intellectual property claims ... (b) under Patents Claims infringed by the making, using or selling of

Original Code, to make, have made, use, practice, sell, and offer for sale, and/or otherwise dispose of the Original Code (or portions thereof).'

24. Mozilla.org (1999), Section 3.4(b): 'If Contributor's Modifications include an application programming interface and Contributor has knowledge of patent licenses which are reasonably necessary to implement that API, Contributor must also include this information in the LEGAL file.'
25. 'A Word About the GNU GPL and Patent Rights: LizardTech has chosen to use the GNU GPL because this license is accepted in the Open Source community.... At the same time, LizardTech does not endorse or agree with all of the underlying assumptions or biases of the Free Software Foundation that are reflected in the GNU GPL. In particular, the GNU GPL is generally antagonistic to patent protection for software.... On the issue of patents, LizardTech believes the approach taken in the Mozilla Public License...is preferable to that in the GNU GPL, because the MPL explicitly provides for contributors to grant patent rights to users of the open source software....' (LizardTech 2001).

Part III

Networks Shaping Open Innovation

10

The Interorganizational Context of Open Innovation

Wim Vanhaverbeke

10.1 Introduction

Open innovation is almost by definition related to the establishment of ties of innovating firms with other organizations. Companies are increasingly forced to team up with other companies to develop or absorb new technologies, commercialize new products, or simply to stay in touch with the latest technological developments.

Firms are working more and more as part of broader networks to create customer value. Those networks are based on the collaborative efforts of specialist companies each providing complementary intermediate goods and services. As Information and Communication Technology (ICT) becomes a powerful technology, it allows those companies to be linked by sophisticated business-to-business information systems. But networking can also imply collaboration with other partners. The set of partners can be quite different depending on the goal an innovating company wants to realize: companies develop relations with universities and research laboratories to explore the technical and commercial potential of new technologies, they establish alliances with or acquire technology-based start-ups, or set up networks with selected suppliers and customers to launch radically new products or services based on new technologies or a new business model. Learning how to create and capture value when companies are highly dependent on each other is still an underexplored area in the network literature. Most firms are used to making decisions within their boundaries taking the external environment as an exogenous variable or as an arena where firms compete with one another. But in networks value is coproduced: the total value created in the network depends directly on how well partners' objectives are aligned to each other and on the commitment of the partners to invest in complementary assets (Teece 1986; Moore 1991). Similarly, in developing systemic technologies, the innovating company depends on the technological skills and commitment of other companies. Most firms do not feel comfortable in these 'open' scenarios where the return essentially depends on the partnering actors.

The three chapters in Part III analyze in greater detail how companies have to team up with other actors in the business system and build interorganizational networks to support open innovation. But firms are not only embedded in their environment by interorganizational networks: they can be part of regionally bounded clusters of competitive firms which, in turn, can be considered as a subsystem of a regional (or national) innovation system. Chapters 12 and 13 apply Open Innovation to interorganizational networks but in two different settings: Maula et al. show in Chapter 12 that, in the context of systemic innovations, innovating companies are highly dependent on complementary innovators forcing them to take an external perspective to resource allocation processes. Vanhaverbeke and Cloodt (Chapter 13) describe how the commercialization of radically new products based on technological developments in the agro-food biotechnology requires the establishment of a value network where central players establish a network of firms with complementary skills and assets to create value for the targeted customer group. Chapter 11 takes a broader perspective starting with an interorganizational knowledge flow and a description of the geography of Open Innovation to derive taxonomy of different interorganizational networks that enable Open Innovation.

10.2 Open Innovation and Different Levels of Analysis

Chesbrough (2003*a*) conceives Open Innovation from the point of view of large, incumbent companies. Although Open Innovation entails by definition the close collaboration with a broad set of potential partners to insource or outsource technologies, the Open Innovation framework has been analyzed so far at the (focal) company level and not at the network level where the benefits of the network for the focal firm are jointly analyzed with those for the collaborating organizations. There are of course good reasons to emphasize the firm level perspective as Open Innovation is always expected to have an impact on a company's bottom line and is based on a business model which is by definition centered on a single firm (Amit and Zott 2001).

Analyzing Open Innovation on other levels of analysis can however broaden the scope and enrich our understanding of Open Innovation. Following Figure 13.2, there are a number of levels at which Open Innovation can be analyzed.

The first level is that of the intraorganizational networks. There exists a considerable literature about intraorganizational networks to stimulate innovation (Nonaka and Takeuchi 1995; Szulanski 1996; Tsai and Ghoshal 1998; Hansen 1999; Foss and Pedersen 2002; Lagerstrom and Andersson 2003).

However, these networks have not been analyzed explicitly within the context of Open Innovation. Since many companies struggle leveraging the commercial potential of innovations that have been developed externally, it is interesting to analyze how firms' internal organization plays a role in improving the assessment and integration of externally acquired knowledge. Internal networks play a crucial role in the way companies get organized to increase the effectiveness of acquiring external knowledge (Hansen 1999; 2002; and Hansen and Nohria 2004). None of the chapters in this volume are focusing on intra-firm level networking, leaving open an interesting avenue for future research.

The second is the firm level: Open Innovation has been explored at this level of analysis in Chesbrough (2003*a*) and in different chapters of this volume.

The third is at the dyad level; that is, considering the interest of two (or more) companies that are tied to each other through equity or nonequity alliances, corporate venturing investments, and so on. This level of analysis has been explored in depth in the academic literature (e.g. Gulati 1995*b*) as well as in the business press (e.g. Kanter 1994; Doz and Hamel 1998). A rich literature exists on how to select partners, how to assess the return and risks of an alliance, how to evaluate the fit between potential partners, and how to structure the cooperative agreement and manage it over time. As Open Innovation is basically about non-arm's-length relations between companies it can take advantage from a dyad level perspective and from the management lessons about alliance management (Lynch 1993; Bamford and Ernst 2002) and external corporate venturing (Keil 2002*a*).

The fourth level of analysis refers to interorganizational networks. Within this approach Open Innovation is no longer studied at the level of a single company or the dyad level. Individual alliances or other non-arm's-length transactions between organizations usually do not account for a company's success, but it is determined by the way the firm integrates its external relations into a coherent strategy and manages them over time. Interorganizational networks provide a durable structure for interfirm relations, which both enables and constrains dyadic interactions. The fact that individual relations between companies are 'embedded' in broader networks also leads to the formation of more complex topologies.

The last level of analysis consists of the national or regional innovation systems. Innovating companies are embedded in a broader institutional setting that can enhance or hinder the innovativeness of the local companies. Academics have debated whether the impact of innovation systems is on the national, regional, or supranational level (Lundvall 1992; Cooke 1992, 1998), but it is beyond doubt that the external, geographically bounded innovation systems play a crucial role for companies' innovativeness. Cooke (2005) explains that Open Innovation is one of the key concepts to explain how regional innovation systems, and clusters within them, have to be organized to be globally competitive.

Chapters 10, 11, and 12 focus on the two last levels of analysis. First, I will explore the interorganizational networks. Next, I will have a closer look at regional systems of innovation and their link to Open Innovation. Interorganizational networks and regional systems of innovation are two levels of analysis that are complementary with the traditional, firm-oriented approach of Open Innovation. In this way, they have the potential to enrich our understanding of Open Innovation.

10.3 Interorganizational Networks

During the postwar period, innovations were managed in what Chesbrough (2003*a*) calls the 'closed innovation' paradigm. Within this view successful innovation requires that firms generate and develop ideas internally, nurture, and market them until they are launched as a new product or business. It is an internally focused logic where the innovating company trusts on internal capabilities to successfully innovate. Interorganizational ties like technology alliances have been around for decades but did not challenge the 'closed innovation' paradigm as long as companies relied mainly on their internal (technological) capabilities to develop new products or services. However, recently this paradigm has been challenged because of the increasing costs and complexity of research and development (R&D), the shortening of the technology life cycles, the presence of increasingly knowledgeable suppliers and clients, the growth of venture capital, and the growing diffusion of leading-edge knowledge in universities and research laboratories around the world. If most of the new knowledge emerges outside the firm as a result of these ongoing trends, a closed innovation approach is likely to overlook the business opportunities from this large pool of external knowledge, while it cannot prevent internally built knowledge from leaking out as entrepreneurial employees leave the company and start their own business with venture capital financing. Companies embracing Open Innovation actively tap into these external technology sources to strengthen their businesses. Similarly, internally developed technology and resulting IP are no longer only valuable for internal use, but the company can also profit from the selective use of its IP by other companies with different business models. Open Innovation thus implies an extensive use of interorganizational ties to insource external ideas and to market internal ideas through external market channels outside a firm's current businesses (Chesbrough 2004; see chapter 6).

There are many types of interorganizational ties. Spin-ins and spin-offs, corporate venture investments, joint ventures, and several types of nonequity alliances are only a few examples. Simard and West (Chapter 11) develop a taxonomy of network ties that enable Open Innovation. They make a distinction between deep versus wide ties and formal versus informal ties. Both authors argue that companies have to build ties that are both wide and deep. Deep ties enable a company to capitalize on its existing knowledge and resources. They are the result of a company's strong network position that allows it to tap into key resources for innovation. Deep ties are enhanced by the geographical proximity of the partners and by building trust in networks. They are appropriate to deepen the strength of companies in their existing businesses. Wide ties on the contrary enable a company to find yet untapped technologies and markets. In contrast to deep ties that are associated with the exploitation of existing technologies, wide ties offer a firm opportunities to explore new technologies. Explorative search is enhanced by ties that span structural holes and link the innovating firm with diverse technological environments by means of different types of ties. Because of this diversity geographical proximity is very valuable. Wide and deep ties, or explorative and exploitative ties, have to be balanced (March 1991). In this way, companies have to combine both deep and wide ties to profit optimally from their external relations (Uzzi and Gillespie 1999*a*).

Simard and West make another distinction between formal and informal ties. The former are agreements based on a formal contract. They are planned channels for knowledge exchange between organizations. However, formal contracts bring people from different firms together who, in turn, establish informal networks. Similarly, existing informal networks lead to more formal arrangements to cooperate.

Simard and West combine deep versus wide ties and formal versus informal ties to get a better understanding of the role of interorganizational ties in Open Innovation. The combination of both dimensions (deep–wide and formal–informal) leads to different types of networks; for example, deep ties are characterized by redundant information overlapping with the existing knowledge base of the companies involved. This suggests that deep networks tend to lead to incremental innovations. Wide networks give a company access to nonredundant information and have as such greater potential for innovation. Informal networks are harder to manage and make it more difficult to control the knowledge flows in and out of the firm. Simard and West developed this framework to understand the role of different networks in innovation management. It is a guideline to identify opportunities for further research about the relationship between knowledge flows, interorganizational networks, and the practice of Open Innovation.[1]

Formal ties have been studied extensively, but the role of informal interorganizational ties is less well understood. How is commercially viable knowledge accessed through informal networks? Different case studies give scattered evidence that informal ties of employees with employees in other organizations or institutions are crucial to understand how new ideas are generated and turned into commercial successes. Hamel (2000) recalls the story of Ken Kutaragi, a Sony Corp. engineer who had been 'outsourced' as engineer to one of the leading game console producers and which ultimately led to Sony's successful PlayStation. However, informal networks might also be too 'closed' to generate the desired information from other organizations. Porter et al. (2005) remark that in biotechnology informal social networks are too tightly centered on star scientists that act as a bottleneck for information sharing. Hence, both formal and informal ties have their advantages and disadvantages and an innovating firm has to balance the mix to optimize the return on open innovation.

Simard and West also point at the management challenges related to the network portfolio. The role of network portfolio management (Ashkenas et al. 1995; Bamford et al. 2003; Parise and Casher 2003; Sarkar et al. 2004; Heimeriks 2005; Hoffmann 2005; Ozcan and Eisenhardt 2005; Reuer and Ragozzino, forthcoming) has not been linked so far to the promises of Open Innovation strategies of innovating companies. We do not know what aspects of the network portfolio significantly can raise (or lower) the effectiveness of Open Innovation. What is the optimal mix of different types of ties in the portfolio? Are portfolios dependent on the industry setting or the business model that the innovating company is pursuing? These are just a selection of yet unanswered questions opening a new avenue for future research.

Defining new metrics for managing Open Innovation is a final topic that Simard and West analyze. They argue that managers have to use new metrics to measure Open Innovation. Measuring interorganizational knowledge flows is an important challenge in realizing that objective. Citing patents of partners give us a first indication of these flows, but patent citations are known to be imperfect measures of interdependence between firms. They do not indicate how much value a company creates from its externally acquired technology. Licensing agreements and royalty payments measure some forms of knowledge used in formal ties, but it is harder to measure knowledge that flows through informal ties. Hence, we need new measures to accurately manage Open Innovation.

10.4 Network Management and Open Innovation

Interorganizational relations and networking are a crucial dimension of Open Innovation. They are implicitly present in the Open Innovation framework when external ideas are insourced to create value in a firm's current business or when internal ideas are taken to the market through external channels, outside a firm's current businesses (Chesbrough 2004). When companies are highly dependent on other organizations for their supply of new technologies or when they need the support of others to bring a new technology to the market, it seems logical that Open Innovation has to put an emphasis on the management of external networks to be successful. However, Chesbrough's work (2003*a*, 2004) on Open Innovation is analyzed at the level of the innovating firms and network management is not treated explicitly. This does not mean that network management is not present in the existing literature about Open Innovation. On the contrary, Chesbrough and Rosenbloom (2002) consider the value network as a function of the business model. The latter describes 'the position of the firm within the *value network* linking suppliers and customers, including the identification of potential complementors and competitors' (p. 534). 'The value network created around a given business shapes the role that suppliers, customers and third parties play in influencing the value captured from commercialization of an innovation. The value network increases the supply of complementary goods on the supply side, and can increase the network effects among customers on the demand side' (pp. 534–5). However, it is not explicitly mentioned whether the innovating company should manage the entire value network, and if so, how it should do this. Chapters 12 and 13 analyze how innovating companies manage the external network to create and capture value. The context is however rather different in the two chapters: Chapter 12 focuses on systemic innovations within the ICT sector, and Chapter 13 applies network management to the commercialization of innovations in the agro-food biotech where companies want to commercialize genetically modified crops that create value for a targeted customer group in completely new ways.

10.4.1 *Network Management and Systemic Innovations*

Maula, et al. (Chapter 12) describe how companies face new challenges when creating systemic innovations that require significant adjustments to be made in other parts of the business system. The benefits of a systemic innovation can only be realized in conjunction with complementary innovations or components. But companies usually cannot wait until new technologies emerge. How can they coordinate the activities of the relevant players so that the development of components and subsystems is mutually aligned assuring the success of the value creating, systemic innovation?

The classic answer to this problem is provided by Chesbrough and Teece (1996). In systemic innovations, innovating companies are exposed to strategic hazards because of their reliance on suppliers or partners. As a consequence Chesbrough and Teece (1996) recommend that companies develop the technology in-house when the required capabilities still have to be developed, and that they ally with caution (because of the strategic hazards) in case these capabilities exist in other organizations. Maula et al. remark that companies face problems to develop systemic innovations in-house because of the growing complexity of technologies and the shortening of the technology life cycles. Since vertical integration is rarely an option, innovating companies have to take a broader, network level perspective to resource allocation.

'Ally with caution' is now translated into the management of an interorganizational network to successfully create a systemic innovation. In that regard, the innovating companies need tools to manage this network; examples are external venturing, research collaboration, and industry consortia. The difference with autonomous innovations is that there must be a collective governance in the case of systemic innovations giving each partner incentives to stick to the network. In autonomous innovations, a firm can team up bilaterally with another company irrespective of its existing network portfolio.

The systemic nature of innovations thus forces innovating companies to manage other actors in the network in a proactive way. Maula et al. argue in this respect that companies can manage these mutual dependencies by creating *foresight* and *shaping* the development of these industries over different time horizons. The foresight process is necessary because the systemic innovation requires that the company monitors the development of multiple innovations simultaneously. Multiple external contacts provide rich information about the development of different technologies which ultimately allows the company to translate technological developments into new business opportunities. The foresight process keeps companies alert to create new offerings that offer the highest potential value for the targeted customers.

The shaping process is in its turn necessary to avoid the strategic hazards related to systemic innovation as mentioned by Chesbrough and Teece (1996). This process intends to influence the resource allocation decision of other companies. With this process a company tries to keep all the partners in the boat offering them a sufficiently large share of the pie.

Both foresight and shaping mechanisms can be analyzed within different time horizons. Timing and differentiation of management tools over time is important to succeed with systemic innovations. Industry leaders can, in an early phase, long before the commercialization takes place, collaborate on research to speed up the technological progress defeating in that way other (groups of) competitors. They also agree on the standardization of technologies, which, in turn, allows them to partition the complexity of the system, enabling other companies to provide pieces that can be easily integrated in the system. During the early commercialization phase, companies get involved with corporate venturing to keep their options for a successful commercialization open. In the full commercialization phase, boundary-spanning activities are changing again as customer and supplier alliances, joint ventures, and mergers and acquisitions dominate the scene. Furthermore, leading incumbents are usually involved in different product generations and they have to cannibalize sales of their existing products when introducing a new generation. The fact that companies have to manage over different time horizons requires that they signal their commitment to a new technology in a credible way to their (potential) partners.

10.4.2 *Network Management and the Commercialization of Novel Product Offerings*

Vanhaverbeke and Cloodt (Chapter 13) apply network level management to the case where radical technologies are commercialized by internal paths to the market (Chesbrough 2003*a*). The difference with Maula et al. (Chapter 12) is that the latter focus on the supply of technologies to develop a new technology, while Vanhaverbeke and Cloodt focus on the commercialization process of innovations that have already been developed successfully in the laboratory. They take the use of biotechnology in agricultural products (or agbiotech) as a particular setting to illustrate how network level management actually works during the commercialization of new technologies.

Taking a firm level perspective on the commercialization of new technologies that radically change the business model of the targeted customers does not reveal all management issues related to the commercialization process. New, genetically modified (GM) food products do not sell automatically, and sales will not take off by establishing loose, arm's-length transactions with other players in the value system.[2] Take for instance the case of a GM tomato that has a better flavor and targets the fresh tomato market.[3] First, commercialization requires that different partners in the value system that own complementary assets make investments that are transaction specific (Teece 1986). These firms will only join the innovating firm when they get some guarantee that these investments will be profitable. Next, new product offerings typically suffer from thin market problems. The innovating company has to develop a 'take off' strategy pulling all relevant actors together in order to get the new product on the market. As a result, radical innovations require a value network perspective where the innovating company (or a small clique of central players) manages the external network with all the actors that are necessary to launch the new product offering. This is in sharp contrast with incremental innovations where a company can rely on existing relations with suppliers, channels and end-consumers. The fact that the innovating company has to team up with different partners in the value system and has to organize this external network indicates that Open Innovation also applies to the commercialization of radical innovations. Managers of the innovating firm have a difficult task to manage the interface between the different links in the value network.[4]

Value creation and value appropriation are central to the commercialization of new technologies. The value network is created in order to create value for a particular customer group. Three examples: The Flavr Savr tomato was targeting the end-consumer who was willing to pay a premium-price for a better tasting tomato. Herbicide-resistant corn allows the farmer to save on spraying costs and time. GM improved cotton reduces the need to blend it with polyester or other materials to strengthen the natural cotton fibers. The targeted customers—the end-consumer, the farmer, and the textile industry in the three examples—are always better off as long as the premium price for the GM-crop is smaller than the extra value they get from the product offering compared to other offerings (i.e. traditional crops). Other actors in the value network, however, are not necessarily better off: logistics can become quite complex when different GM types of cotton have to be transported and stored separately. As a result, balancing the value appropriation among the different actors in the value network requires the active management of a central firm. Besides the task to organize the network to create value from the innovation, this central firm also has to manage the potential tensions between partners about value capturing. This is a difficult task because competition is no longer based on rivalry between single firms but between groups. Different product offerings—and not firms—are competing in the market (e.g. herbicide-resistant corn versus traditional corn). It is a group-based competition (Gomes-Casseres 1994, 1996) where the total value created depends on the quality of the relations between the partners in the value network. The profitability of the companies or the distribution of the total value created depends not only on the traditional bargaining power of each partner (Brandenburger and Stuart 1996). Contrary to the firm-based competition, value appropriation has to be considered jointly with value creating strategies in group-based competition because the total value created depends on the quality of the interorganizational relations. In other words, too much fighting about the share of the pie reduces the total volume of the pie. This subtle interaction between value creation and value appropriation implies that there exists a continuous tension between maximizing joint value creation and firm level profitability. The innovating company has to manage this tension carefully.

How value is created and distributed in the commercialization process of agbiotech innovations illustrates two central ideas. First, the commercialization of an innovation is based on a business model of the innovating firm but its scope is much wider than the firm itself: its path to the market entails the establishment and management of an interorganizational network of partners with different assets and positions in the value system. Thus, although a business model is always centered on a particular firm, it has as a unit of analysis a much wider scope than the firm since it encompasses the capabilities of multiple firms in multiple industries (Amit and Zott 2001: 514). Business models are in this way no longer tied to the boundaries of the firm but can be analyzed in terms of Open Innovation—or is open commercialization a more appropriate term? Second, the previous analysis also suggests that the analysis of competitive advantage can be centered on the value-creating system and not necessarily on the firm or the industry (Normann and Ramirez 1993; Iansiti and Levien 2004*a*, 2004*b*). The source of value creation lies in networks of firms and the configuration of their roles in these networks (Bettis 1998; Dyer and Nobeoka 2000; Gulati et al. 2000).

The commercialization of new technologies also challenges the established theoretical frameworks about value creation (and distribution). In line with Amit and Zott (2001), Vanhaverbeke and Cloodt (Chapter 13) argue that in order to understand the commercialization of new technologies one has to integrate various theoretical perspectives. The commercialization of GM crops as described in Chapter 13 shows that the way how value is created cannot be fully captured by a single theoretical framework. The value creation in agbiotech—but also in ICT as demonstrated by Amit and Zott (2001)—can only be explained when different theoretical perspectives are brought together. First, the commercialization of new technologies is situated at the crossroad of strategic management and entrepreneurship (Hitt et al. 2002): it combines how value is created for buyers who want to pay premium prices with the exploitation of new business opportunities based on the emergence of a new technology. The commercialization of agbiotech products can also be described in terms of the Schumpeterian model of creative destruction since new products based on GM crops will replace traditional products in, as well out of the food industry. Next, the resource-based view of the firm (RBV) is also applicable since the value network brings together different players with complementary resources and capabilities that are necessary to market the new products (Barney 1991). The establishment of a value network is also related to dynamic capabilities (Teece et al. 1997; Eisenhardt and Martin 2000) because it activates, coordinates, and reconfigures these resources in new ways to create value. Value networks are almost by definition related to strategic networks and the relational view of the firm (Dyer and Singh 1998). Finally, value networks cannot be analyzed without entering the question why firms internalize transactions that might otherwise be conducted in markets (Coase 1937; Williamson 1975, 1985). Value networks are hard to analyze from a single firm's transaction cost minimization point of view. The partnering firms are rather interested in the pursuit of joint transactional value (Zajac and Olson 1993; Dyer 1997). Hence, the commercialization of GM crops—or the commercialization of technologies in general—which is essentially an Open Innovation process, calls for an *integration* of various frameworks (Amit and Zott 2001). The recipe of 'open innovation' can only be understood when different ingredients such as transactions, capabilities, value creation and appropriation, and interorganizational networks are linked to each other and integrated in a coherent strategy. The challenge to relate Open Innovation to an integrated approach of the existing theoretical perspectives has just begun. This is a most promising area for future research.

10.5 The Geographical Dimension of Open Innovation

Interorganizational networks constitute one level of analysis for Open Innovation. These networks can be part of larger regional clusters that can be defined as 'geographical concentrations of interconnected companies and institutions in a particular field' (Porter 1998: 78). Clusters, in turn, are part of broader regional or national innovation systems (Lundvall 1992; Nelson 1993; Cooke 2004*b*). Although there is a huge literature stream on clusters and the link between geographical proximity and economic growth, the relation between interorganizational networks and geography is still underexplored.

Clusters and regional innovation systems are important for Open Innovation because the knowledge flows between companies is crucial to Open Innovation. An optimal Open Innovation strategy would exploit multiple ties to multiple types of institutions. Since knowledge flows more readily to closer entities (Jaffe et al. 1993), the organization and institutional embeddedness of geographically networks might be crucial in explaining the differences in effectiveness of Open Innovation in different regions or nations. Simard and West (Chapter 11) identify universities, research laboratories, venture capitalists, focal firms, and other industry specific actors as powerful institutional forces that shape Open Innovation and determine its effectiveness. The institutional changes have also been explored by Chesbrough (2003*a*: chs. 2 and 3). He mentions for instance different factors that erode the strength of the closed paradigm such as the increasing availability and mobility of skilled workers, the emergence of the VC market, and the increasing capabilities of external suppliers. These erosion factors are not necessarily linked to the existence of clusters and regional innovations systems, but they clearly show that Open Innovation is fostered within particular institutional settings.

This is an important observation because there exist huge differences in the knowledge capabilities' of regions depending on the presence and the level of global competitiveness of clusters and regional innovation systems. Since the effectiveness of Open Innovation strategies of companies is strongly related to the presence of regional innovation systems, these regional differences can also explain why some regions are much more successful in attracting multinationals ensuring a steady flow of knowledge workers and entrepreneurs. Examples are Silicon Valley, Helsinki's and San Diego's telecommunications clusters, biotechnology in Boston (Owen-Smith and Powell 2004) and ICT clusters in Cambridge and on the Leuven-Eindhoven axis just to mention a few. Companies depend increasingly on the external supply of knowledge, which is locally embedded in regional innovation systems, and forces them to tap into these local epochs of knowledge. Hence, getting access to local knowledge is of crucial importance in the current knowledge economy. As a result, Open Innovation has to be connected to regional economics in the future.

There is to my knowledge only one author who has explicitly linked Open Innovation to clusters and regional innovation systems. Cooke (2004*a*, 2004*b*) explains in two recent papers how Open Innovation plays a crucial role in the explanation of regional innovation systems. Based on a Penrosian-inspired

(1959/1995) theorization of 'regional knowledge capabilities' as drivers of globalization, he argues that Open Innovation plays a crucial role in the changing spatial structure of industries. He claims that instead of the organization of industry determining spatial structure, the economic geography of public knowledge institutions determines industry organization.

> Thus firms agglomerate around universities or centers of creative knowledge like film studios. Learning was the central attraction where knowledge capital could have rapidly escalating value. Now it is clear that knowledge itself is the direct magnet. The more knowledge-based clusters thrive, the more imbalanced the economy is likely to become spatially and in distributional terms and the more important it becomes to seek ways of moderating this without killing the golden goose. This is an important challenge confronting economic policy-makers everywhere for the foreseeable future. (Cooke 2005: 31)

Although this conceptualization of the link between Open Innovation and regional development might be still in an embryonic phase, it is clear that this is an interesting topic for future research.

Open Innovation has already (although implicitly) been applied to new internationalization strategies for multinational companies. Within this respect, Doz et al. (2001) have developed the notion of the metanational company. They acknowledge that globalization and the distributed presence of leading research institutes around the world reduces the knowledge preeminence of any single location. So, the profits from projecting home-grown advantages (the traditional drive to internationalize) are falling. Instead, metanationals prospect the world for good ideas and technologies. Since valuable knowledge is complex and hard to move, they have to be present in these knowledge centers to sense the new technological or market developments. This is a nice example how Open Innovation can be applied into the context of international management. These ideas are also echoed in a recent book of Hagel and Brown (2005).

10.6 Conclusions

We can draw a number of conclusions from Chapters 11, 12, and 13. First, the 'open innovation'-framework is not only applicable to ICT or to industry settings where network economies play a role. Vanhaverbeke and Cloodt (Chapter 13) illustrate how Open Innovation is also prominently present in the commercialization of agbiotech innovations. Cooke (2004*b*) also provides evidence that Open Innovation is abundant where biotech start-ups and big pharma companies in the development of pharma-based or 'red' biotech applications. Developments in the innovation and commercialization strategies of companies in other industries suggest that Open Innovation is applicable in a growing range of industries.

Second, business models always refer to a particular firm (Chesbrough and Rosenbloom 2002; Chesbrough 2003*a*) but its impact easily spans the firm and even the industry boundaries. Interorganizational networks play a crucial role in that respect. Companies with complementary capabilities or positions in the value system have to be fully committed to cooperate. Creating value cannot be done unilaterally based on the efforts of a single, focal firm, nor can it be done without keeping the different and divergent interests of all collaborating partners in mind. Hence, in order to understand Open Innovation correctly, it has to be analyzed on complementary levels of analysis. Two of these levels, the network and the firm level, play a crucial role in the understanding of Open Innovation.

Third, external network management is one of the new roles for the central firm. There is always one company that operates as the organizer of the network when companies develop a new systemic innovation or commercialize a radically new product (see also Gomes-Casseres 1996).

These focal companies—or industry shapers—establish boundary-spanning activities for two purposes. On the one hand, they design the whole process starting from the idea or business model how the innovation or new product offering has to deliver value: the complexity of the technology requires that a central firm monitors the multiple simultaneous innovations in the case of systemic innovations and the changes required in different parts of the value network in order to deliver value for the targeted customer in the case of the commercialization of GM-crops. On the other hand, they have to make sure that they have an impact on the resource allocation decisions of the other actors in the network. These two processes—industry foresight and industry shaping—are dynamic concepts since a company has to manage its dependencies on other actors by shaping the industry over different time horizons.

Next, Open Innovation has a number of implications for the theory of the firm especially when one is focusing on the need to team up with partners to successfully commercialize new product offerings based on breakthrough technologies. I have suggested, echoing Amit and Zott (2001), that the network perspective on Open Innovation calls for an integration of the various theoretical frameworks such as value chain analysis, transaction costs theory, the relational view of the firm, and the RBV. This is probably one of the most promising areas for future research.

Finally, Open Innovation also implies that innovating companies choose a particular governance mode for their ties with other actors in the network. Maula et al. (Chapter 12) indicate that the appropriate modes of external technology sourcing might change depending on the time horizons a company considers to change the business environment. Simard and West (Chapter 11) provide a taxonomy suggesting that different types of ties are required for different Open Innovation settings. However, the choice between different external sourcing modes of technology still has to be linked to the broad literature stream about the 'buy-ally-make' decisions (Hoffman and Schaper-Rinkel 2001; Roberts et al. 2001; Dyer et al. 2004).

Notes

1. I exclude the geographical dimension of this topic, which will be discussed in Section 10.5.
2. I use Porter's (1985, 1990) terminology: 'A value chain disaggregates the firm into its strategically different activities in order to understand the behavior of costs and the potential sources of differentiation' (1985: 33). 'A firm's value chain is embedded in a larger system stream of activities that I term the value system' (p. 34). The value system links suppliers to buyer firms, channels, and the final customer.
3. This example is based on the Flavr Savr tomato described in Goldberg and Gourville (2002).
4. We will call an organized network of partners in the value system a 'value network'. This concept is not new and has been applied in many business settings (Amit and Zott 2001; Normann and Ramirez 1993; Normann 2001) but the article of Amit and Zott (2001) is the only exception linking the concept to the market introduction of a new technology (ICT in that case).

11

Knowledge Networks and the Geographic Locus of Innovation

Caroline Simard
Joel West

11.1 Introduction

A crucial goal of Open Innovation is to capture external knowledge that flows between organizations, allowing firms to be more successful at innovation than firms that close off such flows. As has been discussed in Part I, one goal of external innovation is to capture the value of internal knowledge transferred to other firms. In other cases, firms find it more efficient and effective to incorporate external knowledge rather than develop it internally. In the overall scope of innovation, most new ideas emerge from outside companies, and those that emerge inside can leave if not quickly captured (Chesbrough 2003*a*; Moore and Davis 2004). However, prior research on Open Innovation has mostly focused on the firm level of analysis and has not emphasized the role of the firm's external institutional and geographic context in shaping the flows of knowledge that the firm can act on in pursuing an Open Innovation strategy.

In this chapter, we consider the context of firm innovation by building upon prior research on interorganizational networks and innovation and draw their implications for Open Innovation. Networks are an inherent part of an organization's institutional environment and, whether formal or informal, are key conduits through which knowledge travels from the environment to the firm. Furthermore, such networks often have a geographical locus, as with the dense networks that form within a regional economy. Despite the availability of global travel and electronic communications, regional networks such as the ties that are the fabric of Silicon Valley continue to play an important role in interorganizational knowledge flows. Government policy often seeks to encourage such innovative knowledge flows through efforts to synthesize or strengthen national innovation systems.

To explore the implications of the network literature for Open Innovation, we begin by reviewing prior research on network organizations and interorganizational networks. We then consider studies on regional and national innovation systems and their impact on knowledge flows between firms. From this, we offer a framework of how firms build networks to support Open Innovation based on key network dimensions that are likely to influence the flow of knowledge to the firm, and conclude with suggestions for future research linking the two bodies of research.

11.2 Knowledge Flows, Networks, and Innovation

Knowledge flows between firms are crucial to many innovations. Such knowledge flows through networks of formal and informal ties, enabling firms to build upon the broad pool of knowledge located outside the boundary of the firm.

11.2.1 *Interorganizational Knowledge Flows*

Knowledge and information are distinct constructs. Information has the potential to be used in a way that creates new knowledge, or adds to or transforms existing knowledge (Machlup 1983; Nonaka 1994). Unlike information, knowledge requires a knower: 'information is a flow of messages, while knowledge is created and organized by the very flow of information, anchored on the commitment and beliefs of its holder' (Nonaka 1994: 15). Thus knowledge flows through and resides in individuals.

Knowledge is often subdivided between tacit knowledge—that which is not articulated or codified—and explicit knowledge (Polanyi 1967; Nonaka 1994). Because tacit knowledge is generally harder for competitors to imitate, it generally has greater competitive value (Nelson and Winter 1982). One problem in operationalizing the distinction between the two types of knowledge has been an absence of bright line tests between the two categories (Cowan et al. 2000; Johnson et al. 2002).[1] However, firms use both tacit and explicit knowledge to increate innovations (Cowan et al. 2000).

Essential knowledge flows along the value chain, between customer and supplier, where highly tacit knowledge may be necessary to use an innovation or for a supplier to refine its offering to meet customer needs. Von Hippel (1988, 2005) also notes that customers themselves often refine innovations or identify new ideas. In other cases, firms work closely with suppliers of complementary products to complete the whole product offering (Teece 1986; Moore 1991). In many cases, firms must organize and lead an entire value network to support their specific innovations (Christensen and Rosenbloom 1995; Iansiti and Levien 2004*a*; See Chapter 12). Universities transfer scientific knowledge, whether through their faculty research or through the education carried in their students (Audrescht et al. 2004).

Conversely, unintended knowledge spillovers occur between firms, as when labor mobility allows a firm's knowledge to 'walk out the door' and end up at a customer, supplier, or competitor. The conduit for desirable and undesirable knowledge flows is through people, both through mobility and through interpersonal interaction between individuals (Kogut and Zander 1993; Spender 1996). This is particularly true for the tacit knowledge held by individuals which is an essential antecedent to creative breakthroughs (Polanyi 1967).

To 'seal' valuable knowledge within their boundaries, firms have used a variety of approaches, including developing new technologies in-house from scratch, surrounding their research and development (R&D) activities with secrecy, closing their boundaries through noncompete and nondisclosure agreements, and acquiring external knowledge through costly vertical integration strategies (Chesbrough 2003*a*). But if, as Chesbrough (2003*a*) argues, most new knowledge emerges outside the firm, then approaches to closing the boundaries of the firm run the risk of overlooking opportunities from the far larger pool of knowledge outside the firm—while failing to prevent the leakage of knowledge that will eventually escape the firm if not acted upon. Open Innovation, in fact, seeks to address both issues, by creating processes for incorporating such external knowledge as well as capturing a return from potential outflows.[2]

11.2.2 *Networks and Innovation*

In contrast to the market and hierarchical forms of Williamson (1975), Powell (1990) identified a third alternative, the network organization. This form considers the identity and enduring reputation of the organizational actor as important, enforcing a norm of reciprocity and interdependence between network members (Powell 1990). Networks are highly flexible (Piore and Sabel 1984) and are embedded in ongoing relationships between social actors, where sanctions are reputational rather than normative. Network forms in the new economy are made of big and small players alike, across multiple industries, and encompass multiple types of ties (Powell 2001; Stark 2001).

Network ties may reflect formal collaboration, such as joint ventures, alliances or R&D partnerships. They may correspond to customer-supplier relationships (such as licensing, contracting, or providing key components) or more lateral alliances to comarket or develop complementary products. Or they may reflect informal ties between individuals, built through past collaborations (which might be sanctioned or unsanctioned).

Like the network form, Open Innovation is a value-creation strategy that is an alternative to vertical integration. In Open Innovation, some firms need to identify external knowledge and incorporate it into the firm; others seek external markets for their existing innovations (see Chapter 5). The pathways of network ties create opportunities for both types of Open Innovation. Accessing a network allows a firm to fill in a specific knowledge need rapidly, without having to spend enormous amounts of time and money to develop that knowledge internally or acquire it through vertical integration. Similarly, networks can facilitate (or result from) efforts to commercialize internal technologies, such as through creation of a spin-off, corporate venture investment in a start-up, or establishment of a joint venture.

Prior research has shown the role of interorganizational ties in enabling firm innovation. Teece (1989) showed how cooperation between companies increases knowledge gain and reduces the inherent waste of duplicated effort. Networks have been found to have beneficial returns on innovation such as increased patenting rates, improvements on existing products, and new product creation, faster time to market, and access of new markets (Powell et al. 1996; Gemunden et al. 1996; Almeida and Kogut 1999; Baum et al. 2000). By providing access to complementary skills, scale benefits, and a broader knowledge base, network ties positively influence firm innovation (Shan et al. 1994; Powell et al. 1999; Ahuja 2000; Baum et al. 2000). Studies further showed that firms involved in multiple types of ties are more innovative than organizations that engage in a single type of tie, since different types of ties can transfer different types of knowledge (Powell et al. 1999; Baum et al. 2000).

Networks are especially well suited to knowledge-intensive industries where joint problem-solving is paramount: networks foster problem-solving and learning mechanisms (Powell et al. 1996). Hence, the Open Innovation phenomenon has been most often identified in technology-intensive industries (Chesbrough 2003*a*), although using networks to tap into external knowledge is potentially relevant for companies in all industries. Innovation-related knowledge is not just limited to technical knowledge, but may also include the knowledge necessary to commercialize an innovation, such as the knowledge of customers, market segments and product applications. Such knowledge may come from customers or other partners in the value chain (von Hippel 1988, 2005; Lynn et al. 1996; Chesbrough and Rosenbloom 2002).

11.2.3 *Formal and Informal Ties*

Organizations and individuals are embedded in networks, and thus both interorganizational and interpersonal knowledge flows are guided by the formal institutionalized and the less visible informal interrelationships of those involved in innovative activities.

Formal ties are contractually agreed upon, planned channels for knowledge exchange between organizations, such as a strategic alliance. These ties are more easily incorporated into an Open Innovation strategy: a firm can identify gaps of internal knowledge and then seek potential partners for collaboration to fill that knowledge without having to build it internally.

While planned and thus part of a firm's strategy, formal ties can also have unexpected knowledge spillover benefits. Formal ties such as licensing agreements and alliances also represent channels for informal knowledge flows, and are 'more open than their portrayal as pipelines suggests' (Owen-Smith and Powell 2004). For example, a joint technology development agreement will not only foster the planned technological knowledge exchange, but also can enable labor movement between the two companies, create access to unforeseen knowledge through informal ties between those individuals developing the technology, and thus create a possibility of tapping into the networks of the respective participants. Formal ties between organizations are embedded in social networks (Gulati 1998), consistent with economic sociology's view that economic action is embedded in social structures (Granovetter 1985; Uzzi 1994). In the Open Innovation context, the challenge for firms is to develop the capabilities to recognize those unanticipated spillovers and capture their potential benefits.

Informal ties provide an important pathway for flows of valuable knowledge—particularly for exploiting unforeseen knowledge opportunities. Informal and unplanned ties can lead to knowledge spillovers (Agrawal and Henderson 2002; Murray 2002), when individuals move between companies or are members of a community that spans multiple organizations. Informal ties may emerge as the consequence of formal ties, or they conversely can open paths to the formation of formal ties (Gulati and Westphal 1999).

Knowledge can flow through people moving between organizations. Labor market mobility is an important source of network ties between organizations in regional economies such as Silicon Valley (Castilla et al. 2000; Cohen and Fields 2000). These informal ties are a source of human and social capital[3] for organizations (Murray 2002; Porter 2004). That is, an individual's stock of knowledge, experience, skills, and connections are brought to an organization at the time of hiring. Knowledge has been shown to flow through career movements (Almeida and Kogut 1999). People moving between organizations in a region is one way through which knowledge is transferred and applied to new contexts, leading to innovation. Almeida and colleagues (2003) concluded that informal knowledge flow mechanisms benefited small firms more than large firms.

In past research, organizations have been characterized as social communities involved in knowledge creation and transfer (Kogut and Zander 1996). Studies report that entrepreneurs typically gain knowledge of an industry in existing companies before founding a firm of their own (Sorenson and Audia 2000). Not only do past career experiences affect an individual's stock of knowledge, but these experiences also affect the social networks that one can draw upon to support subsequent economic activity (Uzzi 1996).

Embeddedness has been shown to favor organizational performance up to a threshold point, where it becomes detrimental by cutting off external sources of information (Uzzi 1996, 1997).

Social networks affect the creation of further intellectual capital by promoting knowledge sharing and innovation (Coleman 1988; Nahapiet and Ghoshal 1998; Sorenson and Audia 2000). Some research has emphasized that groups of individuals share knowledge through a community-of-practice, an informal network of knowledgeable people who share a common work identity and create knowledge flows across organizational boundaries (Brown and Duguid 2000). Thus, an Open Innovation strategy would need to recognize the external knowledge opportunities possible from their employees' embeddedness through informal network ties.

11.3 The Geography of Open Innovation

Open Innovation benefits may be more readily achieved in regional clusters, since the effect of networks on innovation is magnified by geographic proximity; such clusters are defined as 'geographic concentrations of interconnected companies and institutions in a particular field' (Porter 1998). Marshall (1920) first noted regions that are rich in ideas (and thus knowledge) will attract economic activity. Economists have pointed out the benefits of localization on economic growth (Romer 1987; Stuart and Sorenson 2003), such as reduced production and transport costs leading to increased access to markets and economies of scale, specialized labor markets, and the lower costs of accessing information locally (Weber 1928; Rotemberg and Saloner 1990; Krugman 1991; Maskell 2001).

Less developed are the linkages between networks and geography. While many have described the geographical nature of knowledge flows, very few studies have quantitatively measured the effect of geography on such flows and on innovation. A few studies show that knowledge flows more readily to closer entities (Jaffe et al. 1993), whether through organizations or through individual labor mobility (Almeida and Kogut 1999). Owen-Smith and Powell (2004) stress that interorganizational networks act as a signal of membership in a local community of knowledge. Recent research confirms that membership in a regional community increases innovation benefits (Bunker Whittington et al. 2004). This regional network effect applies both to high-tech and other industries such as apparel (Uzzi 1996, 1997), shoes (Sorenson and Audia 2000), knitwear (Lazerson 1995) and wine (Benjamin and Podolny 1999).

Networks have been a key building block of the formation of regional economies in high-tech (Saxenian 1996) and in biotechnology (Owen-Smith and Powell 2004, forthcoming); start-up firms in those regions have long recognized that colocation enables them to tap into necessary knowledge.

High-tech regions can be viewed as multidirectional knowledge flows (Brown and Duguid 2000), or as Porter and colleagues (2006) put it, 'the intersection of multiple networks is the wellspring of technology clusters.' Networks of colocated organizations are necessary to construct a regional social structure of innovation and the knowledge flows that lead to innovative activity (Owen-Smith and Powell forthcoming).

11.3.1 *Key Institutions*

Regional innovation is enabled by the knowledge exchanges among a diverse set of institutions and organizations. An optimal Open Innovation strategy would exploit multiple types of ties to multiple types of institutions, as each type of tie and institution favors the flow of different pieces of knowledge.

Universities. First among institutions known for creating basic knowledge is the university. While fully integrated firms once were renowned for their basic research abilities, the Open Innovation framework calls for tapping into other institutions with that basic research capability. Universities have been shown to be a central creator of such basic knowledge in regional economies (Teece 1989; Rosenberg 2000; Kenney 2000*b*; Miner et al. 2001).

High-quality research universities produce knowledge spillovers through such formal interfaces such as commercialization initiatives (patenting and licensing), industrial parks, and informal flows of students entering the labor market (Saxenian 1996). As research institutions with a culture of knowledge sharing, universities tend to generate more knowledge spillover effects in regions than other organizational forms (Dasgupta and David 1994; Owen-Smith and Powell 2004). However, increasing attempts by universities to profit from their research are potentially reducing those spillovers (see Chapter 8).

Venture Capitalists (VCs) are another important source of regional knowledge since they are actively involved in the creation of start-up companies (Gompers and Lerner 1999; Hellmann 2000; Kenney and Florida 2000; Kortum and Lerner 2000; Leslie 2000). With their ties to multiple start-up companies, VCs can help identify needed knowledge and potential synergies that are beneficial to both established companies and start-ups. VCs' knowledge base is geared toward commercialization of innovation and act as connective agents in a regional economy (Owen-Smith and Powell 2004).

VCs are a 'powerful institutional force' that are inherently focused on commercialization of technologies, converting ideas into products, and hence can be a crucial partner in an Open Innovation model (Chesbrough 2003*a*). Firms create informal ties through joint participation in advisory boards, trade associations, and other indirect collaborations. Formal ties to venture capitalists can be created in a variety of ways, such as creating formal ties through joint investments in start-ups or spin-offs. Firms can also create captive venture capital divisions to access external knowledge and commercialize firm technologies, as with Intel Capital (Chesbrough 2003*a*) or Qualcomm Ventures (Simard 2004).

Focal Firms. Another frequently cited force of knowledge creation in a regional economy is the presence of a highly successful start-up that acts as a breeding ground for knowledge creation and further ventures. In Silicon Valley, Fairchild Semiconductor and Hewlett-Packard are often depicted as key generators of future start-ups (Lecuyer 2000). In Helsinki's telecommunications cluster, Nokia has been identified as the 'star organization' attracting other multinationals to the region and ensuring a steady flow of knowledgeable workers and entrepreneurs (Porter and Solvell 2000);[4] for San Diego's telecommunications cluster, Linkabit played a Fairchild-like role in generating spinoffs while Qualcomm was the star organization (Simard 2004). Hence, companies in a cluster may gain some innovation benefits by favoring network ties to a local 'star' organization over less known companies. Star organizations may fluctuate over time; recently, Google has replaced HP as a 'star' organization in Silicon Valley acting as a major attractor for knowledge and talent.

Each industry may have its own institutions that lead to location innovation benefits. In biotechnology, for example, public research institutes may be an important source of knowledge (Owen-Smith and Powell 2004). According to the context, other key government entities may include the military, which provided both markets and knowledge spillovers for the development of clusters in semiconductors (Leslie 2000) and wireless communications (Simard 2004). Other organizational forms such as law firms and consultants can also act as important sources of knowledge or bridges to other organizations (Suchman 2000; McKenna 2000) and vary in their organizational form and spatial distribution depending on the type of industry (Kenney and Patton 2005).

11.3.2 *National Innovation Policies*

In a broader geographic scope, policymakers have sought to identify and systematize policies that enable the creation and incorporation of innovation within a national economy. Variously referred to as 'national systems of innovation' (Lundvall 1992) or 'national innovation systems' (Nelson 1993; Montobbio 1999), contemporary research on such national innovation policy has attempted to link between-country differences in innovation outcomes to differences in their respective supporting institutions. The studies focus on the role of nation-state in enabling (or constraining) innovation activities, focusing on institutions that facilitate collaborative innovation such as university and government-sponsored research, as well as many of the same spillover issues as the regional innovation literature. The work often attempts to identify policy proscriptions that will allow a national policy body to improve innovation creation and flows. Thus, understanding the differences between innovation systems (as well as the antecedents of such differences) would help us to anticipate national differences in the degree and nature of Open Innovation. Such understanding would also help us understand the relationship between changes in innovation systems and changes in Open Innovation.

In some cases, the policy linkages are overt, as with direct government subsidies for industrial research, or indirect subsidies through government procurement of military or other goods. Such research benefits both the direct recipients and related firms through spillovers to civilian applications (Nelson 1993; Steinmueller 1996; Bresnahan and Malerba 1999). In this case, the government acts as what Chesbrough (2003*a*) terms an 'innovation benefactor,' creating external sources of innovation without attempting to appropriate the full returns of such innovation. However, spillovers from military projects are often accidental, as in the shift from military to commercial technology in San Diego's 'Wireless Valley' (Simard 2004).

Other research has sought to identify the role played by nongovernmental institutions to explain national differences in their ability to exploit new technological opportunities, based on the flow of tacit knowledge and organizational learning (McKelvey 1991; Lundvall 1992; Mowery 1996). Early studies attempted to isolate specific 'national' patterns of innovation common across all high-tech industries in a given country. So the studies edited by Nelson (1993) show that industries with high up-front R&D costs tend to be found in large, affluent countries—except for those smaller countries (e.g. Sweden, Israel, Korea) with disproportionately large defense industries. The more successful firms have been exposed to stronger competition, typically but not always in their home market (Porter 1990; Nelson 1993).

However, a key limitation is that these studies have assumed that between-country differences in innovation institutions are more important than within-country ones. Other studies have noted the importance of firm-specific factors to explain the relative success of national industries (Dertouzos et al. 1989; Chandler 1990; Nelson 1993). Mowery and Nelson (1999) combine the two approaches with the concept of 'industrial leadership' to encompass both firm and industry effects.

Of course, in a globalized environment, many firms source technology and seek customers across national boundaries. Still, home market customers play an important role in developing the innovative capabilities of firms (Porter 1990). And labor markets remain one of the few innovative inputs that are imperfectly traded across national boundaries, due to lingering labor market protectionism (Rodrik 2000).

Thus we would expect to find several key national factors to explain the differences in the application of Open Innovation. Some countries will have a larger supply of innovation spillovers available to firms (whether due to scale or innovation sponsorship). Countries will differ (due to industrial structure) in the number of firms ready to incorporate such spillovers, with a Japan quite different from the Netherlands or Sweden. Finally, countries differ in the role played by start-up companies and thus the importance of venture capital firms, which as Chesbrough (2003*a*) notes, often serve to disseminate innovative knowledge within an industry.

11.4 Building Networks to Support Open Innovation

What conditions would increase the likelihood and effectiveness of Open Innovation strategies? Network attributes have important effects on firm performance (Beckman and Haunschild 2002). However, prior research suggests that firms cannot assume that 'the more network ties, the more innovation'.

Here we suggest three factors to consider when using networks as the interface to obtain knowledge in an Open Innovation strategy. Firms need to build ties that are both wide and deep. At the same time, they must also make sure that the value of the knowledge flowing into the company is greater than the value that knowledge outflows provides to potential competitors.

11.4.1 *Deep Ties*

Gulati (1999; Gulati et al. 2000) argues that a firm's position in a network provides 'network resources' that are difficult to imitate and thus potentially provide enduring competitive advantage. If a firm is to obtain innovation advantage through its network position, then its position not only needs to be unique, but it must also tap into key sources (and markets) for innovation.

One way that such uniqueness can be created is through a deep embeddedness in a key technology or market. Firms may do so by locating in densely populated networks, by building their own value networks, and by strengthening the ties within their networks through building trust. Repeated interactions breed trust in networks (Gulati 1995*a*).

Geographic Embeddedness. If the effects of network ties on innovation are enhanced by proximity, then firms may decide to establish a physical presence in regions that are repositories of knowledge in their specific industry. High-tech firms (such as IBM and Microsoft) have opened branch offices in Silicon Valley to tap into regional knowledge. Intel has created research laboratories near key research universities to facilitate knowledge transfer between the firm and university researchers (Chesbrough 2003*a*). Firms that lack geographic proximity to key innovation networks instead must build their own networks, as in switched amplification (see Chapter 2) or materials science (see Chapter 3).

Increasing Tie Strength. Networks of innovation are often based on repeated interactions between firms, and thus depend on trust—particularly in regional clusters where firms and people develop a local reputation based on past interactions. Network forms rely on trust as a coordination mechanism (Powell 1990).

Limited research has been done on trust and interorganizational relationships. Trust is an important coordination mechanism of networks (Powell 1990; Uzzi 1997). Empirical evidence suggests that interorganizational trust, which is more institutionalized, is longer lasting than the interpersonal trust inherent in informal networks. Trust is crucial in reducing the risks associated with interfirm tie formation (Nooteboom et al. 1997). Repeated interactions through interpersonal ties can lead to a more institutionalized interorganizational trust, where organizations come to recognize each other as long-lasting partners and can engage in knowledge exchange ties rapidly (Zaheer et al. 1998).

At the same time, organizations must consider a balance of strong and weak ties when considering their Open Innovation strategy. Strong ties benefit from more institutionalized trust and are likely to be more quickly and easily activated, yet weak and bridging ties provide access to new information which is paramount to innovation. There is an inherent trade-off between trust and novelty, safety and flexibility (Gargiulo and Benassi 2000). In turbulent environments, Powell and Smith-Doerr (2005) argue that the linkages are not driven by loyalty but by the need to stay informed, and that proximity leads to greater trust in tie formation. However, Erickson (2005) concludes that the trust between two firms built through past interactions may be reduced through major changes in their respective network roles.

Limitations. Overembeddedness happens when firms rely too much on repeated interactions with the same partners; when these partners are themselves linked through strong ties, the network becomes closed to external information and starts having access to only redundant information, leading to the stifling of innovation (Uzzi 1997). Indeed, some research suggests that spatial concentration leads to conformity in firm behavior and less innovation (Sorenson and Audia 2000). Regional clusters, while known for their innovative capacity, run the risk of becoming closed to outside knowledge and becoming overembedded.

11.4.2 *Wide Ties*

Weak Ties. One way of countering the problem of overembeddedness is to form some weak ties. Since Granovetter (1985) posited the 'strength of weak ties', significant attention has been given to the power of arm's-length ties. Based on occasional rather than frequent interactions, these ties offer more pathways to new information, because they provide access to different networks and thus different sources of information. Informal professional affiliations such as common organizational affiliation are such weak ties that can be acted on in an Open Innovation model. Weak ties can act as a counterforce to the overembeddedness problem. Little research has applied Grannovetter's (1985) weak ties argument to formal interorganizational network ties, but there is some evidence that firms who combine a mix of strong and weak ties gain more information benefits (Uzzi and Gillespie 1999*b*).

Exploiting Structural Holes. Another strategy to avoid becoming overembedded is to exploit structural holes, the gaps between otherwise disjoint networks. Burt (1992) shows that forming ties to nonredundant, nonconnected others leads to more information benefits. Acting as a bridge between diverse actors enables the firm to tap into the knowledge contained in multiple networks (McEvily and Zaheer 1999).

Diversity of Ties and Institutions. Central to an Open Innovation strategy is to maintain diverse types of ties to a diverse set of institutions. There is a delicate balance between exploration and exploitation ties (March 1991; Koza and Lewin 1998). Exploration in organizational learning involves searching for new opportunities and developing new product or technological development through alliances (Rothaermel and Deeds 2004), whereas exploitation involves capitalizing on existing knowledge and resources. Exploration alliances have been found to predict the future occurrence of exploitation alliances (Rothaermel and Deeds 2004). As the measure of success for Open Innovation is commercialization, the occurrence of exploitation alliances could be used as a dependent variable in the Open Innovation literature.

Each firm has its own appropriate mix of institutions, but these might include universities, other firms with complementary knowledge, government institutions such as research institutes, firms more geared toward commercialization such as venture capitalists, and potentially other professional firms such as law firms. However, it is not enough to connect to a diverse set of partners: firms pursuing Open Innovation also need to utilize diverse types of ties. Formal ties may encompass joint research, commercialization agreements such as licensing, or marketing agreements; informal ties may include labor movements, regional communities of practice, and past common organizational affiliations. When considering the mix of variables—strong versus weak ties, connectedness, and structural holes—research has yet to identify either the optimal mix of variables or the process for achieving this mix.

One value of tie diversity is that innovation often happens through the recombination of sometimes-unforeseen knowledge elements (Smith-Doerr and Powell 2003), which can be enabled through collaboration between companies. Access to heterogeneous knowledge through networks has increased benefits by increasing chances for recombination leading to innovation (Hargadon and Sutton 1997; Pelled et al. 1999). However, some point out that heterogeneity can come at the cost of trust (Hambrick et al. 1996), and there may be a threshold where decreasing returns occur when too many diverse ties are maintained, if they overwhelm the firm's ability to recognize the relevant knowledge in each (Beckman and Haunschild 2002) and to tie them together to create innovation.

While a firm seeks a diverse set of ties, it has only a limited set of resources to manage these ties. These limits are particularly important for firms creating global network ties for both its inputs and outputs. Research on multinational ties finds a negative relationship between alliance diversity and performance (Goerzen and Beamish 2005). For transnational alliances this diversity may be too complex to manage and lead to decreasing returns (Goerzen and Beamish 2005) and structural holes had no beneficial impact (Ahuja 2000), suggesting that tie diversity is most valuable when coupled with geographic proximity.

While so far the research on networks and innovation suggests that firms should concentrate its resources on forming and capturing knowledge from regional network ties, knowledge and markets in the new economy are increasingly globalized, so that successful specialized firms need to tap into knowledge and markets scattered across the globe to rapidly deploy innovative applications (Doz et al. 2001). Organizations are part of far-reaching and diverse ecosystems that hold distributed knowledge which is key to one firm's innovation capacity, and thus 'the crucial battle is not between firms but between networks of firms. Innovation and operations have become a collective activity' (Iansiti and Levien 2004*a*: 11).

Some firms must manage innovation ties at both the regional and global level due to the nature of their institutional environment. For example, the importance of compatibility standards force telecommunications firms to balance regional supply ties with multinational ties to help them promote their technology in new markets. New research is attempting to measure regional versus global effects: one study suggests firms are less successful if they attempt to maintain centrality in both their regional and global networks, and thus for optimal performance must choose whether to focus on local or global innovation ties (Bunker Whittington et al. 2004). We suggest that the most appropriate balance between local and global ties in an Open Innovation strategy may depend on the nature of the firm's institutional environment, and thus that the institutional environment needs to be included in analyzing and explaining a firm's practice of Open Innovation.

Technological environments. Different industries have different institutional environments and require different types of tie formation. Hence, in biotechnology, where new knowledge creation and commercialization is heavily based on basic science, R&D ties are the main 'ticket entry' through which later commercialization benefits are realized (Powell et al. 1996). By contrast, in industries operating in a technological environment characterized by network externalities (Katz and Shapiro 1985), there are different strategies of innovation (Sheremata 2004) and hence different patterns of tie formation. These markets are driven by interoperability standards and the provision of complementary products, as when Qualcomm built ties to promote its technology through standardization bodies and to attract complementary products, which enabled its subsequent licensing business model. However, Qualcomm's business models depended not only on its ties but its intellectual property (IP) strategy (Simard 2004). Thus, the use of Open Innovation may also depend on the available IP regime both for the industry and desired market (see Chapter 6).

11.4.3 *Maximizing Returns from Knowledge Outflows*

Prior research on Open Innovation has underemphasized the importance of a firm's institutional environment in designing strategy. The network research has, conversely, examined how firms together form an ecosystem of knowledge flows but has said less about how these may be incorporated into strategy at the firm level. How, then, might we combine the network and Open Innovation perspectives to develop a richer view of the external factors affecting a firm's Open Innovation strategy?

Location Matters. The first implication for Open Innovation is that location matters. In some industries and technological environments, forming ties with and establishing a physical presence in a region where important knowledge resides will be key. Thus, a firm may decide to open a branch close to a partner or competitor that to attempt to establish knowledge spillover benefits, as when large telecom firms established a presence in San Diego to tap into Qualcomm's CDMA knowledge. Firms may also locate in proximity to an elite university where partnerships with faculty and the hiring of top students can become crucial for the firm to keep abreast of cutting-edge scientific knowledge in a field, as has been documented in the biotechnology industry (Porter et al. 2006).

The Learning Race: Maximizing Returns from Spillovers. In any firm, key knowledge will spillover from a firm to its customers, suppliers, partners, and competitors. Strategies and mechanisms that enable inflows of key knowledge—such as building a broad and deep network and locating in a dense cluster with high labor mobility—can also enable a comparable outflow of knowledge. Even in formal alliances for learning and sharing information, the complementary stocks and deficits lead alliance partners to a 'learning race' (Hamel 1991; Gulati et al. 2000), whereby one organization tries to maximize its learning from the other and minimize the amount learned by the other while trying to retain trust. Khanna et al. (1998) show that firm learning expectations predict resources allocated for learning, and thus learning success.

The response of the traditional innovation model is to clamp down on such flows, by segregating access to knowledge, locating away from dense networks of suppliers and competitors, and attempting to minimize job turnover. This has been the inherent approach of large US, European and (especially) Japanese multinational companies. But such an approach also cuts a firm off from finding markets for its technologies and often impedes the flow of inbound innovation as well. But if firms are unwilling or unable to be part of a network, they may be a disadvantage compared to those firms that gain knowledge and increased innovation capacity by belonging to such networks.

Another approach (as recommended by Chesbrough 2003*a*) is to adopt an IP strategy that allows and encourages the outward flows, but maximizes the economic returns that accrue from commercial application. Instead of using trade secrets to keep the ideas within the firm, a firm would aggressively patent its ideas and disseminate them widely, assuring a stream of patent royalties should those ideas be adopted (see Chapter 4). At the same time, such strategies are increasing the cost of inbound flows of external innovation, as when universities seek to profit from publicly sponsored research that they once would have allowed to spillover to local firms (see Chapter 7).

Where patents are ineffective, firms can develop polices to license their tacit knowledge and thus actively participate in the success of its spillovers. Such policies both accelerate commercialization of the innovation and also provide the recipient with an advantage over potential rivals. For example, when Xerox PARC declined to exploit key inventions and was faced with the likely defection of key scientists seeking to commercialize these inventions, it developed a range of policies to allow Xerox to participate in the commercial success of any spin-off companies (Chesbrough 2002).

Firms may also differ in their knowledge-sharing intensity with different partners. That is, some collaborations or alliances can be identified as particularly crucial to a firm's innovation. In that case, the firm may decide to maximize knowledge exchange by establishing more open knowledge-sharing routines in order to maximize absorptive capacity (Dyer and Singh 1998). Dyer and Singh argue that knowledge transfer and absorption are maximized by processes that maximize social and technical interaction between the firms, such as sending employees at the other firm and repeated interactions.

Finally, approaches to maximizing the returns to spillovers need to recognize the role of both formal and informal ties carrying knowledge away from the firm. Business models are more likely to be successful if they acknowledge the existence of informal ties and spillovers that cannot be stopped, by assigning a price to essential knowledge that can be protected and is an essential complement to the free spillovers. For example, Information technology systems vendors widely disseminate knowledge about building complements that increase the value of the firm's products, but aggressively protect the information necessary to build competing implementations via trade secret, patent, and often copyright law (West 2006).

Building an Open Innovation Network. In Figure 11.1, we consider the trade-offs between two dimensions of network ties identified above: deep versus wide ties and formal versus informal ties.

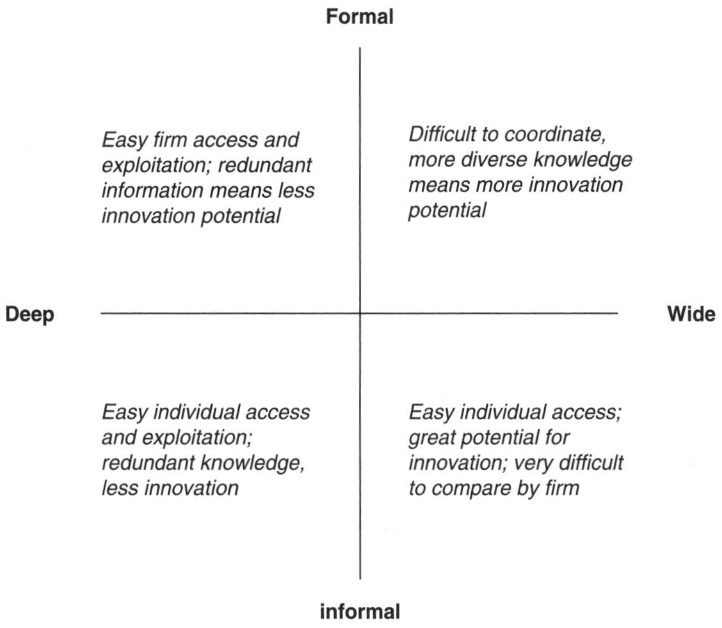

Formal

Easy firm access and exploitation; redundant information means less innovation potential

Difficult to coordinate, more diverse knowledge means more innovation potential

Deep **Wide**

Easy individual access and exploitation; redundant knowledge, less innovation

Easy individual access; great potential for innovation; very difficult to compare by firm

informal

Figure 11.1. Nature of interfirm ties enabling Open Innovation

Deep networks are easily activated and the knowledge contained in them easily captured, however the knowledge contained in these networks is likely to be redundant with knowledge already possessed by the organization. This trust and access to knowledge is further enhanced by geographic colocation. The potential for the networks to increase innovation is thus comparatively small; one hypothesis would be that deep networks tend to lead to incremental innovation as opposed to radical innovation. When ties are deep and informal, they also provide the potential for easy access to information but add another challenge for the firm that needs to recognize and act upon information hidden in the fabric of employees' social lives. Such informal ties, while an extremely important part of the knowledge flowing into and out of the firm, would be difficult to predict and incorporate into an explicit Open Innovation strategy.

Wide ties provide the benefit of access to nonredundant information and thus a greater potential for innovation, but without the trust inherent in deep ties. Wide ties are also hence more difficult to manage, particularly in capturing and recombining these sometimes disparate information elements into new knowledge. The coordination and trust difficulties are further compounded when there is an absence of geographic colocation. Wide networks of informal ties have high potential value for knowledge creation, but pose significant challenges in managing the inward and outward knowledge flows to maximize firm benefit. Again, a major role for informal ties makes it difficult to predict, capture, and plan the role of such ties, but this does not mean that they can (or should) be ignored.

The need for firms to balance the need for deep and wide ties parallels the need identified by Tushman and O'Reilly (1996) to balance short-term and long-term technological change. They contend that firms require an 'ambidextrous' capability to cope with incremental and radical innovation. Consistent with Tushman and O'Reilly, we would expect that wide ties would be necessary to cope with new technological trajectory (Nelson and Winter 1982), while deep ties would be needed to strengthen innovative capabilities within a given trajectory.

11.5 Implications for Future Research

Open Innovation is about harnessing knowledge flows across firm boundaries (Chesbrough 2003*a*). The channels for these repeated flows are interorganizational networks, constituted from a diverse range of possible ties. Each tie may vary in strength, the enabling mechanism, the level of analysis, and the direction of knowledge flow that it provides. And the portfolio of network ties managed by firms may differ in the breadth and depth of the knowledge they collectively provide, and in the geographic locus of the network partners.

Thus the study of the role of network ties in innovation is implicitly (if not explicitly) one that relates to potential Open Innovation. Here we identify opportunities for future research about the relationship of knowledge flows, interorganizational networks, geography, and the practice of Open Innovation.

11.5.1 *Understanding Informal Ties*

Studies of networks in innovation have emphasized the role of formal ties at the organizational level, but the role of informal ties is less well understood. These informal ties may be those that arise from formal alliances and other ties (and thus reflect an unmeasured confound), or they may be those ties utilized by a firm's employees in a way that may not be a visible part of the firm's strategy.

Similarly, while research on Open Innovation has emphasized formal institutions, the framework should also consider how commercially valuable knowledge can be accessed through informal networks. Firms can and do exploit informal knowledge flows, by hiring the best possible sources of knowledge—individuals with not only strong backgrounds but from companies or industries on which the organization wishes to gain knowledge. Firms seeking to capture external innovation through informal ties will seek to employ not only the ones with the most knowledge in specific areas, but also the ones with past career affiliations to firms that act as repositories of knowledge in specific areas.

The benefit of formal and informal ties comes from inbound flows of commercially valuable knowledge. But the existence of a tie is not a guarantee of knowledge transfer; a key moderator is the level of trust by the disclosing party. Trust may also play other roles in interorganizational networks, such as in a willingness to form ties and the ability to interpret tacit knowledge to unlock its latent value. And in at least some forms of networks (such as interactions with universities, open source communities, or other nonprofits), efforts to realize commercial value from knowledge flows can potentially reduce the trust that enables such flows.

At the same time, both formal and informal ties have their costs—the direct costs of managing the ties and as well as the potential indirect costs if the knowledge provided obtained by the firm is less valuable than that which flows out to competitors. The trade-offs are likely to differ greatly according to institutional context, depending on the social fabric of the industry or geographical cluster in which the firm is located.

For example, recent research suggests that formal networks in the biotechnological industry may be more open and more conducive to innovation than informal networks which are more closed. In the case of biotechnology, informal social networks tend to be clustered around star scientists who act as a bottleneck for information sharing (Porter et al. 2006).

So what are the industry, regional, firm, and individual factors affecting both a firm's efforts to create a mix of formal and informal ties, and also the value of that mix for Open Innovation? Are there commercialization benefits that extend across industry and institutional contexts? Or is the relative role of such ties primarily due to a firm's technological, economic, and geographic context?

11.5.2 *Managing the Network Portfolio*

Rather than a single tie, the interorganizational networks of innovative firms will include a portfolio of complementary ties. Firms thus must determine what individual ties best support their innovation strategies, what interaction effects they are (positive or negative) between the various ties, and how to maintain and improve the overall portfolio.

The individual ties can vary across numerous dimensions: formal versus informal, strong versus weak, local versus national versus international, and individual versus firm level interactions. Within the within the formal versus informal dimension, a range of mechanisms for creating ties exist, including formal R&D alliances, arms'-length licensing, or (on the informal side) harnessing employee coworker networks. Each tie can also support inbound knowledge flows, outbound flows, or some combination thereof.

There are additional issues to consider when valuing a combination of multiple ties—whether the whole is more or less valuable than the sum of the parts. Firms can choose to develop wide or deep ties, and a high or low level of diversity in tie dimensions and mechanisms. Prior research has implied there are trade-offs and a possible U-shaped relationship for each.

There is also potentially an interaction between firm size, level of integration, and use of external ties. As part of an Open Innovation strategy, a small firm is likely to build deep and lasting ties to integrate its particular business model into a larger value network. However, large firms—particularly vertically integrated ones—may be tempted to develop in-house (or acquire) its own deep knowledge in areas that play an important role in supporting its business model; this would fit the fundamental idea of a core competence, as discussed in Chapter 3. One would expect both types of firms to use weak ties to find new knowledge that they didn't even know they needed—but these hypotheses are all testable propositions.

Thus, there are numerous unresolved questions regarding the role of these network portfolios in promoting Open Innovation, including balancing the trade-offs on each dimension, the influencing of external factors in determining the available tie options, and the optimal tie mix (moderated by internal and internal factors) to maximize knowledge flows that support innovation.

There is also the question of the direction of causality. Chesbrough (2003*a*) focuses on examples where firms have successfully implemented Open Innovation strategies. But does each firm have an endogenous set of choices for building its network portfolio? Or are the tie options (particularly for younger and smaller firms) sufficiently constrained that the network portfolio drives the innovation strategy? Are there particular aspects of the network portfolio that would significantly raise (or lower) the effectiveness of Open Innovation as part of the firm's innovation strategy?

11.5.3 *Geography and Innovation Networks*

Considerable research has shown that geographic proximity facilitates network formation. Such proximity can identify partners for formal ties such as agreements to license technology or supply key components. It can also allow firms to better utilize the value of informal ties, as when a biotech firm hires the alumni of the local research university both to identify potential partners at the university and provide entrée for future collaboration.

Regional clusters can provide an ideal setting to study Open Innovation: start-up firms in technology-intensive industries cannot spend the time and resources to build their own fully integrated innovation funnel as the old model of innovation implies. Rather, these companies can rapidly form network ties to institutions and firms with complementary knowledge in order to bypass the innovation funnel and be first to market.

At the same time, firms cannot limit their search for innovation sources or markets to a subset of desirable partners. So it remains an open question whether firms embedded in regional networks practice more Open Innovation than those more geographically distant, or whether other factors determine the openness of innovation.

At the opposite extreme, metanational firms increasingly seek to capture specialized knowledge in different parts of the world (Doz et al. 2001). Are Open Innovation practices across national boundaries different from those within a nation-state? Do factors that would attenuate tie strength—for example, measure of cultural distance such as language (West and Graham 2004)—also apply to tie formation or knowledge flows within ties? Do such factors have a greater impact on informal than formal ties?

Finally, there are interaction effects for both regional and global influences on open networks. Are regional innovation ties more important for early stage industries (or those with rapid rates of technological change or new firm formation) than for more mature, slowly changing industries. Conversely, for industries with globally dispersed specialized knowledge, does Open Innovation success depend on a competency in creating, maintaining, and utilizing such cross-national innovation networks (cf. Dedrick and Kraemer 1998).

11.5.4 *Measuring Innovation Creation and Flows*

Understanding the role of external innovation and opportunities to commercialize internal innovations requires, in turn, an understanding of the firm's interorganizational knowledge flows. Measuring such flows remains a challenge, whether they are to be used as an antecedent, mediator, or outcome of the firm's level of innovation.

Patent data is often used as a measure of both innovative output and (through citation analysis) of the relationship between individual inventions. Such data is readily available, corresponds to a population of a particular type of innovation (patented invention), and use of rigorous statistical techniques. One important impact is that they provided an externally relevant measure of invention influence through citations of prior art.

However, as Gallini (2002: 138) notes, 'patent counts are an imperfect measure of innovation.' For example, the patent propensity of some industries is comparatively rare, while in other industries patents are used for defensive purposes.

More fundamentally, patents measure technological invention, the outcome of a process of knowledge generation. Open Innovation draws the distinction between a technology and realizing the commercial value of that technology, as mediated by the business model (Chesbrough and Rosenbloom 2002; Chesbrough 2003*a*). Assuming that the latent economic value of all patented inventions can be realized assumes away the role of business strategy, complementary assets, and all the other factors identified by Teece (1986) to appropriate the value from a technology; we know from prior research that firms can and do differ dramatically on such dimensions.

Ideally, Open Innovation research would both measure technological innovation (such as through patents) as well as the commercialization of that innovation. Examples of the latter would include annual licensing revenues, new product development, and market share of new products; many of these measures have been used, although there are often very difficult to obtain for a wide range of firms in a given industry.

Finally, Open Innovation presumes knowledge flows between firms. Patent citation counts have been used as one measure of such flows, but as Jaffe et al. (2000) report, they are only partially correlated to self-reported knowledge flows, which suggests at least one measure is an imprecise measure. In other cases, network studies often assume that knowledge is flowing through ties without investigating the type and content of knowledge in these ties (Simard 2004). Measures exist for some forms of knowledge utilization across formal ties—such as licensing agreements and royalty payments. But flows across informal ties are inherently harder to measure, and without such measures it would be impossible to analyze the relative importance of formal and informal ties—as well as the antecedents of such knowledge flows (such as industry or firm characteristics) and their consequences (i.e. whether the flows lead to innovation). Such processes could be studied through comparative case studies.

Notes

1. A 'bright line' test is one that provides 'an unambiguous criterion or guideline especially in law.' (*Merriam-Webster's Collegiate Dictionary,* 11th edn.), analogous to the bright lines displayed in a spectrograph.
2. We do not mean to suggest that networks have completely supplanted vertical integration. Examples where the latter remain desirable include controlling downstream markets for innovation (Chesbrough and Teece 1996), or obtaining an upstream supply of crucial innovation (Podolny and Paige 1998).
3. Social capital is understood here not in the sense of civic participation theorized by Putnam (1993), but instead as the structural and relational assets created by interpersonal relationships (Tsai and Ghoshal 1998).
4. Instead of 'star' organization, Feldman (2003) allows for multiple 'anchor tenant' firms, analogous to shopping malls; her study does not examine the case of anchors entering, exiting, or changing in relative importance.

12

Open Innovation in Systemic Innovation Contexts[1]

Markku V. J. Maula
Thomas Keil
Jukka-Pekka Salmenkaita

12.1 Introduction

The competitive consequences of different types of innovation have been one of the central themes in research on technological innovation (e.g. Mansfield 1968; Abernathy and Utterback 1978; Tushman and Anderson 1986; Henderson and Clark 1990). Studies have repeatedly shown that incumbents ability to create and to adapt to nonincremental innovation differs starkly from their ability to deal with incremental innovation (Henderson and Clark 1990; Christensen and Bower 1996). Several types of innovations have been identified that create particular challenges for incumbents. Tushman and Anderson (1986) argue that *radical innovations* that change core technical concepts and their linkages create adaptive challenges for incumbents. Henderson and Clark (1990) extended this research by showing that innovations that change the linkages between core concepts, so called *architectural innovations*, create similar challenges for incumbents. More recently, Christensen and Bower (1996) yet extended this class of difficult-to-handle innovations by their concept of *disruptive innovations* that initially address new customer groups and focus on different performance characteristics. These three groups of innovations have in common that they require organizational arrangements for innovation that differ from processes and arrangements suitable to create or adapt to incremental innovations. The findings further suggest that firm behavior differs (and needs to differ) depending on the nature of innovation an incumbent is faced with in its operating environment.

While the above classifications of innovations and the different classes of non-incremental innovations have added to our understanding of the innovation process, they are yet incomplete. In this study we focus on *systemic innovation* (versus autonomous innovation), that is, innovations that require significant adjustments in other parts of the business system they are embedded in (Teece 1986; Chesbrough and Teece 1996; Teece 1996; De Laat 1999). Due to the fact that systemic innovation processes frequently span beyond the boundaries of the firm, they often entail the coordination of different parts of the value network and entail Open Innovation organization models of innovation activities (Chesbrough 2003*b*). In systemic innovation processes firms need to coordinate as well with producers of complementary products and in many cases even with direct competitors to ensure the viability of the innovation, rather than coordinating solely with suppliers and customers as is frequently the case in closed innovation models. While systemic innovation processes are widely practiced in industries such as telecommunications or information technology (IT), the processes how incumbents and new entrants achieve this coordination and ultimately how they jointly create systemic innovation are ill-understood in academia.

Although some prior studies have examined innovations that could be characterized as systemic, these studies have not systematically analyzed the management of proactive creation of the entire commercialized system of innovations. The studies have frequently taken as given the long and evolutionary development process of the complementing innovations, such as the development of the petrochemical industry to provide fuel for the combustion engines of automobiles or the development of the production and distribution of electricity to enable electric light to displace gas lamps (Abernathy and Clark 1985; Utterback 1994; Hargadon and Douglas 2001). However, in many cases such complementing innovations are vital for the commercial success of the actual value creating innovation, and in current business environment companies rarely have the luxury of time to wait for the emergence of such essential complementary resources for decades. The question is how can companies lead the development of systemic innovations proactively?

In this chapter we make several contributions to the innovation literature. First, we contribute to a better understanding of Open Innovation by showing how systemic character of innovation acts as an underlying driver for Open Innovation processes. Second, our chapter shifts the focus of attention from the firm level to the value network level. In particular when faced with systemic innovation, resource allocation processes have to stretch beyond the boundaries of the firm and need to take complementary innovators into consideration. Proactive creation of systemic innovations requires a dynamic interplay between complementary innovators including incumbents, start-ups, and research institutions through various collaboration mechanisms including external corporate venturing modes, research programs, and industry consortia, a variety of mechanisms and innovation processes common in Open Innovation. Third, we develop a dynamic framework of resource allocation mechanisms. By analyzing multiple time horizons, we are able to show

how different boundary spanning mechanisms enable the firm to dynamically manage resource allocation. Fourth, we show how in systemic innovations, industry leaders frequently play architect roles (Chesbrough 2003*b*) in steering other participants in the value network. To play this architect role, industry leaders add industry shaping activities to the strategic foresight processes that often drive resource allocation decisions in autonomous innovation. Ensuring the commitment of complementary innovators through signaling own commitment (e.g. Spence 1973) is an important part of resource allocation process in systemic innovation contexts.

The rest of the chapter is structured as follows. Sections 12.2 and 12.3 briefly summarize prior literature on systemic innovation and make the case that traditional resource allocation approaches are insufficient for systemic innovations. Section 12.4 develops our framework for the management systemic innovation processes and necessary resource allocation processes. Section 12.5 discusses conclusions.

12.2 Systemic Innovation

We want to start our introduction to systemic innovation with an example. For third generation mobile telephone systems to be successfully commercialized, at the minimum, third generation (3G) network infrastructure needs to be developed by mobile network equipment providers and deployed by mobile operators, mobile phones need to be developed by mobile terminal producers (only some of which are the same companies that produce infrastructure) that are capable to interact with this new infrastructure and make use of the functionalities that it provides. Finally, applications and services need to be developed by mobile operators and third parties that make use of the new functionalities of both handsets and infrastructure so that mobile operators end-users have an incentive to purchase the handsets and use the infrastructure creating a business case for mobile operators to invest in deploying the infrastructure and distributing the handsets. If either of these building blocks (as was the case in 3G deployment in Europe) is delayed or functions below expectation, frequently the producers of other parts of the system cannot move forward with introducing their part of the system facing significant economic consequences.

Conceptually, such systemic innovations were introduced to the literature of innovation management as a category of innovations requiring specialized complementary assets for successful commercialization of the innovation in question (Teece 1986). Extending this initial limited conceptualization, systemic innovation has been defined as an innovation whose 'benefits can be realized only in conjunction with related, complementary innovations' (Chesbrough and Teece 1996). During the past decades and years we have seen that such innovations have become increasingly common including the Internet, the 3G mobile telephony, Linux, Java, Symbian, and many others. Systemic innovations thus have to be distinguished from autonomous innovations that 'can be pursued independently from other innovations' (Chesbrough and Teece 1996). In the following we will assume that both of these refer to the requirements that can be observed at least in retrospect for full-scale commercialization of the innovation, defining an innovation as an invention that has been fully diffused to the marketplace (Garcia and Calantone 2002).

Systemic properties of innovations have been subject to very limited discussion in the literature. The systemic characteristics of innovations have been identified to impact selected business dimensions of innovative activity, and examples of such discussion include the literature on platform leadership (Gawer and Cusumano 2002), Complex Product Systems (CoPS) (Miller et al. 1995; Hobday 1998), industry standardization questions (Kano 2000; Keil 2002*b*), and Open Innovation (Chesbrough 2003*b*). However, the existing research does not clearly identify the distinction between individual autonomous or systemic innovation and the broader system. The characterizations tend to classify innovations as linked to either one firm or one product or technology category, forcing the analysis to extend to a more complex environment in terms of organization and dynamics of innovations. Literature that has directly focused on systemic innovations has largely focused on whether systemic innovation should be managed within a single firm by vertically and horizontally integrating complementary innovations or whether these innovations are better created through markets. Teece (1986) as well as Chesbrough and Teece (1996) have argued that systemic innovations should be typically managed in an integrated fashion to avoid the substantial difficulties in coordinating the innovation activities of multiple players in the market place. However, this view has been seriously challenged in some contexts (e.g. De Laat 1999) with the argument that many contemporary systemic innovations are just too big and complex even for the largest integrated companies to manage alone. While integrating systemic innovation economizes on the cost of coordination and provides control benefits, it is frequently infeasible since even the largest firms lack the financial resources let alone technological and market capabilities to create the simultaneous complementary innovations necessary for successful systemic innovation. Empirical evidence supporting the integration argument is rather inconclusive, limited (Teece 1996) and at least partly contradicting. Observations concerning for example the telecommunications and Internet technology industries since the mid-1990s present several examples of highly systemic technologies pursued through various types of collaborative efforts by a number of firms (Kano 2000; Keil 2002*b*).

The systemic nature of innovation makes companies increasingly dependent on others. Because in complex systemic innovations, vertical integration is rarely an option, the innovation processes become increasingly collaborative processes. Innovating companies are dependent on complementary innovators. The move from internal innovation processes to collaborative open processes forces companies to take a wider perspective to resource allocation processes and to adopt new governance modes to carry out activities related to the creation of systemic innovations. In systemic innovation, companies need new tools for foresight and shaping to manage the business environment of the corporation over different time horizons. This increases the role of tools such as external venturing, research collaboration, and industry consortia.

12.3 Strategic Resource Allocation Processes

Resource allocation processes in innovation processes have received continued attention in the strategy and technology and innovation literature. Starting from the general resource allocation model developed by Bower (1970), this literature has evolved through a cumulative body of research over a thirty-year period (e.g. Bower 1970; Burgelman 1983*a*, 1983*b*; Christensen and Bower 1996; Noda and Bower 1996; Gilbert 2002*a*, 2002*b*). The Bower-Burgelman process model, named after the seminal contributions of Bower (1970) and Burgelman (1983*a*), views resource allocation as part of a larger strategic management process conceptualized to consist of multiple, simultaneous, interlocking, and sequential activities that take place on the frontline, middle, and top management level of the organization. Through bottom-up processes of defining technological and market forces and championing strategic initiatives in a sociopolitical impetus process and top-down processes of structural and strategic context determination, the organization arrives at strategic decisions and in particular at resource allocation decisions. The central feature of the Bower-Burgelman model is that strategic initiatives emerge predominantly from the activities of frontline managers and then compete for resources and top management attention. In later work, Burgelman (1994) shows how the process of the emergence and selection of initiatives can be understood in an intraorganizational ecology perspective as a variation, selection, and retention framework. Noda and Bower (1996) further extended the model by showing how iterative processes of resource allocation lead to escalation and de-escalation of strategic commitments. More recently, scholars have identified limitations of the Bower-Burgelman model for some innovation types and have developed additional perspectives to explain how corporations can address resource allocation for different types of innovations. For instance Christensen et al. (1996) argue that resource allocation along the lines of the Bower-Burgelman model fails for disruptive innovations. Disruptive innovations differ from incremental innovation in that they lower product performance along traditional metrics, but find an untapped need with a new set of applications, find a broader set of new and initially different customers, who value these attributes and applications, and creates a significant change in the underlying business model of the firm, often lowering gross margins or changing the basic drivers of firm profitability (Christensen and Bower 1996; Christensen 1997; Gilbert 2002*a*, 2002*b*). Resource allocation in line with the Bower-Burgelman model frequently fails to support such innovations since they do not fit the financial and operating criteria required to sustain the core business. The problem for incumbent firms is that when disruptive proposals are considered, analysis based on established performance criteria reveals the new opportunity as inferior when compared with other potential opportunities that sustain the existing business. Gilbert (2002*a*, 2002*b*) complements this perspective by arguing that organizations frequently need to employ different and changing cognitive frames for disruptive innovations. Yet others have suggested separating the development of disruptive innovations in new venture divisions with separate resource allocation processes to enable them to survive in organizations (Tushman and O'Reilly 1997).

The research on disruptive innovation suggests that different innovation types might require differing resource allocation logics. For corporations to be able to develop radical or disruptive innovations, the prescription has been to establish separate new venture divisions to insulate the immature disruptive ventures from the pressures of the core businesses and to thereby create space for the long-term development of more explorative ventures that are critical for the long-term competitiveness of the firm. However, we argue here that these prior resource allocation models optimized for allocating the internal resources of the corporation are insufficient for systemic innovation. For corporations to be able to create systemic innovations, yet further development is needed in the resource allocation processes. In prior research into the resource allocation processes, the resource allocation deals with allocation of internal resources, that is, employees, machinery, financial resources of the focal company. However, for innovation processes that require multiple simultaneous innovations in independent companies, such a perspective is too narrow. In systemic innovations, partners and external developer communities make up a significant resource pool working on developing different components of the systemic innovation (West 2003; Franke and von Hippel 2003; Hertel et al. 2003; von Hippel and von Krogh 2003). These developer communities and external partners are critical for the success of the innovation in question but are not under direct control of the focal corporation. Attracting and retaining the commitment of these external resources is the key to proactively building systemic innovations. The difference between resource allocation processes in autonomous innovations versus systemic innovations is highlighted in Figure 12.1. For creating disruptive innovations, the recommendations from prior literature have included the advice to establish

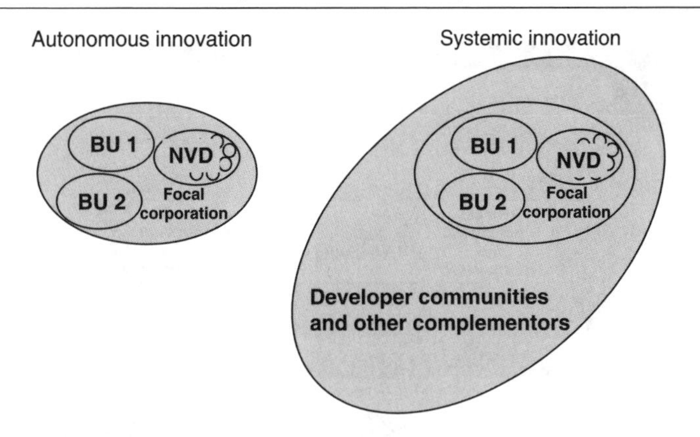

Figure 12.1. Impact of systemic innovation on resource allocation process

separate new venture divisions. However, optimizing the allocation of internal resources in a corporation may lead to suboptimization when viewed from the perspective of creation of systemic innovations. In systemic innovation, resource allocation is not only about own resources. For successful proactive management of systemic innovations, resource allocation processes have to take the external resources into account. The allocation of internal resources, and credible commitments to do so in the future, influence the allocation of external resources that form a major part of the total pool of resources that are needed to create systemic innovations.

Resource allocation processes that do not consider these external resources and provide mechanisms to steer the resource allocation in these partner companies and communities outside the boundaries of the focal firm run the risk to lead misallocation of resources and to the ultimate failure of the systemic innovation. In Section 12.4 we will develop some arguments as to how boundary spanning mechanisms can support internal resource allocation mechanisms and processes in systemic innovation.

12.4 A Framework for Proactively Managing Systemic Innovations in Industry Leading Companies

In systemic innovation, innovating companies are dependent on other complementary innovators and need to learn to proactively lead the development of systemic innovations. Systemic nature of innovations will increase the need and alter the means of corporations to manage their dependencies of their environment by creating foresight and shaping the development of their industries over different time horizons. This calls for Open Innovation processes and dynamic resource allocation processes that consider both internal and external resources available to further the systemic innovation in question. In these processes, in particular, the linkages between different activities need

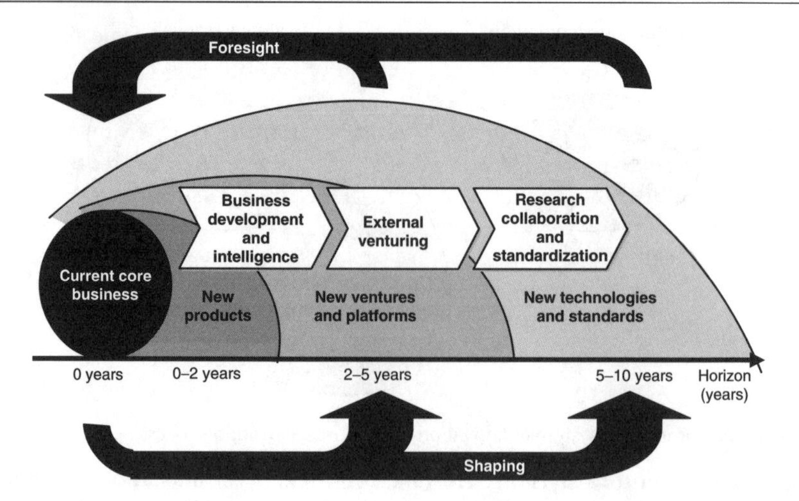

Figure 12.2. Tools for foresight and shaping to manage the business environment of the corporation over different time horizons in industries depending on systemic innovations

to be understood to capture the dynamics of systemic innovation. Our framework, as depicted in Figure 12.2, highlights these aspects that are needed for proactive creation of systemic innovations in large corporations.

The framework consists of three building blocks. First, similar to the top-down and bottom-up processes of the Bower-Burgelman model we identify two main processes: foresight and industry shaping. Foresight and industry shaping are fundamental goals relating to resource allocation that firms aim to accomplish when interacting with firms in their environment. Our second building block is the time dimension. By focusing on three time dimensions we develop a dynamic resource allocation model that spans beyond technology creation. In systemic industries, a focus on the technology creation process alone falls short of understanding systemic innovation processes, since the drivers of resource allocation shift as a firm moves from early technology development to full technology commercialization. We argue that the whole process from early research to product development and introduction, needs to be observed to understand how industrial organizations react to environmental changes and shape their environment. Our third building block consists of tangible mechanisms that firms can utilize to accomplish foresight and industry shaping goals during different stages of the emergence of a systemic innovation.

12.4.1 *Foresight and Industry Shaping as Resource Allocation Processes*

In systemic innovation, boundary spanning serves two main purposes, to provide foresight and to shape the environment. In Figure 12.2, these two processes are depicted as two arrows spanning the different time horizons. Firms draw information about the evolution of technologies and markets but also about the resource allocation decisions of other firms by linking to multiple actors in their environment. The information derived is similar to the information that derives from strategic initiatives in the Bower-Burgelman model in that it allows the firm to define the technological and market forces and so influences internal resource allocation decisions. We refer to this process as *foresight processes* (the upper arrow in Figure 12.2). The complexity of systemic innovation means that firms need to monitor the development of multiple simultaneous innovations. Since internal resource allocation decisions need to be adjusted according to the evolution of these innovations, intense interaction with suppliers, customers, partners, developer communities, and competitors are needed to provide rich information about the development of different elements of the systemic innovation. External contacts formed through research consortia, alliances, and other forms of cooperation provide a way for companies to gain information (Powell et al. 1996; Gulati 1999), to gain understanding and make sense of this new information (Weick 1995), and ultimately to provide foresight into emerging new business opportunities.

The foresight process can be understood as passive in that it provides input into the firm internal resource allocation process but does not change the decisions of actors in the firm environment. Thus foresight does only provide limited coordination among the firms creating systemic innovation. A second process similar to the structural and context determination process for internal resource allocation is needed that provides for stronger coordination to avoid the risks that Chesbrough and Teece (1996) discuss. We refer to this process as *shaping process* (the lower arrow in Figure 12.2). Firms need to proactively influence the evolution of technologies and markets and the resource allocation decision of others in their environment. Shaping can take several forms. First, shaping can take place through providing actors in the environment with financial incentives. Focal firms frequently provide financial incentives to customers or suppliers to support a new technology or to develop complementary products or services. Other financial incentives might include investment into producers of complementary technologies. For instance INTEL has been investing in recent years millions of dollars into firms that create products based on its INTEL64 architecture (Leamon and Hardymon 2000). IBM, on the other hand, has invested significantly in companies that have built complementary innovations to enable the corporate use of the Linux platform.

Additional incentives can include sharing of firm proprietary resources, technologies, or access to information. For instance, Texas Instruments has provided small and medium sized firms with access to advanced architecture information and with development support that committed to adopt its DSP technology. Sun Microsystems has sponsored the Java Platform (Garud et al. 2002). Second, shaping frequently takes place through participation in standardization processes run by industry associations or standardization organizations. Important aspects of third generation wireless technologies or the Bluetooth short-range wireless standard were standardized in open standardization long before these technologies were ready to be introduced (Keil 2002*b*). Finally, shaping can also take less formal form through signaling intentions, for instance, through press announcements or in informal communications with customers suppliers, partners, or competitors.

12.4.2 *Time Horizons and Boundary Spanning Mechanisms*

To understand the mechanisms that firms can use as foresight and industry shaping mechanisms it is useful to analyze three distinct time horizons in the emergence of a systemic innovation. Since systemic innovations require multiple simultaneous innovations, coordination needs to start well before the innovation is introduced to the market place. Early technology development and standardization of technologies underlying systemic innovations frequently predate their broad commercialization by five to ten years. During this early technology development stage, the firm faces the challenge to understand the development of multiple technologies and influence their evolution and standardization. During this early stage of evolutions, focal firms frequently rely on precompetitive cooperation mechanisms that usually take the form of research collaboration. These might include research consortia, cooperation with universities, and participation in academic research such as presentation at a conference or publication in scientific journals. Through these research cooperation mechanisms, focal firms can gain important information about the evolution of complementary technologies but also influence early technology decisions by other actors in the value network. For instance, participation in research consortia provides firms with access to knowledge spillovers from other firms participating in the consortium (Katz 1986). These spillovers can therefore either substitute for internal research and development (R&D) thereby reducing the cost of R&D or provide knowledge complementary to a firms own knowledge base through skill sharing (Sakakibara 1997). R&D cooperation frequently takes place either in government supported consortia such as SEMATECH in the US or VLSI in Japan or is related to basic research during the precompetitive phase of technology development. During this early stage of technology development, focal firms might as well be willing to make knowledge publicly available to influence others to follow a certain technology path. For instance, the joint article on Semantic Web in the Scientific American written by Ora Lassila of Nokia with Tim Berners-Lee of W3C and James Hendler of University of Maryland has had an instrumental role in bringing the Semantic Web closer to the reality by giving a joint direction for a variety of diverse research efforts. For systemic innovations that require multiple complementary innovations also the standardization of technologies frequently occurs during this early phase through formal and informal standardization bodies. For instance the standardization of 3G wireless technologies took place during the second half of the 1990s several years ahead of the release of any commercial products (Gandal et al. 2003). Participating in these standardization bodies provides the focal form with important foresight on future standards but also allows it to influence the choice among multiple standards candidates.

During early commercialization, the emphasis in tools for foresight and shaping changes from research and technology oriented mechanisms to exploratory commercialization through corporate venturing and other mechanisms. During this intermediary phase that often precedes full-blown commer cialization by two to five years, the focal firm participates in early commercialization efforts, however, without committing large-scale resources. Early commercialization of systemic innovations usually takes place on a small scale and faces a significant risk of failure. The early commercialization efforts might involve complementary innovations, or might require experimentation with product configurations, customer groups, or business models. To control these risks and enable experimentation, the boundary spanning corporate venturing mechanisms that firms can utilize in this context include corporate venture capital investments in, and alliances with start-ups involved in early commercialization, as well as small-scale acquisitions.

In recent years, a large number of companies has utilized some form of corporate venture capital investments (Maula and Murray 2001; Chesbrough 2002; Dushnitsky and Lenox 2005) to participate in the early commercialization of emerging technologies. In this form of venturing, the corporation imitates the behavior of traditional venture capital firms by participating with an own investment fund in the private equity market or by participating in a fund managed by a traditional venture capital firm. By coinvesting in start-ups the corporation gains access to information about the commercialization efforts of these ventures allowing it to avoid resource allocation to failing products, business models, or customer groups. Corporate venture capital investments also help to shape the emerging value network forming around the systemic innovation. Investments provide important resources for the start-ups such as finance, knowledge, or access to distribution channels. In addition, these investments provide legitimacy to the start-ups as well as the technologies or business models these pursue. Venturing through alliances includes nonequity alliances and joint ventures to explore new business opportunities. Similarly to corporate venture capital investments, alliances and joint ventures can provide foresight into emerging technologies and support the creation of complementary innovations. In an acquisition, an independent venture or an external venture is internalized by acquiring a majority position in the venture. Acquisitions can provide control over emerging critical complementary innovations.

During the full commercialization and development stage, the emphasis of the organizational mechanisms changes once more. During full commercialization, the focal firm focuses on learning about competitive threats through business intelligence, countering competitive moves by competitors and on creating efficient value networks through shaping mechanisms. Business development mechanisms such as supplier and customer alliances, joint ventures, and large-scale acquisitions play an important role in providing business intelligence and developing the business system supporting the systemic innovation. Alliances and joint ventures during this stage of development are frequently used to create efficient supply chains, increase commitment compared to arm's-length market transactions and to lock-in important suppliers, customers, or providers of complementary products. Being embedded in a network of relationships with other supporters of a systemic innovation reduces the risk of opportunistic behavior and provides for a mechanism of social control (Uzzi 1997). The enfolding network of partners further provides a focal firm with improved access to information (e.g. Koka and Prescott 2002). Acquisitions are frequently used to gain control of firms that own important assets in the business system, to alter the power balance between actors in the business system, and thus to manage interdependencies in the focal firm's environment (Pfeffer and Salancik 1978).

12.4.3 *The Industry Leaders Challenge: Managing Credible Commitments to Create Systemic Innovations*

The initiation of systemic innovation often poses particular problems to industry leaders. Since systemic innovation requires the simultaneous development of multiple complementary innovations, business systems often require leadership by a small group of firms that can function as anchors for coordination. Following Chesbrough (2003*b*), we refer to these firms as architects (see also, Chapter 3). Chesbrough (2003*a*) describes architects as firms that develop the technical architectures of systems thereby partition the system's complexity, enabling that other companies can provide pieces of the system while guaranteeing that these pieces can be integrated. Innovation architects provide leadership to a systemic innovation by establishing the system architecture, communicating it, and enabling others to support and further develop it. This role is frequently played by large firms that can provide the emerging innovation legitimacy. The architect stands to gain significantly from the innovation since they frequently produce important components of the systemic innovation or play a central role in its commercialization. However, contrary to small- and medium-sized players that can solely focus on the emerging systemic innovation, innovation shapers face a different set of challenges. Being industry incumbents, architects frequently have vested interests in technologies that are to be replaced by the emerging systemic innovation. For instance, virtually all of the central players in the emerging 3G business system have been involved in second generation mobile systems. Therefore, architects need to balance driving the emerging systemic innovation with maintaining their existing business base.

A critically important perspective particularly for architects in this expanded resource allocation equation is the perspective of credible commitments (cf. Williamson 1983). Incumbents can use their own internal resource commitments to signal their commitment to specific systemic innovations and thereby attract external complementary innovators behind them. Because of their dependence of the evolution of the innovation, the external complementary innovators constantly monitor and assess the commitment of the leading promoters to the specific systemic innovation. The signals created by the resource allocation of the leading promoters of the innovation may have big effect on the commitment of the external complementary innovators. For instance, a wrong move by SUN could easily impact the number and commitment of Java developers. The same challenge applies to Nokia concerning Symbian developers, IBM concerning Linux developers, and Microsoft concerning .Net developers. Attracting and retaining the commitment of external resource pools is a critical issue for builders of systemic innovations and should be considered in internal moves made by these corporations.

A visionary leader cannot be credibly committed to all alternatives. Yet also visionary leaders need to explore and manage different time frames and prepare for multiple possible futures. Simultaneously when sponsoring a disruptive new technology for the future, large corporations need to sell their existing products. The need to manage different time horizons may create conflicting signals and create confusing signals for the complementary innovators. The challenge is how to orchestrate so that your customers are focused on what you are offering now and so that simultaneously developers are focusing on what you are pushing forward for the fifth year (or even second or third year for rapidly changing industries) from now. Corporations need new tools to manage these external resources with external venturing and research collaboration becoming increasingly valuable. The credible commitments constraint limits the moving space of corporations, which forces them to think carefully what tools they can use for different purposes. For example, IBM influenced strongly the emergence of Linux by making supportive and legitimizing CVC investments in Linux companies in 1998–1999 while at the same time they were selling fully competing products (Young and Rohm 1999; Väisänen et al. 2003; Maula et al. 2003). Too early use of wrong tools such as establishing a business division would have distracted the attention from the core business and created false signals and expectations. It is only now when setting up an internal business makes sense and does not cause too bad a reaction in other partners. As seen during the recent years, Linux has grown to be a multibillion business for IBM.

12.5 Discussion and Future Research

In this chapter, we have focused on the impact of the systemic nature of innovation on the resource allocation processes. We have argued that prior resource allocation models have focused on developing the optimization of internal resources for example, to enable the creation of disruptive innovations with the use of separate new venture divisions. However, when viewed from the perspective of creating systemic innovations, these models fail to address important challenges faced by firms. We have argued that factors exist, which make it challenging for incumbent corporations to internalize the whole process of developing a systemic innovation, that is, to create the new technology, standardize it, find relevant business models, and commercialize the technology. Creation of systemic innovations often requires the use of external resources. Therefore, models focused solely on optimizing the use of internal resources may lead to suboptimization when viewed from the perspective of creation of systemic innovations. Systemic character of innovation leads to Open Innovation processes.

12.5.1 *Contributions for Theory*

This chapter makes three important contributions. First, we contribute to a more detailed understanding of systemic innovations and their implications. We argue that systemic character of innovation is an important underlying driver of Open Innovation processes. In particular, this chapter contributes to a better understanding of the constraints and challenges that incumbent corporations face in the creation and commercialization of systemic innovations. We also contribute to a more in-depth understanding of why corporations employ a variety of governance mechanisms—internal research and development, alliances, and venture capital—in the various stages of a systemic innovation process. Our arguments suggest that in systemic innovation these Open Innovation mechanisms (Chesbrough 2003*b*) are necessary complements to ensure the successful creation of systemic innovations.

Second, our chapter makes contributions to the general theme of this book—Open Innovation. The shift from closed to Open Innovation is an empirical phenomenon readily observed in many industry contexts. Our argument that the systemic nature of innovation in industries such as information technology or telecommunications forces firms to broaden their innovation activities beyond their internal resources provides a theoretical underpinning why these changes occur. We further contribute to the Open Innovation concepts by identifying how architect firms (Chesbrough 2003*b*) accomplish the process of establishing their systems solution, communicating it, and persuading others to support it and develop it in the future. In particular, this chapter details important requirements for the architect role for a specific type of innovations systemic innovations that are particularly well suited for Open Innovation.

Third, our chapter contributes to theories of resource allocation in firms. While the Bower-Burgelman model (Bower 1970; Burgelman 1983*a*, 1983*b*) has been the dominant model of resource allocation in the past twenty years, recently limitations and boundary conditions for the model have been identified. Our paper contributes to this discussion by suggesting that in systemic innovation, models of internal resource allocation and the optimization of internal resource allocation are insufficient and might in some cases even hamper firm success. When, as in systemic innovations, a major share of resources necessary for success is located outside the corporate boundaries, these external resources have to become an important part of the resource allocation process. Our discussion further suggests that the processes through which external resource allocation is accomplished vary dramatically from internal resource allocation since the focal firm lacks direct control of these resources and needs to devise indirect steering mechanisms.

12.5.2 *Implications for Practice*

Our results also have important implications for practice. We propose that in order to create and commercialize effectively systemic innovations, collaboration between incumbents, start-ups, and other stakeholders should be encouraged. Our findings also show how collaboration already in the early stage of research can have critical impact on the speed of convergence in the creation of systemic innovations. Public–private partnerships and industry consortia can play a major role in facilitating the development of the new technological system. Companies need to manage their future business environment over different time horizons. Depending on the time frame, different tools can be used to create foresight of the development and to shape the development of future systemic innovations. Our framework presented in Figure 12.2 provides an overview of the tools for creating foresight and shaping to manage the business environment of the corporation over different time horizons in industries depending on systemic innovations.

Given the increasing role of systemic innovations not only in the information and communications technology industries but increasingly also in more traditional industries, understanding of the systemic innovation process and the functioning of various policy measures in supporting this type of innovative activity is becoming increasingly important. Although there is very little prior research on systemic innovations and even less on the policy measures for supporting them, several factors suggest that policy actors may play an increasingly important role in stimulating innovative activity in a systemic innovation.

First, because of typically relatively weak intellectual property (IP) protection regime in systemic innovations, innovative activity may be susceptible to different types of market failures when contrasted to traditional R&D activity. On the other hand, policy measures may have disproportionally important effects in spurring the development of important systemic innovations as evidenced for instance in the development of the Semantic Web technology significantly spurred by EU and DARPA programs that accelerated and integrated a fragmented set of research efforts. In large systemic innovations individual innovating companies rarely have incentives to invest sufficiently until they are ensured that the overall systemic innovation will materialize. This easily leads to underinvestment and delays in one or more critical components of the systemic innovation thereby delaying the creation of the whole systemic innovation. Therefore, government R&D programs can have an important role in kick-starting and aligning the development of complementary innovations and thereby facilitate faster development of systemic innovations.

Third, many systemic innovations have important societal effects because systemic innovations commonly alter the way people and organizations work and live. The Internet is one example of such systemic innovations. Fourth, systemic innovations have significant implications also for companies who need to take a more proactive approach to research collaboration and venturing on longer time horizons to secure the future path of growth. Therefore, the increasing role of systemic innovations and adoption of proper tools for managing them are likely to have impacts on the level and type of collaboration between government, universities, and firms. Because of the special characteristics of the innovative activity in the creation of systemic innovations and the resulting different systemic and market failures, the potentially important role of policy measures in spurring and steering the development of systemic innovations as well as the important economic and societal effects of systemic innovations, it is important for policymakers to create a thorough understanding of the characteristics of systemic innovations and applicable policy measures.

12.5.3 *Future Research*

The arguments made in this chapter suggest several potential avenues for future research. First, our arguments have suggested on theoretical underpinning why we are observing a shift from closed to Open Innovation. We suggest that linking specific innovation types with the closed and Open Innovation processes might help to further our understanding of the applicability and boundaries of Open Innovation processes.

Second, our arguments have focused on the challenges for industry leaders that play the role of an industry architect. While the creation of systemic innovations is frequently steered and to some extent controlled by these industry leading firms, the question arises what role other firms in the systemic innovation process can play and what the unique challenges are, for these players. For instance, future research could investigate how producers of complementary products influence the evolution of systemic innovations for instance to position their complementary innovation as centrally as possible in the system. An additional interesting question would be to which extent small- and medium-sized firms can play the role of architect steering the evolution of a systemic innovation. Many of the mechanisms discussed in this chapter will be unavailable for small- and medium-sized firms. At the same time, alternative Open Innovation processes such as open source development (Kogut and Metiu 2001; Grand et al. 2004; see also Chapter 6; von Hippel and von Krogh 2003) might open alternative avenues for small- and medium-sized firms to steer Open Innovation processes but these avenues would need to be investigated further.

Third, our arguments raise important research avenues regarding resource allocation. While we have provided a framework for resource allocation in a systemic innovation context, future research should further develop and detail the mechanisms of resource allocation available to firms in an Open Innovation model. Studies should for instance investigate how these mechanisms are integrated into the internal decision-making. Future research could also investigate how firms reconcile the at-times-conflicting demands, time horizons, and resource allocation mechanisms between internal and external resource allocation and thereby further develop the Bower-Burgelman model (Bower 1970; Burgelman 1983*a*, 1983*b*) to incorporate external resource allocation.

12.6 Conclusion

To sum up, innovations are increasingly systemic. This makes companies increasingly dependent on external parties and changes the resource allocation equation because a majority of the potential relevant resources are located outside the boundaries of the corporation. Successful visionary industry leading corporations are able to tap and leverage these external resources to create systemic innovations by keeping the credibility of their commitments in mind and using tools such as external venturing and research collaboration to create foresight of and to shape the business environment over different time horizons. Systemic innovations require Open Innovation processes.

Note

1. We acknowledge with appreciation the financial support from The Research Programme for Advanced Technology Policy (ProACT) of the Ministry of Trade and Industry and the National Technology Agency of Finland, Tekes.

13

Open Innovation in Value Networks

Wim Vanhaverbeke
Myriam Cloodt

13.1 Introduction

New technologies such as information and communications technology (ICT), biotechnology, or new materials are becoming an increasing powerful driving force generating competitive advantages and commercial success for companies in a wide range of industries. The complexity of new technologies often goes beyond the capabilities of individual companies (Hagedoorn and Duysters 2002) and forces innovating companies to cooperate with other firms and organizations to reduce the inherent uncertainties associated with novel product markets. Simard and West (Chapter 11) have analyzed the role of different types of networks in Open Innovation. In Chapter 12 Maula et al. have shown that central firms have to take the role of a network orchestrator in the case of the development of systemic innovations. In this chapter we focus on the role of interorganizational networks in the *commercialization* of new product offerings based on the technological breakthroughs in the agricultural biotechnology (or agbiotech).[1]

Have a look at Figure 1.2. Insourcing of externally developed technologies is crucial for the innovativeness of a company. The funnel in Figure 1.2 might give the impression that interorganizational networks only play a role in the research and early development phases. This chapter shows that innovating companies also have to set up and manage interorganizational networks to commercialize their innovations successfully. These networks are different from the networks that firms establish to tap into external technology sources in the early stages of the funnel.[2] They have received less attention than the research and development (R&D) networks in the academic literature, but are nevertheless important because they are directly responsible for the market success and profitability of new technologies.

We take the commercialization of new products based on agricultural biotechnology (agbiotech) as an illustration. To deal with business opportunities that are enabled by technological developments in agbiotech, economic actors with different assets and competencies have to be linked together into interorganizational networks to create jointly, value for targeted customer groups (Normann and Ramirez 1993; Brandenburger and Nalebuff 1996; Parolini 1999; Normann 2001).

We call these networks 'value constellations' following Normann and Ramirez (1993): they are interorganizational networks linking firms with different assets and competencies together in response to or in anticipation of new market opportunities. Value constellations create value for a target customer group by means of a business model translating technological developments in agbiotech into new, commercially viable products. A central firm sets up a value constellation through acquisitions, licensing agreements, nonequity alliances, joint ventures, contracting, and other types of relations that go beyond arm's-length relations. Partners in the network are biotech firms, seed producers, chemical companies, farmers, but also manufacturing companies, retailers, and so on, depending on the end application. Such constellations, however, bring with them significant strategic and organizational challenges, about which there is very little prior knowledge.

Interorganizational networks have many links with the idea of Open Innovation. One reason why firms are teaming up with each other is technological complexity. Chesbrough (2003*a*: 53) argues that: 'The cascade of knowledge flowing from biotechnology... is far too complex for any one company to handle alone... so companies have to identify and build connections to excellent science in other labs'. These are networks that are established to absorb externally developed knowledge, they can be situated in the left side of Figure 1.2. Interorganizational networks are also established when the innovating company is not capable to reap the business opportunities stemming from agbiotech on its own. New business models that are radically different from those underlying existing (or competing) product offerings force the innovating firm to set up a value constellation with different partners to successfully launch the new product. These networks are situated at the right side of Figure 1.2.

Value constellations are also interesting from a theoretical point of view. When networks of companies become crucial in understanding how value is created and captured, different theoretical perspectives have to be brought together to understand this phenomenon. First, value constellations have a lot in common with the relational view of the firm (Dyer and Singh 1998) and the maximization of transaction value (Dyer 1997; Zajac and Olson 1993). Value constellations challenge us to rethink value creation not from a single firm's point of view but as the outcome of the interplay of the network partners. Moreover, value appropriation can no longer be analyzed in terms of the negotiation power of individual firms as too much fighting among the participants for a share of the pie reduces the volume of the pie. There is also a clear link to the resource-based view of the firm since organizations have to combine their internal resources with those of their partners to generate value. Value constellations are a particular type of interorganizational networks, where value creation and distribution, external resource or knowledge sourcing, interorganizational ties, and network governance call for an integration of various frameworks (Amit and Zott 2001; Gomes-Casseres 2003).

The chapter is structured in the following way. First, we explore some evolutions in the agbiotech and analyze how companies establish different value constellations that are enabled by the agbiotech. Section 13.3 explains how value constellations create value. When firms launch radically new products based on new technologies (e.g. agbiotech) they use (new types of) business models that determine how value will be created and appropriated, how external resources have to be sourced by establishing different interorganizational ties with different partners. We also focus on the external management tasks of the central firm ensuring that the potential value of the network is maximized. Section 13.4, examines how value constellations can improve our understanding of Open Innovation. Three topics are analyzed. First, we analyze how central firms choose between different governance modes when they establish relations with partners. Second, value constellations are a nexus for the integration of different theoretical perspectives since they are a nexus where value creation, non-arm's-length transactions, external resource sourcing and interorganizational networking are welded together. Third, we suggest that Open Innovation has to be analyzed at different levels of analysis and that interorganizational networks are important to improve our understanding of Open Innovation. The chapter concludes with some final observations and possible avenues for further research.

13.2 Agricultural Biotechnology and New Business Opportunities

In this section we describe how companies are profiting from new business opportunities that emerge in the wake of the technological breakthroughs in the agbiotech. We show how the same technology—genetically modified (GM) crops—leads to completely different ways to create value depending on the business model and targeted customers.

Before the advent of agbiotech, agriculture was a mature and slowly growing business, characterized by a standardized production of commodity-like food and feed based on arm's-length transactions between different players in the value creating system. Competition was based on price and economies of scale were crucial.

The first generation of GM-crops was designed *to reduce farm production costs or improve crop yields*. The most popular examples are pest-resistant and herbicide-tolerant crops. Pest-resistant crops (e.g. Bt Corn)[3] generally lower insecticide use and require fewer trips across the field. The advantages for farmers are time and cost savings (both on pesticides and energy) and using less pesticide is beneficial for the environment. These GM-crops are a challenge for chemical companies producing pesticides: sales of traditional pesticides drop sharply and price competition intensifies as a result of the recurring introduction of new pest-resistant crops. Entering the business of insect-resistant crops is one way to substitute for the reduced insecticide sales.

In conventional weed control, farmers used preplant herbicides before or at planting combined with the use of selective post-emergence herbicides to fight weeds. The introduction of herbicide-tolerant seeds—resistant to a particular, firm-specific, and patented herbicide—provides broad spectrum weed control[4] with one herbicide that can be applied over the crop at any stage of growth. Farmers can spray crops with a single herbicide during a much longer period with productivity increases as a result. They also profit from time and cost savings because herbicide-tolerant crops require no more tillage or at least less than in conventional weed control. Herbicide-tolerant crops are particularly interesting to chemical companies because it allows them to offer seeds and herbicides as a bundled good to farmers.

These enhancements of the agronomic traits did however not change the commodity nature of crops like corn, canola, and others. It affected only a small part of the existing value creating system, and the business model stayed strictly focused on commodity-like agricultural products. For farmers, competition is still based on price and efficiency gains remained the most important competitive driver. As a result, the introduction of the first generation of agricultural-related biotech products was intended to strengthen the position of agricultural firms by productivity gains but without restructuring their business model. Agrochemical companies and seed companies were the only ones that have to revisit their business model to some extent.

Contrary to the first generation of biotechnology products, many of the innovations under development in the agricultural biotechnology focus on *value-enhanced or output traits*. These GM-crops are designed for specific needs of end-users in different industries. Some are focusing on agriculture-related industries (mainly processing industries), but some applications are targeting turnarounds in industries that were previously not related to the agriculture industry. We first discuss the GM-crops focusing on agriculture related industries.

Different GM-crops under development are delivering better or healthier food and feed. Corn, canola, and soybeans are genetically modified to improve quality, digestibility, and taste. Some GM-tomatoes remain firm longer and retain pectin during processing into tomato paste, which generates advantages throughout the entire value chain; the farmer can get a better tomato to the factory, the processor can reduce waste (less rotten tomatoes), and he has lower energy costs (less need for evaporation) during puree production, and the end-consumer is offered quality at cost savings. Another example is the improvement of the fiber quality of cotton, such as polyester-type traits to make the fabrics sturdier and produce fibers with superior insulating qualities. Some companies envision making wrinkle-resistant or fire-retardant qualities of cotton. These developments will eventually have a major impact on the processing of cotton and the textile industry.

Biotech also transforms the agricultural industry into an important upstream industry for many industrial sectors that were previously not related to agriculture. Biotechnology has the potential to make major inroads in industries in which agricultural products have previously been absent. Nutraceuticals for instance are foods or food components that reduce risk of certain health problems. Other examples—just to name a few ones—are edible vaccines, cheap drug development, biodegradable polymers, or new energy sources.

These examples illustrate that agbiotech enables companies *to set up new value creating systems* in *different industries*. The initial primary target of plant biotechnology was to improve the production of plant-derived food. Biotechnology, however, enabled agriculture to shift from the production of commodity-like food and feed to high-priced, specialized plant-derived products that can be applied in a wide range of industries. Agbiotech is introducing a broad range of new products that not only redefine today's key agricultural markets, but also create business linkages to other previously nonrelated markets and industries including pharmaceuticals, animal health, chemicals, and a broad range of industrial markets (Shimoda 1998; Enriquez and Goldberg 2000).

Consequently, agriculture and the entire agribusiness have entered a period where the pace of biotech innovations intensify speedup change and where new business models challenge existing ones.

13.3 Value Constellations: Organizing for Value Creation and Distribution

Innovation can be defined as the conception of an idea up to the introduction of an invention into the market (Ernst 2001). Technology has no value as long it is not commercialized in some way. To create and extract economic value from new technological developments, each firm needs a suitable business model, which operates as a mediator between technology development on the input side and economic value creation on the output side (Chesbrough and Rosenbloom 2002).

A business model can be defined as

> the set of which activities a firm performs, how it performs them, when it performs them as it uses its resources to perform activities, given its industry, to create superior customer value . . . and put itself in a position to appropriate value. (Afuah 2004: 9)

Creating and capturing value from early stage technologies can in most cases only be realized if the innovating company links other parties like customers, suppliers, complementors, and competitors to the commercialization process. Establishing a 'value network' with partners to market a new technology, '... shapes the role that suppliers, customers and other parties play in influencing the value captured from commercialization of an innovation' (Chesbrough and Rosenbloom 2002: 534).

We will argue in this section that the value constellation concept offers a coherent framework to understand the formation of the current interorganizational networks and merger and acquisition wave in the agribusiness. Following Figure 13.1, we will first focus on how value is created in value constellations. Second, we analyze how different players can capture part of the value created in the value constellation. Third, we analyze how a central firm has to set up and manage value constellation to realize the potential business opportunities. Both value creation and value capturing can only be realized if a central company acts as an orchestrator and manages the 'value constellation' (Iansiti and Levien 2004*a*). Fourth, we discuss briefly set-up strategies for value constellations. Finally, we focus briefly on the role of governments.

The examples of new product offerings generated by agbiotech illustrate that new technologies provide opportunities to set up value constellations to introduce new product offerings based on new business models and to compete against existing ones with new business models. New technologies can be disruptive (Christensen 1997; Christensen and Raynor 2003) but it is not the technology itself but the business model behind the application of that

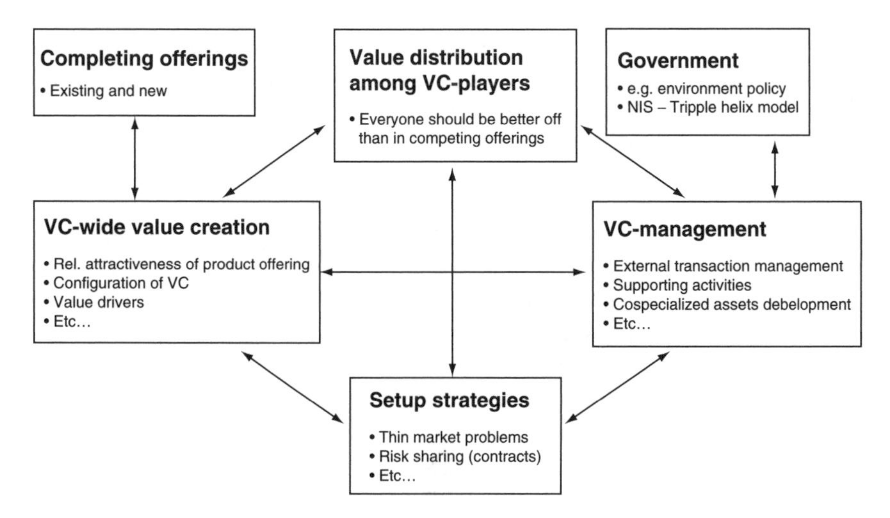

Figure 13.1. Analyzing value constellations

technology that gives it its disruptive power. Similarly, Ramirez and Wallin (2000) and Normann (2001) argue that value migration from incumbents to new entrants occurs when the latter enter the market deploying a business model that imposes new rules for competition. Hence, not the technology as such but the business model grafted upon technological innovations opens up new business opportunities (Chesbrough and Rosenbloom 2002; Chesbrough 2003*a*).

In Sections 13.3.1 and 13.3.2 we enter two topics related to the value creation in value constellations. First, we give a brief overview of the value drivers that play a role as sources of value creation in the establishment and organization of value constellation. Next, we analyze how a central company can create value from combining its capabilities with that of its constellation partners.

13.3.1 *Value Drivers*

A new business model has to identity different *sources of value creation* or 'value drivers'. We identified four different types of value drivers that enhance the value-creation potential of agbiotech.[5] The first value driver which is prominently present in agbiotech is *efficiency*. The first applications of 'agbiotech' focused on efficiency gains for farmers. Both insect-resistant and herbicide-tolerant seeds are developed to improve farm productivity. GM-seeds are bought at a premium price vis-à-vis the traditional seeds but a farmer also saves on fuel and insecticides, and profits from time savings because of reduced tillage, spraying, and so on. Farm productivity also improves because operational risk can be reduced (e.g. less damage by insects or fungi; strawberries that resist frost). Productivity enhancements can of course also be realized in the processing industry: for example, GM-tomatoes that retain longer their pectin, save on wastage and energy consumption for the processor. Finally, cost efficiency; for example, dramatically reduced capital costs to produce different drugs in previously unthinkable ways.

Efficiency enhancements should be considered *relative* to the traditional ways of breeding and producing agricultural products (see box in top left-upper corner of Figure 13.1). Competing offerings are always the benchmark to evaluate new offerings, but it is a moving target as companies in the traditional value creating system can retaliate or change strategy. Pesticide and herbicide prices dropped sharply as a reaction to the plummeting sales following the introduction of GM-crops with agronomic traits. Retaliation and strategic moves of companies that are part of a competing value creating system, should always be taken into account. As a result, companies considering setting up a new business model should not consider the *actual* behavior of competitors but their *potential reactions* of incumbents and new entrants vis-à-vis the new business model. Focusing on niche applications to avoid head to head competition with incumbents may be one strategy to avoid retaliation (Yoffie and Cusumano 1999).

A second type of value drivers is *convenience*. Edible vaccines based on GM-crops can be administered in a more convenient and cost-efficient way than traditional vaccines. Insect-resistant and herbicide-tolerant crops also increase convenience for farmers as they reduce the need for tillage and spraying.

Agbiotech's most important value driver is its *'enabling'* property. It enables targeted customers to do things that they were not able to do before (Normann 2001: 74). Especially the value-enhanced crops offer lots of possibilities. Agbiotech is developing nutraceuticals reducing the risk of particular health problems. It may develop new drugs that are too expensive to make by traditional production methods. Textile fabrics may become better insulated or sturdier without additional manufacturing processes. New types of plastics that were not accessible by standard chemistry may be produced by GM-plants, and polymers may become fully biodegradable.[6]

Finally, agbiotech creates value by bundling complementary goods—Amit and Zott (2001) call it *'complementarities'*. GM-crops 'bundle' into seeds complementary goods that have previously been offered separately: pesticides and seeds are now bought as a one stop purchase of insect-resistant seeds, herbicide and GM-seeds purchase are necessarily two sides of the same coin, and so on. Customers value 'bundled' complementary goods when their costs are lower than when they are delivered separately or when the performance of the 'bundle' is better than when customers have to bundle the products themselves.

In sum, companies have a broad range of sources to create value from agbiotech. In the following paragraphs we analyze how companies create value in value constellations.

13.3.2 *Value Creation*

Innovation-based value creation for a targeted customer group is at the center of Open Innovation in general and value constellations in particular. However, value creation is also at the center of business strategy. Porter (1985, 1996) has argued that value is created by a 'value creating system'—a vertical chain extending from suppliers in upstream industries to buyers of products or services: 'Gaining and sustaining competitive advantage depends on understanding not only a firm's value chain but how the firm fits in the overall value system' (Porter 1985: 34). However, the value system is not crucial in the further analysis to understand the competitive positions of companies. In a value system every company occupies a particular position within the value system and adds value to the inputs before passing them to the next actor in the chain. Relationships between firms (suppliers, distribution channels, substitutes, and so on) are usually restricted to arm's-length transactions where price negotiations play an important role. The value system can therefore be decomposed into bilateral transactions between companies.[7] In other words, analyzing the whole value system does not offer any additional insights: it is only the outcome of the bilateral transactions between firms in the value system.

The 'value constellation'-concept (Normann and Ramirez 1993) is related to the value system in that they both are focused on delivering value for the targeted customer group. However, the value constellation approach offers a different view of how value is created by the participating firms. First, within value constellations not the individual companies but different products or services compete for the time, attention, and money of the customers. Gomes-Casseres (2003) calls this collective competition: competition is at the level of product offerings that the participating firms are producing together. Second, actors in the value-creating system produce value together through rethinking their roles and interrelationships. Therefore, value creation is not just adding value step after step but reinventing it by means of a reconfiguration of the roles and relationships among actors of the value-creating system (Ramirez and Wallin 2000). Competitive advantage of a constellation is not only based upon the resources of its participants but also how they are assembled, structured, and managed within the constellation. Finally, within this logic, networking and the overall structure of the constellation become central to explain how companies gain and sustain competitive advantages.[8]

Within collective competition—that is, competition between value constellations—the competitiveness of the product offering is determined by the firm-level resources and competencies that are aggregated at the constellation level. These group-level resources determine the relative value of the constellation's products vis-à-vis other products in the market. The value of product offerings has to be expressed in relative terms because the price customers want to pay is also affected by the price of competing products. Moreover, one should also take competitive dynamics into account because competitors could retaliate or develope substitutes with a better price–performance ratio.

Aggregating resources of firms into a value constellation is however not sufficient to explain the potential of a constellation to create joint value. Resources have to be effectively combined and governed effectively at the constellation level. Gomes-Casseres (2003) discusses four factors that are crucial to ensure that constellations are effectively governed.

> A unifying vision is important to bring disparate partners together. A corollary of this is that competition among members erodes the cohesion of the constellation (Hwang and Burgers 1997). Leadership is important in making collective decisions and in disciplining constellation members. Group size is a self-evident factor: the larger the group, the harder it is to manage, all else being equal. (Gomes-Casseres 2003: 331)

In sum, value creation in constellations is determined by the (*a*) the resources it assembles; (*b*) the way how it can combine and govern them,[9] and (*c*) the value of competing products and the competitive reactions of other competing firms and constellations.

13.3.3 *Value Appropriation by Different Actors in Value Constellations*

In Section 13.3.2 we have analyzed how value constellations create value. That value also has to be distributed among the different participants (including the targeted customers). According to Brandenburger and Nalebuff (1996) the total value-created in a value-creating system equals the sum of values appropriated by the different actors. Amit and Zott (2001: 515) extend this approach '... by positing that total value created through a business model equals the sum of values appropriated by all the participants in a business model, over all transactions that the business models enables'. However, the total value that can be captured by the participants of the value constellation does not tell us anything yet about the distribution of that value among the participants.

Brandenburger and Stuart (1996) argue that value appropriation by the different players depends on their bargaining power. Gomes-Casseres (2003) explores two different sources of bargaining power. One strand of literature emphasizes the role of the position of firms in the network as an important determinant of their power—crucial concepts are network centrality, structural holes, and participation in multiple networks (clique overlap) (Nohria and Garcia Pont 1991; Burt 1992; Lorenzoni and Baden-Fuller 1995). Others underline the role of scarce resources that companies bring to the value constellation (Pfeffer and Salancik 1978; Brandenburger and Nalebuff 1996; Ghemawat et al. 1999). Future research has to analyze how important those factors are in companies' ability to extract profits from the constellation.

The bargaining power of the individual companies can however only partially explain how value is extracted from value constellations. Value appropriation in a value constellation has to be *considered jointly* with the value-creating strategy at the constellation level because the quality of the collaboration of the participants and the value sharing among them both determine how much value the constellation as a whole can create. Moreover, all participants should profit from its participation in a value constellation. The strength of the value constellation is determined by (*a*) the extra value created in comparison with competing value systems and (*b*) the commitment of the different partners in the value constellation. The latter is in turn the result of the (financial) benefits each one can reap compared to alternative value systems (e.g. farmers will not purchase insect-resistant seeds when they are better off with traditional seeds and a cheap herbicide). Hence, it will be necessary to calculate the benefits along the value constellation and to ensure *that each part of the constellation gets a return so that every participant stays committed.*

Therefore, life science companies,—although they might have monopoly power thanks to their intellectual poperty rights (IPR)-protected innovations—share the value created in the value constellation with the other economic players in order to ensure that GM-plants gain rapid market penetration. Traxler and Falck-Zepeda (1999: 95) calculated for the use of Bt Cotton in the period 1996–1997 that

> US farmers received the largest single share of benefits ranging from 42% to 59% of total surplus generated. The combined share of Monsanto and the seed firms ranged from 26% to 44%. The main conclusion of our study is that even under monopoly conditions, the innovator is only able to extract a portion of the surplus it creates. The monopolist must provide farmers with an adoption incentive by setting a price that makes the new input more profitable than existing options.

These findings are in line with previous studies (Griliches 1957; Teece 1986).

'Fair' value distribution in a value constellation is important, because some actors are automatically better off in the new constellation compared to the old value-creating system, but others might be worse off and have to be compensated to get/stay committed to the value constellation. In a value constellation there are always customers that are explicitly targeted as groups that potentially can benefit from agricultural biotechnology. These target groups differ from application to application. New GM-crops and derived products may be beneficial for farmers (e.g. herbicide-tolerant crops), downstream industries (e.g. oil processing industry) or the end-customer (e.g. fresh tomato market). However, focusing on those players that can directly benefit from a new GM-crop is not enough. A value constellation can only be successfully established if *all* players that are necessary for a smooth working of the value constellation are better off than in competing business systems.

Most 'agbiotech' innovations are designed to affect only part of the value-creating system: a nice example is the 'agronomic trait' crops that, in monetary terms, only affect agrochemical companies, seed companies, and the farmers' community. In this case it is tempting to leave downstream sectors and end-customers out of scope. However, the public opposition to transgenic plants (especially in Europe) indicates that (perceived) value is a concept with many more dimensions than the 'economic value' that can be measured as cost reductions, quality improvements, or other product characteristics for which the end-customer wants to pay a premium price. Agronomic trait crops created benefits for agrochemical companies, seed companies, and farmers but they have not provided end-customers with food that is significantly cheaper, safer, and tastier. End-customers have become critical because they do not benefit from agbiotech and are confronted anyway with their potential (but unproven) environmental and health hazards. Escalating public opposition has a serious economic impact on the GM-crops; some of them are nowadays sold at a discount because exports to Europe cannot be guaranteed.

As a result, farmers and downstream industries have little incentive to grow and process them.[10] In short, this story about public opposition shows that *all* actors, who might be affected one way or another by the value constellation, should be committed to it. Focal firms in agbiotech have ignored their critics or have been too defensive in the past; nowadays they are beginning to engage in public dialogue and to teach the public.[11]

This is only one example illustrating the rule that all actors have to be better off in the new value constellation compared to the existing value-creating system(s). Value-enhanced crops pose new and increasingly difficult challenges compared to agronomic trait crops. First, the number of actors is larger as life science companies target downstream industries and even the final-customer (e.g. the fresh tomato market). As a result, many different actors—or even the whole value-creating system from upstream industries to end-customers—are affected by the new business model and they all have to be convinced to get committed to the value constellation. The higher the number of actors, the more difficult it becomes to distribute the value created and to manage the value constellation. Second, value-enhanced crops require huge adjustments from a number of key players. End-customers have to learn how to appreciate and take advantage of these crops. This, in turn requires that retail business has to introduce new types of branded products. Processors must learn how to leverage quality enhancements. Elevators must learn how to effectively segregate output trait crops and how to optimize identity-preserved supply chains. Hence, the larger the adjustments and investments required for the commercialization of value-enhanced crops, the tougher the management challenges will be to compensate companies that might be worse off in a new business system compared to the old ones.

13.3.4 *Value Constellation Management*

Creating and capturing value from life science applications neither happens spontaneously nor is it the result of an adaptation process of firms to changes in the business environment. It requires a central firm that explores the potential of life sciences to create value for customers in radically new ways and shapes the external environment accordingly (Normann 2001, Iansiti and Levien 2004*a*). Take the Flavr Savr tomato as an example to understand what actions the central firm has to take to launch a new GM-crop on the market. This GM-tomato developed by Calgene more than a decade ago had a better flavor and targeted the fresh tomato market.[12] Its commercialization required that different actors in the value system joined Calgene in its efforts to launch the new tomato: the central firm had to manage carefully its relations with seed companies, farmers, packers, retailers, and end consumers. These other players own complementary assets that are crucial for the commercialization process. Typical examples are complementary R&D, manufacturing processes, logistics, and distribution channels.

Arm's-length transactions between the innovation firm and the other actors are in most cases not a viable option because investments the partners have to make are sometimes (co)specialized and thus transaction-specific (Teece 1986). Vertical integration (or ownership of assets) is one possible way to overcome transaction cost problems. However, vertical integration is only applicable to very specific transactions since the commercialization of GM-crops may affect whole value-creating systems (see Flavr Savr example) including companies that are many times larger than the innovating company.[13] As a result, market transactions and asset ownership (integration) are not appropriate to commercialize radical innovations that require the redesign of value creating systems.[14] The central firm has to control and make the most of critical capabilities that reside in other firms by establishing a value constellation. In a value constellation the central firm brings together players with disparate assets and competences (Normann 2001; Gomes-Casseres 2003). This implies that the company has to set up an interorganizational network and manage the constellation by means of mergers and acquisitions, strategic alliances, licensing agreements, contracting, and other types of relations that go beyond arm's-length contracts. This 'critical capability sourcing' is not unique for the commercialization of new technology applications but has also been explored within the context of 'strategic sourcing' (Iansiti and Levien 2004*a*; Gottfredson et al. 2005).

Value constellations imply that interdependency becomes crucial in business: the performance of the innovating company is increasingly dependent on the influence it has over assets outside its own boundaries. Although value constellations are important in current business practice, there is yet no comprehensive framework that provides a general guideline how to manage it successfully. Following Iansiti and Levien (2004*a*) there are in our opinion two important issues for a central firm to manage a value constellation First, it has to structure and manage the constellation so that the potential of the constellation to create joint value is maximized. Second, it has to make agreements with other participants to *share this jointly created value*.

How much value there will be created depends on the design of the constellation. Gomes-Casseres (1996, 2003) has shown that the collective competitiveness of the participants depends on the size of the constellation, its technological capabilities, market reach, unifying vision, leadership at the core, and absence of internal competition among the participants. These different factors all have an impact on the competitive strength and growth potential of a value constellation. In that regard, Iansiti and Levien (2004*a*) also emphasize that the central firm actively has to nurture the constellation to manage potential tensions between participants and to discourage competitors to match the strength of the value constellation.

Second, the focal firm has to make a number of arrangements with other participants in the value constellation so that *everyone is better off* compared to competing offerings (see also Section 13.3.2). Companies will only join and stay in the constellation if participation offers a higher expected net return compared to competing offerings. This implies that the firm has to share the added value with others to spur adaptation of the GM-crops. But it also implies substantial *support for and compensation of* actors that have to invest in new (transaction specific) assets when they intend to join the value constellation.

We take the elevators (logistics systems) as an illustration. Segregation or identity preservation is necessary to deliver value-added crops to downstream industries but they also imply extra logistical costs. Identity preservation for example is crucial in the production and distribution of nutraceuticals and agriceuticals because health and environmental hazards require a fully separated and dedicated logistics system. In line with the traditional commodity crops, elevators' asset configuration and logistics of commodity grain handling have been based upon volume-based shipping and mingling of grains. Logistical redesign focusing on segregation and identity preservation is necessary and will lead to additional direct expenses and considerable long-term investments. Maltsbarger and Kalaitzandonakes (2000) found that operational costs and switching costs for elevators are not trivial. Moreover, segregation becomes prohibitively expensive for very low threshold levels of contamination. Therefore, a logistic system based upon segregation is one of the potentially 'weak' links in the value constellation. The focal firm has to strengthen this 'weak link' by supporting elevators in their efforts to come up with logistic systems that can ensure segregation or identity preservation.

Finally, there are a number of setup strategies (see fourth block in Figure 13.1) that have to be performed by the focal company. A recurring problem during the initial phase of a new value constellation is the 'thin market' problem: buyers—downstream industries—may be discouraged by an erratic or insufficient supply while farmers face a market that is too thin to support large enough premiums (when they risk to have excess supply). In that case, guaranteeing pull through demand and contracting to mitigate farmers' risk may be a convenient way to get through the initial phase of the value constellation. Pull through demand may require that the focal firm integrates vertically into some downstream industries.

In the case of agbiotech the central firm is likely to be a large corporation with deep pockets because of the considerable investments both in tangible and intangible assets necessary to set up a value constellation.[15] Usually they have a stake in a particular industry that might be affected by 'agbiotech'. These companies might be incumbents, defending their traditional turf such as agrochemical companies, or they may be new entrants in those industries, taking the biotechnology as an enabler to enforce an entry strategy and changing the strategic game in a particular industry.

13.3.5 *The Role of the Government*

Biotechnological innovations have always been tested intensively by regulatory agencies before they could be commercialized. The government is an important player within this context and it has the power to decide about the fate of new value constellations that are enabled by 'agbiotech'-innovations.

In Europe, governments have been preoccupied with the regulatory hurdles accompanying GM-labeling. This preoccupation to protect the end-consumer is legitimized as GM-crops with agronomic traits mainly benefit farmers and agrochemical industry. The commercialization of value-enhanced crops may fundamentally change the role of governments. Benefits for end-consumers will become tangible: nutraceuticals reduce the risks for particular diseases, drugs will become more effective and cheaper, and new industrial applications will become available. In defending end-customers' interest, the governments 'regulatory' task may become a highly complex one.

Similarly the government can embrace 'agbiotech'-innovations as a strong tool in realizing environment protection targets. Bio-fuel, clean energy sources, biodegradable plastics, and so on, are innovations that can be stimulated by tax-initiatives differentiating between the prices of traditional, polluting products and those offered by means of GM-crops (or biotechnology in general). Hence, the role of the government can chance from a cautious and defensive regulator into a more proactive initiator of initiatives that are designed to reach higher consumer surplus and to realize policy targets in the realm of health care and environmental protection.

13.4 Using Value Constellation to Improve Our Understanding of Open Innovation

Value constellations have a number of possibilities to better understand Open Innovation. We chose three topics to explore. First, the choice of the central firm for a particular governance mode for the relations with its partners. Second, value constellations as a nexus to combine and integrate different theories of the firm. Third, the need to analyze Open Innovation at different (but nested) levels of analysis. We only make a quick exploration of these topics and provide some interesting research questions for future research.

13.4.1 *Choice Between Governance Modes in Value Constellations*

Moving from a business model in the midst of well-defined, mature businesses to one that tries to capture the potential of the life science sector does not happen spontaneously. First, it requires new business model architectures developed by key companies (Normann and Ramirez 1993; Normann 2001). Next, the development of agbiotech-based business models requires that economic actors with different assets and competencies are tied together into a value constellation. Finally, a central firm has to choose the appropriate governance mode for its relations with each constellation partner. In theory, coordination between the partners can be accomplished by choosing any of the options ranging from external market-based contracts to the vertical integration of complementarities within the firm, and any collaborative arrangement in between.

The choice for an appropriate governance mode in value constellations has not yet been analyzed in a comprehensive framework. Nonetheless, it is at the core of Open Innovation since the shaping of these external relations will determine the success of the commercialization of new technologies. Chesbrough and Teece (1996), Chesbrough (2003*a*), and Pisano (1990) describe how user firms have different options when they want to source externally developed technologies. The question of how to shape the relations with partners to commercialize the innovation is not within the scope of their research.

Some researchers have analyzed how companies choose between different governance modes to shape the relations with partners to commercialize an innovation. However, almost all publications focus on dyadic relations with one partner (see Pisano 1991; Hoffman and Schaper-Rinkel 2001; Roberts and Liu 2001; Almeida et al. 2002; Dyer et al. 2004; Grant and Baden-Fuller 2002, 2004). In addition, these publications have focused on make-buy-ally decisions from a transaction cost perspective. Barney (1999) takes a different perspective and states that a firm's internal capabilities affect its boundary decisions. Recently, Jacobides and Winter (2005) argue that transaction costs and capabilities should not be considered separately but are intertwined in the determination of the vertical scope of the firm. We suggest from the analysis of the value constellation that the choice for a particular governance mode (including integration) cannot be analyzed from one theoretical perspective and is determined by the role of the different partners in the constellation.

Seed firms provide a nice illustration. Seeds were previously little noticed commodities, but they changed into highly valued, strategic assets with the coming of agbiotech because they incorporate the intellectual capital of biotech companies (Bjornson 1998). As a result, in the second half of the nineties seed companies were acquired at extravagant price earning ratios by agricultural, chemical, and pharmaceutical companies. Those companies controlled large biotech research budgets and had promising technology applications in the pipeline, but they lacked access to the seeds that could incorporate their patented know-how and the seed distribution systems that could give them the possibility to reach the highly fragmented agriculture sector.

Coordination modes between the partners also vary depending on the level of control and coordination that is required to ensure that quality, technological specifications, and product specifications can be delivered to the targeted customer. In the case of nutraceuticals and drugs tight controls from farmer to end-user are of utmost importance because of the quality control of the product and the potential health (and environmental) hazards. Similarly, value-enhanced products also involve serious producer risks for farmers: to mitigate that risk specific contracts between life science companies and farmers will become more fashionable.

In short, we believe that the choice of the governance mode of the interorganizational ties in the value network should be analyzed from their role in the value constellation. These choices maximize the joint value created by the network partners and assure that the created value is shared among them so that each of them is better off than when they would leave the constellation. Hence, the analysis of the determinants of these make-buy-ally decisions in value constellations is in our opinion an interesting topic for future research.

13.4.2 *Value Constellations as a Nexus for the Integration of Different Theoretical Perspectives*

Value constellations are also interesting from a theoretical perspective. They are established to create and extract value, they consist of a set of transactions, they combine resources and capabilities of different partners, and are by definition a specific class of interorganizational networks. Value creation, transactions, resources, and networking are the four constituent dimensions of value constellations. Moreover, they have to be considered *jointly* to understand how firms can create and appropriate value within constellations. Consequently, the role of value constellations cannot be sufficiently addressed by 'one-dimensional' theoretical frameworks that emphasize the role of only one of these dimensions.[16]

How value constellations can be analyzed along the value-chain analysis (Porter 1985), the transaction cost view (Williamson 1975, 1985), the resource based view (i.e. Wernerfelt 1984; Barney 1986, 1991), or the relational view (Dyer and Singh 1998) goes beyond the scope of this chapter and is obviously an interesting topic for future research. However, we can point to some of the limitations of these theoretical frameworks arguing that they offer only a partial explanation of value constellations and that we are in need of an integrative theoretical framework.

The value-chain analysis (Porter 1985) analyzes value creation and appropriation at the *firm level* and is very valuable in examining value constellations and Open Innovation. However, in constellations value creation is determined by the cohesion and internal structure of the value constellation as a whole, not by the performance of the individual participants. Competitive advantage and competition are no longer determined at the firm but at the constellation level (Gomes-Casseres 1994). Future research will have to take the complex interplay between competition and cooperation into account to explain value creation in value constellation (see also, Brandenburger and Nalebuff 1996).

The resource-based view (RBV) postulates that a unique bundle of resources and capabilities may lead to value creation and sustainable competitive advantage. We have mentioned before that value constellations bring together and integrate resources and capabilities that reside in different partners. A central firm may/should have some control over these resources but it owns only part of them, since most resources are owned by the partners in the value constellation. Iansiti and Levien (2004*a*, 2004*b*) call this a keystone strategy: 'By carefully managing the widely distributed assets your company rely on ... you can capitalize on the entire ecosystem's ability to generate ... innovative responses to disruptions in the environment' (Iansity and Levien 2004*b*: 74). The RBV is thus crucial in our understanding of Open Innovation because it emphasizes the bundling of unique resources. However, in value constellations the innovation firm should have control over the required resources but should not necessarily internalize them.

In constellations value is created through sequences of transactions between the participation companies. Transaction costs economics is concerned with the choice of the most efficient governance form for a particular transaction. As we have seen, the choice for the appropriate governance mode of the relations between constellation partners is crucial for the optimal functioning of the constellation. However, are transactions in value constellations optimized by minimizing transaction costs?

Some scholars emphasize the importance of maximizing transaction value rather than minimizing transaction costs (Ring and van de Ven 1992; Zajac and Olson 1993; Dyer 1997; Madhok 1997). This is an interesting approach because the structural form of a transaction is derived from the value that can be created within the broader context (of a value constellation). Because both the transaction-value analysis and value constellations are focusing on joint value maximization and on the process of value creation and distribution, there is a good match between both approaches. However, value constellations differ from the transactional-value approach in that value constellations bring the resources of many partners together while transaction value has been analyzed mainly on the dyad level.

The relational view of the firm (Dyer and Singh 1998) offers another theoretical angle to analyze value constellations. This approach recognizes that a firm's critical resources may extend beyond its boundaries and that the economic performance of an individual firm is often linked to the network of relations in which it is embedded. There is a link between the configuration of interorganizational networks and value creation '... and the locus value creation may be in the network rather than in the firm' (Amit and Zott 2001: 513).

It is obvious that the relational view of the firm is an interesting theoretical framework to explain value constellations, but most publications about interorganizational networks have tried to explain competitive success by network positions of network members and structural properties of networks (see amongst others Stuart and Podolny 1996; Powell et al. 1996; Gulati 1998; Rowley et al. 2000; Stuart 2000). Others explain rent generation by the scarce resources networks bundle (Brandenburger and Nalebuff 1996; Eisenhardt and Schoonhoven 1996; Tsai and Ghoshal 1998; Gulati 1999). Hence, if research about interorganizational networks intends to capture the logic behind value constellations it has to integrate value creation and appropriation, resource bundling, and network structure as different dimensions of one and the same strategy.

We conclude that value-chain analysis, transaction costs economics, network theory, and RBV are certainly useful in explaining value constellations but a quick analysis shows that we need an integration of these various frameworks to come up with a complete picture. Interorganizational networks that create value by means of transaction-based bundling of resources and competencies can only be understood when different approaches are integrated (Madhok and Tallman 1998), Hence, value constellations and Open Innovation may become a nexus to combine these different theoretical perspectives in the future. We are not the only ones that point at the need to integrate theoretical frameworks. Amit and Zott (2001) come to the same conclusion in there study about e-business models and Gomes-Casseres (2003) concludes that there is no comprehensive framework explaining the competitive advantage in alliance constellations.

13.4.3 *Open Innovation Research at Different Levels of Analysis*

Our study of the value constellations indicates that Open Innovation has to be investigated at different layers that are nested. Figure 13.2 gives an overview of these layers.

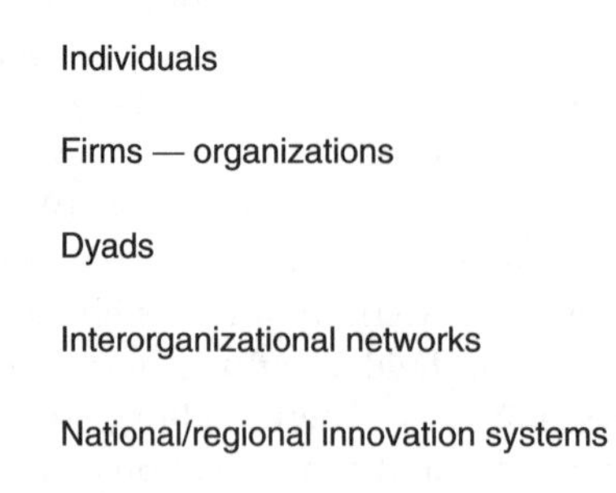

Figure 13.2. Analyzing Open Innovation at different units of analysis

In the past Open Innovation has been analyzed at the firm level and in particular from the technology-user point of view (Chesbrough 2003*a*). The analysis of interorganizational networks (Part III) suggests that an approach with different levels of analysis can deepen our understanding of Open Innovation. The first level is that of the individuals who set up informal intra- and interorganizational networks. This approach has been explored in Chapter 11 and Chapter 10, but has not received much attention in previous Open Innovation research. The next level is the firm level which has been analyzed in Part I of this book.

Next, one can consider Open Innovation from the dyad level, that is, the perspective of two companies that are tied to each other through equity or nonequity alliances, corporate venturing investments, and so on. The dyad perspective takes into account the perspective of the two organizations that are involved in an Open Innovation relationship. As Open Innovation is basically about non-arm's-length relations between companies it can take advantage from a dyad level perspective analysis of strategic alliances (Bamford et al. 2003; Lynch 1993) and external corporate venturing (Keil 2002*a*). Typical research questions at this level of analysis are how to select partners, how to assess the return and risks of an alliance or external venture, how to evaluate the fit between potential partners, and how to structure the cooperative agreement and manage it over time.

The next level of analysis refers to interorganizational networks. The different chapters of Part III contain three main messages in our opinion. First, a network perspective is necessary as a complementary approach of Open Innovations. Key innovating companies do not profit from Open Innovation only by deliberately in- and outsourcing intellectual property (IP) with different external partners. Key innovators have also to set up and manage interorganizational networks both to develop new technologies (Chapter 12) and to exploit technology-based business opportunities (this chapter). External network management becomes crucial when Open Innovation moves beyond bilateral insourcing of externally developed technologies. Key players in Open Innovation have to orchestrate the network of partners that are crucial to develop or to exploit particular innovations. They have to look for interesting partners, lead and nurture the network, minimize tensions between partners, and instill a unifying vision.

Second, when Open Innovation is realized through extensive collaborative networks competition is no longer between individual firms but between groups of firms. Group-based competition is different from firm-based competition (Gomes-Casseres 1994; Brandenburger and Nalebuff 1996; Bamford et al. 2003).

Third, external networks are likely to change substantially when a new venture shifts from the idea generation phase to the commercialization. Chapter 12 explores the need for external networks when a company is involved in the development of systemic innovations (left-hand side of Figure 1.2). This chapter focuses on value constellations that are necessary to commercialize innovations (right-hand side of Figure 1.2). The dynamics of these networks have to our knowledge never been studied in depth.

The last level of analysis (see Figure 13.2) consists of the national innovation systems. This level has been discussed in Chapter 11 and goes beyond the scope of this paper. It is, however, important to mention that the establishment and management of interorganizational networks can be spurred or hampered by the innovation system in which it is embedded.

In sum, we suggest that a multilevel perspective can deepen our understanding of Open innovation and that interorganizational networks play an important but yet underresearched role in explaining Open Innovations. In our opinion, there are ample opportunities for future research in combining Open Innovation with external network management, collective competition, and the dynamics of networks that accompany an innovation in its journey from idea generation to a profitable business.

13.5 Conclusions and Suggestions for Future Research

In this chapter we have shown that market-based transactions within the agricultural industry are increasingly replaced by a complex network of relations between the relevant economic players. Moving from a business model that is appropriate for mature businesses to one that tries to capture the business opportunities related to the emergence of the life science does not happen spontaneously. It requires a purposeful rethinking and shaping of the business model by a central firm (Normann 2001) and economic actors with different assets and competencies have to be tied together into a value constellation. Not only the competencies of the participating firms but also the way how the constellation is structured and managed determines the collective competitiveness of the latter.

What did we learn from this chapter about Open innovation? In our opinion there are five ideas to take away. First, agbiotech is interesting as technology field to test whether Open Innovation can be generalized. Open innovation has been studied predominantly within the context of the information and communication technologies (ICT) (Chesbrough 2003*a*). Although agbiotech provides a complete different setting, Open Innovation is also applicable to this technology field. Future research has to examine whether Open Innovation also applies in others contexts. Our findings about agbiotech indicate that Open Innovation should not be confined to ICT as Graham and Mowery (2003) propose.

Second, we have focused on value constellations, that is, interorganizational networks that are established to create and capture value based on new business models. They could be considered as a mirror image of the innovation networks established to insource externally developed technologies. Innovation networks (the example of systemic innovation networks has been discussed in Chapter 12) are situated at the left-hand side and value constellations at the right-hand side of Figure 1.2. Value constellations are different from 'early-stage' innovation networks: they are established to commercialize an innovation together with partners that own critical resources and are tightly linked to the underlying business. Emphasizing the commercialization stage is interesting because most Open Innovation research has been focused on external technology sourcing and networking with technology providers and innovative, upstream companies. Value constellations are to a large extent oriented towards customers and other downstream players. The study of value constellations shows that the 'openness' of Open Innovation also applies to the commercialization phase. This is not a new idea (see Gomes-Casseres 1994; Normann and Ramirez 1993) but it has not yet been integrated into the broader 'Open Innovation'-picture.

Third, innovation networks and value constellations could be considered as two snapshots that obscure the truly dynamic nature of 'Open Innovation'-networks. Firms continuously change these networks depending on the development stage of the venture. Understanding these dynamics is important both from a theoretical and managerial point of view. We hope future research will explore the dynamics of these networks.

Fourth, value constellations are from a theoretical point of view interesting constructs. They are built to create and capture value and thus have a lot in common with the value-chain analysis. They are also related to transaction cost economics because the central firm has to choose the appropriate governance mode for the relations with its partners. Next, value constellations bring together critical resources that are owned by different companies and have therefore a lot in common with the RBV of the firm. Finally, value constellations can be analyzed in terms of the relational view of the firm. As a result, value constellations are a nexus for the integration of different theoretical perspectives.

Finally, the analysis of interorganizational networks in general and value constellations in particular reveals that research about Open Innovation should be multilayered. Open Innovation from the (user) firm perspective only provides a partial view. Figure 13.2 shows that there are at least five possible layers. Each dimension opens a new perspective on Open Innovation. Since the different layers are nested Open Innovation has to be explored simultaneously at different levels. Therefore we hope that future research will explore Open Innovation at the individual or unit level on the one hand and at the network and innovation system level on the other hand.

Notes

1. Agricultural biotech is that part of the biotechnology that is dedicated to develop genetically modified (GM) plants.
2. This is of course only true for radical or disruptive innovations (Christensen and Raynor 2003).
3. Bt Corn stands for genetically altered hybrids that contain a naturally occuring soil bacterium, Bacillus thuringiensis, that kills the European corn borers. New GM-seeds that resist other pests will be introduced during the next years.
4. Roundup was already introduced in the seventies as a burndown treatment, used to kill all existing vegetation before planting (Carpenter and Gianessi 1999).
5. The four taxonomy of the value drivers is adapted from Amit and Zott (2001) who analyzed the value drivers in e-business. Although the categories have the similar labels, value drivers in agbiotech differ significantly from those in e-business. These differences are an interesting topic for future research because they have a strong impact on the way value constellations are structured.
6. Sustainability of the competitive advantage is not guaranteed as long as strategic countermoves of actors in competing value creating systems are possible. Companies considering to developing crop-based, biodegradable polymers should know for instance that Du Pont already developed fully hydro/biodegradable polymers based on the oil-based polyethylene terephthalate (PET) polyester technology: being biodegradable is not a sales argument that is unique for plastics generated in bacteria or plants.
7. The traditional value system approach has already been challenged in the past by several authors (Brandenburger and Stuart 1996; Stabell and Fjeldstad 1998; Ramirez 1999; Ramirez and Wallin 2000).
8. The configuring of roles of different economic players within new, technology-based value-creating networks has not received substantial attention from scholars. Notable exceptions are Taylor and Terhune 2001; Amit and Zott 2001; Brynjolfsson and Urban 2001; Chesbrough 2003*a*; Bamford et al. 2003; Gomes-Casseres 1996, 2003.
9. We come back on the constellation leadership in Section 13.3.4
10. This situation is likely to change when value-enhanced crops with direct benefits for end-consumers will be introduced in the coming years.
11. Starting the public dialogue and teaching the public are examples of 'value constellation'-management activities (see Section 13.3.4).
12. This example is based on the Flavr Savr tomato described in Goldberg and Gourville (2002).
13. Vertical integration is valuable in very specific circumstances (e.g. acquisition of seed companies) as we will see in Section 13.4.1.
14. This is in sharp contrast with incremental innovations or sustaining innovations (Christensen and Raynor 2003) where a company can rely on existing relations with suppliers, channels and end-consumers.
15. Focal companies are not necessarily large companies with deep pockets. Amit and Zott (2001) indicate that a number of start-up companies have successfully entered

the e-business 'industry'. Normann and Ramirez (1993), Slywotsky et al. (2001), Ramirez and Wallin (2000), and Parolini (1999) focus on new value constellations that emerge from a bright business idea that is usually not linked to technological innovations. They argue (and illustrate with a series of case studies) that small companies have the potential to change the rules of the game into their advantage.

16. The need for a multidimensional approach is echoed in Amit and Zott (2001) and Gomes-Casseres (2003).

Part IV

Conclusions

14

Open Innovation: A Research Agenda[1]

Joel West
Wim Vanhaverbeke
Henry Chesbrough

14.1 Introduction

Chandler (1977, 1990) recounts how the key technologies of the early and the mid-twentieth century were developed by industrial research departments within the large diversified enterprises of US and Europe. Such diversification, along with vertical integration from research and development (R&D) through distribution, provided these firms with competitive advantage over smaller and newer rivals through economies of scale and scope.

Among these leading firms, Chesbrough (2003*a*) argues that the strategy brought with it a certain mindset:

> It is a view that says *successful innovation requires control*.... This paradigm counsels firms to be strongly self-reliant, because one cannot be sure of the quality, availability, and capability of others' ideas: 'If you want something done right, you've got to do it yourself'. (Chesbrough 2003*a*: xx)

However, from his study of US industry practice at the end of the twentieth century, Chesbrough concluded that this model was reaching its limits. Among other factors, he identified the increased mobility of knowledge (through labor mobility) and availability of venture capital to create new firms to capitalize on such knowledge. In a parallel explanation for shifts away from the Chandlerian model, Langlois (2003*a*) identifies the increasing interfirm modularity and subdivision of labor (particularly in high-tech industries) as obviating the need for vertical integration.

In contrast to this 'Closed Innovation' model, Chesbrough argued:

> Open Innovation is a paradigm that assumes that firms can and should use external and internal ideas, and internal and external paths to market.... Open Innovation combines internal and external ideas into architectures and systems whose requirements are defined by a business model. The business model utilizes both external and internal

ideas to create value, while defining internal mechanisms to claim some portion of that value. (Chesbrough 2003*a*: xxiv)

Based on his study of firms practicing Open Innovation, Chesbrough concluded that industrial R&D was undergoing a 'paradigm shift' (in the sense of Kuhn 1962) from the closed to the open model. This is in the spirit of Donald Stokes' (1997) concept of Pasteur's quadrant, where empirical practice preceded the development of the underlying theories that later explained those practices. It also draws heavily from earlier work on industrial evolution (e.g. Nelson and Winter 1982), absorptive capacity (Cohen and Levinthal 1990) and the impact of spillovers on industrial R&D (Rosenberg 1994). A more complete discussion of prior research is provided by Chesbrough (Chapter 1).

Open Innovation is both a set of practices for profiting from innovation, and also a cognitive model for creating, interpreting, and researching those practices. Some of these practices are not new. For example, for more than fifty years government funding agencies and nonprofit foundations have funded scientific research, performing the role that Chesbrough (2003*b*) termed the 'innovation benefactor'.

As a new way of conceptualizing innovation, Open Innovation relaxes many of the assumptions presumed in the Chandlerian model, both in the external supply of innovation to be incorporated into a firm's offerings, as well as the potential demand outside the firm for its internal innovations. However, this does not mean that any innovation model is feasible, any more than the rise of the Internet meant that any 'e-strategy' was profitable. Experimentation within the Open Innovation paradigm has the constraint of establishing a business model for creating or using an innovation, a constraint that may have been obscured by the cross-subsidies often seen with vertical integration.

If the practice of innovation is changing because new forms of innovation are economically feasible, then this offers opportunities for researching and explaining those new practices. The Open Innovation paradigm offers propositions for how such innovation should work, and the earlier chapters in this book have identified how other examples of innovation practices fit within the Open Innovation paradigm.

However, the limited amount of empirical research since the earlier book (both inside and outside this volume) means that there are many unanswered questions about Open Innovation—and thus a concomitant number of research opportunities. To identify many of these opportunities, here we survey the potential scope of Open Innovation research, identifying both unanswered research questions and also the issues researchers will face in addressing these questions.

First we consider how Open Innovation might be studied along five different levels of analysis, and the implications of related research at each level. We then suggest ideas for research methods and data for studying Open Innovation, including ways to establish the limits and boundaries of the Open Innovation paradigm. We conclude with an invitation to other academic scholars to join in shaping and pursuing this research agenda.

14.2 Levels of Analysis

To date, most studies have examined Open Innovation at the firm level, for two reasons. First, innovation is traditionally conceived as the outcome of deliberate actions of a single firm, and thus R&D competition has also been stylized as an innovation race between two or more firms. Second, the value of a technical invention is realized only through a business model of a firm (Chesbrough and Rosenbloom 2002). While business models may span the boundaries of a firm or even an industry, 'a particular firm is the business model's main reference point. This is why one can refer to a business model as "firm x's business model." ' (Amit and Zott 2001: 513–14).

However, neither the practice of nor research on Open Innovation are limited to the level of the firm. Innovations are created by individuals or groups of individuals, usually within organizations, so the subfirm level of analysis is particularly salient in understanding the sources of innovation (cf. von Hippel 1988). At the same time, firms are embedded in networks, industries, and sectors; thus, to understand a firm's business model—particularly the value created and captured from an innovation—it is essential to consider these level of analyses. Finally, Open Innovation is practiced within the context of a given set of political and economic institutions, including regulation, intellectual property law, capital markets, and industry structure. As most (but not all) of the prior research on Open Innovation has focused on the US system, an examination of Open Innovation in the context of other National Systems of Innovation could more clearly identify both the prerequisites for and limits of Open Innovation, and make explicit the linkages between these institutions and practice.

To encourage future research, we now consider Open Innovation using these five levels of analysis, from the individual to the nation-state. At each level, we consider the inflows and outflows of innovation, as well as the associated policies and enabling industry practices (Table 14.1).

14.3 Individuals and Groups

Innovation begins with the efforts of one or more individuals. In the Closed Innovation paradigm, such efforts are within the firm, that is, by company employees, and certainly such individuals play a crucial role in Open

Table 14.1. A framework for classifying Open Innovation research

	Inflow	Outflow	Policies	Enabling practices
Individual	*Creative Commons, music sampling*	Chapter 5 *Blog, open science*		
Organizational	Chesbrough (2003) Chapters 3,4, 5,12,13	Chesbrough (2003) Chapters 4, 5,12	Chapter 9	Chesbrough (2003) Chapters 4, 5,12
Value network	Chapters 11,12,13	Chapter 3	Chapter 8	Chesbrough (2003) Chapters 11,12,13
Industry/Sector	Chapters 3, 7, 8 *Open Source software, pharmaceuticals*	Chapters 7, 8 *Biotech*	Chapter 9	
National institutions	Chapters 7, 8	Chapters 7, 8	Chapters 6, 7, 9	Chapters 8, 9

Prior and *potential* research on Open Innovation

Innovation as well. Under either paradigm, firms want their R&D workers to be productive, using some combination of intrinsic and extrinsic motivations.

However, as Chesbrough (Chapter 2) notes, under Open Innovation there are the additional requirements of avoiding both 'Not Invented Here' and 'Not Sold Here' biases towards the creation and use of innovations. Research is needed to establish how these new requirements affect the incentives and organization of R&D workers. If firms are to be agnostic about the sources and uses of innovation, how can this be reflected in their compensation, recognition, and other motivational techniques? Are other changes to the group or organizational dynamics necessary to support Open Innovation? Is hiring for Open Innovation different than for Closed Innovation, either because it requires more external scanning or because it shifts firm competencies from innovation creation towards system integration?

How are the Open Innovation challenges different for not-for-profit organizations (notably universities) that seek to motivate individuals to generate, appropriate (e.g. patent) and transfer innovations so that they have commercial value, both to the university and for private industry? If scientists differ in their activities in these areas, how much is due to individual differences in attitudes and needs, and how much is due to organizational factors such as incentives and cultural norms? In biomedical research, Zucker et al. (1998) and Bercovitz and Feldman (2003) have attempted to identify some of the individual factors that motivate individuals to create and commercialize innovations, while in their survey of Italian academic inventors, Baldini et al. (2005) identify perceived impediments to such innovation. But research needs to not only explain differences in creation, but also differences in use; for example, do the interpersonal ties of academic researchers affect the use of their innovations by private industry?

In some cases, Open Innovation will also entail utilizing individuals outside the firm to supply (or apply) key innovations in the firm's business model. If these individuals are motivated using financial returns, then the issues faced by the central firm are similar in principle to those faced in dealing with innovation inflows from and outflows to corporate partners. However, there are potential search and transaction costs associated with the lack of scale—are firms used to dealing with 10 or 100 corporate innovation partners able to manage 1,000 (or 1 million) consumer innovation partners?

A potential source of innovations for many products comes from those individuals that use the product in their work or home life. Such users may innovate for their own direct utility, as has been established by the pioneering work of von Hippel (1988, 2005). Users may let their innovations go undiscovered, or may seek to profit from them; in other cases they have rational reasons for freely revealing their incremental contribution to the firm that supplied the relevant technology (von Hippel 2005: 77–91). Research on both licensed and free spillovers from users thus is an important potential research area for Open Innovation. Research could also consider how user needs and requirements are factored into the search for external innovations—or the ways in which technology suppliers create or market such external innovations.

How can such sources of external innovation be encouraged? Prior research on user-based innovations (Franke and von Hippel 2003; von Hippel 2005) has shown how differences among user needs spawn user innovations, and how such innovation can be enabled through product design (such as providing 'toolkits' for user innovation). But are there factors that explain the differences in the ability of firms to utilize user-generated external innovations? The lead user research (e.g. Lilien et al. 2002) has focused more on trying to establish the value of the external innovations, comparing internal and external innovations while holding firm capabilities constant.

Individual motivations are not limited to such direct economic or utilitarian gain. West and Gallagher (Chapter 5) identify examples where software innovations are donated by individuals, often far beyond any direct utility. They point to prior research on extrinsic motivations (particularly among student users) such as external signaling of skill and availability to prospective employers. But is this a significant source of external innovations in other contexts? Or, absent direct utility, would such donations be confined to those with low opportunity cost (such as students or retirees)?

Similarly, while West and Gallagher suggest that the intrinsic satisfaction of creative expression helps attract donated innovation for role-playing computer games, are there other examples where such donations happen? Collaboration mechanisms exist for other forms of creative expression such as blogs, wikis, music sampling, and cumulative creation through Creative Commons (Lessig 2004; von Hippel 2005). However, little if any research has been done on how such creativity is translated through a business model into commercially relevant innovations.

More broadly, the commonly cited examples of shared creativity lie within a broad class of information goods, for which the Internet and relevant software tools enable collaborative production across time and space. If such collaboration were to generalize beyond information goods, what sorts of identification, coordination, and distribution mechanisms would be required? Will the necessary tools (or skills) be available to individual innovators, or only under the umbrella of firms, universities, and other organizations?

14.4 Implications for Firms

The implications of the Open Innovation model for the firm were discussed at some length by Chesbrough (Chapter 2). Here we will reprise some of the most important questions identified in that chapter.

The internal, vertically integrated model of innovation from Chandler that preceded Open Innovation featured one important attribute that Open Innovation lacks. The earlier model generated many new long-term discoveries and inventions, primarily in the central R&D laboratories of large firms. In the Open Innovation model, it is not obvious whether such a wellspring of inventions will continue or not, because it is less clear that there will be a return to the firm's investment in those more basic research activities. If commercial firms do not realize a return on their innovative activities, they will tend to underinvest in innovative activities that are either highly risky (e.g. basic research) or that are easily imitated by free-riding competitors.

An important area of future research is thus to understand the incentives within the firm for generating the new discoveries and inventions that will supply the 'seed corn' for future innovation activities. The Open Innovation model relies upon a specialization of innovation labor (with institutions such as universities playing a more central role) and to intermediate markets (where specialist technology suppliers compete to supply new discoveries to others who commercialize them) to partially or wholly provide the seed corn for new innovation. It is an open and researchable question whether these latter mechanisms supply adequate motivation for individuals and organizations to do the hard work of discovering fundamentally new knowledge. This would include the effect on the aggregate supply of such knowledge, but also whether the new mechanisms change *which* actors are providing that supply.

Even if individual-level effects are overcome, restructuring firms to avoid the Not Invented Here syndrome directly impacts the purpose and organization of corporate R&D activities. Open Innovation subtly shifts the role of internal R&D from discovery generation as the primary activity to systems design and integration as the key function. This builds upon the recent book by Prencipe et al. (2003) on systems integration, and will imply a need for changes in the norms and reward systems in most organizations.

In the Closed Innovation model, firms invested in internal R&D to create new products and services, and lived with the 'spillovers' as an unintended byproduct of the process. These spillovers were regarded as a regrettable but necessary cost of doing R&D. In the Open Innovation approach, firms scan the external environment prior to initiating internal R&D work. If a technology is available from outside, the firm uses it. The firm constrains internal R&D work to focusing on technologies that are not widely available, and/or those in which the firm possesses a core advantage, and seeks advantage from constructing better systems and solutions from its technologies. A testable hypothesis from this new model is that Open Innovation firms may generate fewer spillovers. This hypothesis, if supported, would have offsetting implications for the firm. On the one hand, fewer spillovers may mean that there is a higher yield for R&D spending, encouraging the firm to sustain its commitment to R&D. On the other hand, the lack of spillovers from other firms may deprive the firm (or the industry in which the firm competes) of organic growth opportunities for discovering future technologies.

A related question is that of time horizon for innovation activity under the Open Innovation approach. Research in the Rensselaer Polytechnic project team that studied radical innovation documents the long time frame and convoluted path to market of many (ultimately) successful radical innovations (Leifer et al. 2000). Open Innovation utilizes the company's business model to frame its research investments. O'Connor (Chapter 4) argues that this may shorten the time to market for more radical innovations, thus making the pursuit of radical innovation more sustainable. Would such acceleration result in more radical innovations being undertaken (which might lengthen the overall time horizon for the investments within the R&D portfolio), or would it be used to reduce the overall time spent on a portfolio of innovation? Does it imply a faster time to market for whatever R&D projects lie within the firm at any point in time? What role, if any, do longer term research investments play within the R&D portfolio of an Open Innovation-minded firm? Is Open Innovation more relevant for explorative technology projects compared to exploitative ones (March 1991)?

A more subtle, second-order research question emerges from this potentially shorter time horizon. If projects move faster through the R&D system, does this result in more incremental innovation output? Or does a higher metabolic rate result in the faster incorporation of new knowledge, and in more (re)combinations of technologies in a given period of time? If so, this higher metabolism of knowledge might offset to some degree the issues of overly incremental innovation and the potential loss of the seed corn research noted above. Conversely, Fabrizio (Chapter 7) suggests that as universities seek to profit from their research—rather than allowing free spillovers—both the cost and administrative overhead may slow the pace of cumulative innovation.

A third issue is the control of spillovers by firms practicing Open Innovation. Since spillovers are managed as possible sources of new revenue and new market identification and development in this model, do we see different outcomes for these spillovers? Is there a higher rate of commercialization of spillovers among firms operating within the Open Innovation paradigm? What internal barriers exist that inhibit the greater use of external paths to market for spillovers? Is there a corresponding source of inertia that parallels the Not Invented Here syndrome, as it pertains to utilizing spillovers outside the firm (i.e. do internal business units seek to prevent the use of internal technologies by outside organizations, including potential competitors)?

A fourth question is, what circumstances motivate firms to embrace Open Innovation approaches as part of their R&D efforts? Is adoption of Open Innovation primarily an industry level phenomenon, or do we see significant variation in adoption within industries? If the latter, what firm characteristics are associated with differential adoption rates within an industry? Do large firms differ from small firms in their adoption of Open Innovation, as the initial work of Christensen and colleagues (forthcoming) suggests? Or do firms with relatively greater investments in internal R&D differ from those with little or no investment in R&D? Does Open Innovation provide a way for technology laggards to close the gap with technology leaders, or will Open Innovation reinforce the specialization and scale advantages of the existing leaders?

To do such research, we would need to define what 'adoption' of Open Innovation means. If vertical integration is one extreme, and the fully component model (such as diagrammed in Chapter 6) is the other, how would we classify the intermediate (or hybrid) strategies? If we talk about Open Innovation in terms of degree, then is there a tipping point (such as for the attitudinal and cognitive issues identified earlier)? Would we expect to see a gradual increase in Open Innovation practices over time, or (as with the adoption of other R&D best practices) a clear demarcation between the pre- and post-adoption periods?

A fifth issue turns on the management of intellectual property (IP) under Open Innovation. If firms are utilizing external technologies more frequently, they may need to engage in greater inlicensing activity. Do we see such an increase? How do firms identify potentially useful external technology sources? How do sellers manage the Arrow Information Paradox[2] in offering technology to a buyer? How are the risks of technology hold-up managed in inlicensing discussions? If firms are utilizing external channels to markets for spillover technologies, how are those risks managed? Do firms change the governance of their management of IP as they engage in these transactions more frequently?

Sixth, companies that explore new (disruptive) technologies must often identify a new or adapted business model to create value (see Amit and Zott

2001; see Chapters 5, 12, and 13). What explains a firm's ability to identify the new business models necessary to commercialize disruptive innovations? How is this ability developed or grown? Can it be explained by its technical knowledge or ability to get market feedback, such as through the 'probe and learn' process (Lynn et al. 1996; Brown and Eisenhardt 1998)? Can the capability be developed in isolation, or only through trial and error, developing a firm-specific model of business model development? In developing these business models, how do firms manage the conflicts between the goals of the internal business unit and the external partners?

Seventh, while previous chapters have examined interorganizational networks, Open Innovation also increases the salience of *intra*organizational networks. If firms vary in their ability to access and leverage external sources of technology—as suggested by Gassmann and von Zedtwitz (2002*b*) and Laursen and Salter (2005)—it is quite likely that the heterogeneity of firms to learn and profit from these relationships is largely determined by the internal organization of these firms. To state it more directly, the effective management of externally acquired knowledge likely requires the development of complementary internal networks (Hansen 1999; 2002; Hansen and Nohria 2004) to assess and integrate the externally acquired knowledge. This suggests an important research area: to link the internal networks of the firm to the external use of ideas and technologies outside it. A related insight is that internal reorganization is also necessary to support the formation and sustenance of other organizational capabilities, such as corporate venturing, intrapreneurship, and creating 'newstream' organizational units (Dougherty 1995; Vanhaverbeke and Peeters 2005).

A final very interesting question is whether the widespread use of Open Innovation would change the nature or relevance of 'core competences' for firms. Christensen (Chapter 3) argues that core competences in particular technological fields will give way to a more fluid innovation model, where firms become skilled at incorporating others' specialized technologies, rather than necessarily developing their own. A complementary view is that Open Innovation provides a much broader market for firms' core competences, enabling them to support other companies' businesses and technologies. This could make core competences more valuable, rather than less so. So Open Innovation could separate out core competencies into two broad categories: those related to creating technological innovations, and those related to sourcing or integrating such innovations.

14.5 Interorganizational Value Networks

While Open Innovation research has emphasized the activities of the firm, the innovation sourcing between (at least) two companies implies research opportunities in studying dyads of innovation partners, as well as the interorganizational networks constructed from these dyads and the value networks associated with the value creation from a specific technology.

At the dyadic level, we would expect that the search, negotiation, contractual, implementation, and support phases of Open Innovation would be better understood if researchers simultaneously captured the perspective of both the technology supplier and technology user. For example, Dushnitsky (2004) has shown that success of corporate venture capital can only be understood when the incentives of the technology start-ups are also taken into consideration. Similarly, the growth of markets for technology (selling and licensing technologies) can only be understood when the hurdles for both licensors and licensees are analyzed (Arora et al. 2001*b*).

There are many other potential research questions at the dyadic level. How do two companies find each other to codevelop a technology? How can this search process be improved? Among possible variables—such as transaction costs, the role of tacit or codified technology, complementary assets—which will moderate the benefits both parties see in an external technology sourcing agreement? When do start-ups see corporate venture investments not as a threat but an opportunity to grow? How do firms partners overcome their differences to build trust and a durable alliance (see Chapter 11). How can perceived threats be managed in order to allow both technology supplier and user to profit from the Open Innovation process?

While studying Open Innovation at the dyad level would augment prior research at the firm level, research is also needed on Open Innovation in interorganizational networks, which are more than just the sum of the component dyads. As earlier chapters have shown, researching innovation networks (both within and between networks) is essential to understanding Open Innovation, just as research on corporate innovation (within and between firms) is essential to understanding vertically integrated innovation.

The chapters of Part III explicitly focus on how interorganizational networks help explain Open Innovation, but in other chapters, the network level is implicitly present. Fabrizio (Chapter 7) shows how Open Innovation cannot be conceived without considering the networking between innovating firms on the one hand and universities and research laboratories on the other hand. West and Gallagher (Chapter 5) explore how communities of practitioners and firms create a symbiotic relation that leads to an explosive growth of open-source software.

These chapters offer linkages between our understanding of interorganizational networks and Open Innovation, but many other topics are yet to be researched. What role do interorganizational networks play in Open Innovation? Is the contribution of each player best explained by their competencies, roles, or structural position in the network? For example, Gomes-Casseres (1996) has shown for various alliance networks in the Information and Communication Technology (ICT) sector that network dynamics has to be explained by the interaction of network participants with different roles, assets, and value-chain positions.

If we assume the value of the network to Open Innovation, how can the focal innovator attract and coordinate all the resources necessary to bring a new, technology-based product to the market? Is it necessary to manage the entire value network, or merely a small clique of central players? How is this network creation and management process different for technological discontinuities (cf. Utterback 1994) or market-disruptive innovations (Christensen 1997)? In either case, how does the focal firm both gain the knowledge necessary to manage the network and communicate that knowledge to the network? How does it convince potential partners to join the network during early periods of high-technological and market uncertainty?

Interorganizational networks play a crucial role in the different steps of the innovation process (research, development, and commercialization) as has been illustrated by the cases of systemic innovations (see Chapter 12) and the commercialization of agricultural biotechnology (agbiotech) breakthrough innovations (see Chapter 13). At the R&D phase, how should the innovating firm select the appropriate partners? How is the selection process affected by the potentially disjoint domains of the partners' respective knowledge? In the commercialization phase, how does the focal innovator provide adequate guarantees to partners that they will earn a return on their cospecialized assets (cf. Teece 1986)?

Meanwhile, the firms within a value network are linked through a business model which unlocks the value latent in a technology (Chesbrough and Rosenbloom 2002; Chesbrough 2003*a*). To the degree that the innovation requires a network to realize that value, we would expect differences between networks in their realized value based on the alignment of partner activities. Prior research posits that alignment of network members is orchestrated by the firm that architects the business model (Chesbrough 2003*b*; Iansiti and Levien 2004*b*).

However, we do not have research comparing the effectiveness of various architectural strategies, nor the factors explaining variance in such effectiveness. Nor do we know whether alignment is explained by economic incentives provided to participants—how the value created by the network is shared among the participants—or whether variance in relational or structural aspects of network coordination also measurably affects the value realized. At the same time, any research would have to consider the focal firm's three conflicting goals of maximizing total created value, capturing value for itself and allocating value capture among the network members. To understand this, we would need to study both successes and also networks that failed in one of these three dimensions, such as networks where firms created value but failed to provide enough value to attract complementors?

Even where Open Innovation is enabled by a value network, there are questions about how that network is coordinated and maintained. What are the most important management tasks of the network orchestrator? When and how is governance shared across the network? How do the partners manage conflict within the network, whether between competitors, buyers and sellers, or complementors? How do they overcome the threat of opportunism due to lock-in situations, investments in specialized assets, and structural embeddedness?

There are also structural questions about the shape and size of these value networks. Using a biological metaphor, Iansiti and Levien (2004*b*) argue that certain companies play a crucial role as the keystone of a business ecosystem. But do we expect the value created (and captured) by a firm in an Open Innovation network to be completely explained by the firm's functional role in the network? Even if we assume an optimal keystone position, where in the network are the secondary opportunities for value capture—near the hub of the network, or at the rapidly evolving periphery? Where is the knowledge created by the network most readily accessible? What role does tie strength play in accessing that knowledge (cf. see Chapter 11)? What is the optimal size and density of network evolution for value creation and innovation?

Some networks are less open to new participants than others. For example, industrial groupings such as Japan's *kigyo shudan* or Korea's *chaebol* (Fruin 2006; Steers et al. 1989) tend to buy within their group rather than from outsiders. Hagel and Brown (2005) argue that closed networks need to become more open to develop the necessary specialization and deepening of the innovation capability of the participants. So research could test whether closed networks have performance disadvantages where specialized or deep knowledge is required, and what form of 'openness' provides value over others. If openness has economic value, then research would also be useful to identify the levers of interorganizational change for making an existing ecosystem more open.

Meanwhile, if companies are embedded in networks, then we would expect this to change the nature of competition between such companies: rival firms may not be competing individually but instead as part of groups of networked firms competing against other groups. In this case, the performance of companies in this setting no longer depends only on the internal capabilities of a firm but also on the overall performance of the network to which they belong (Gomes-Casseres 1996). How does group-based competition affect our understanding of Open Innovation? When considering success measures, what are the respective contributions of a firm- and network-level analysis? Should the impact of group-based competition be differentiated along different phases of the technology development—precompetitive and competitive settings—as has been suggested by Duysters and Vanhaverbeke (1996)? How does exclusive network membership shape the dynamics of Open Innovation. Beyond obvious factors such as the costs and risk diversification of participating in multiple networks, are there other moderating factors that make exclusivity more or less attractive?

Thus, we believe Open Innovation practice will be intimately linked to how firm innovation activities are mediated in networks, both interorganizational and (as mentioned earlier) intraorganizational networks. But we also want to note the opportunities to study the impact of individual-level networks, which although they are more likely to be based on informal ties, also play a crucial role in channeling knowledge flows between firms (see Chapter 11). Prior research has suggested that the relative importance of organizational versus individual (and formal versus informal) ties on of some industries (e.g. biotech with star scientists) varies based on the nature of industries or even economic regions, as with Saxenian's (1994) contrast of California and Massachusetts technology start-ups. Since most Open Innovation research has focused on the firm, there remains a research opportunity to identify the antecedents and consequences of individual-level (and informal) network ties upon Open Innovation.

14.6 Industry or Sector

A more traditional level of analysis for the strategic value of innovation is the industry level. Prior innovation research has considered both the differences between the population of firms within a given industry or sector, and also the differences between industries (sectors). Because many of the intraindustry differences have been considered earlier in this chapter, here we consider to what degree differences between industries—as well as changes to a given industry over time—might affect the application of Open Innovation. We also consider the converse, how the use of Open Innovation might change the structure of one or more industries.

As a starting point, prior research has established that the nature, value, and organization of innovation varies between industries and within a given industry over time. For example, it is generally accepted that during the past two decades patents and university research have played a greater role in innovation for biotechnology and pharmaceutical industries than for consumer electronics.

A common characteristic across a given industry or sector is the degree of appropriability available to firms; the role of appropriability (through IP rights) in Open Innovation was the primary focus of Part II. The conventional view is that greater appropriability leads to an increased willingness of innovators to offer internal innovations for others to use (Chesbrough 2003*a*; see Chapter 6).

However, from a large-scale cross-sectional study, Laursen and Salter (2005) concluded that openness was associated with a moderate level of appropriability. Meanwhile, both Fabrizio (Chapter 7) and Simcoe (Chapter 8) identify potentially negative impacts of high appropriability upon the cumulative and decentralized aspects of Open Innovation. This suggests that the relationship between appropriability and Open Innovation is more than a simple linear causal relationship, and thus further research is needed to identify potential moderators of the effect of appropriability upon Open Innovation. This might be done by a comparison of multiple industries, but could also be accomplished by comparing the same industry across multiple appropriability regimes—whether between countries, or in the same industry as it changes over time, as with the increasing use of software patents (cf. see Chapter 9).

But there are other important differences between industries that we could relate to the prevalence or nature of Open Innovation. For example, what is the relationship between R&D intensity (as a percent of sales) and Open Innovation? Are external sources of innovation more highly valued in industries with high levels of R&D? Or is external R&D more likely to be used by low R&D intensity industries where firms lack internal R&D capabilities, and thus are more dependent on external innovation suppliers?

Similarly, questions could be raised about industry concentration: the cases presented by Chesbrough (2003*a*) suggest that more concentrated industries are more likely to vertically integrate. However, as with the example of IBM, such industries might also have more internal innovations generated that can be outlicensed to competitors and other industry partners. Establishing the direction and magnitude of this relationship is an opportunity for empirical research.

We would also expect that other characteristics of an industry—such as stage of the technology life cycle, rate of technological change or growth—would also affect the practice of Open Innovation. The work of Zucker and Darby (1997; Zucker et al. 1998) suggested that using external innovations was a crucial mechanism for firms to deal with rapid technological change in the earliest phase of the development of biotechnologies. Would this broadly apply to other infant or rapidly changing industries, or only those directly linked to university basic research? Are there other factors that mediate or moderate the division of the innovation labor, such as the degree of technical modularity (cf. Langlois 2003*b*; Sanchez 2004) or the specialization of firms within the industry? The adoption of digital components in the audio industry support the proposition that both factors increase the use of open innovation (see Chapter 3; Christensen et al. forthcoming).

Since Caves and Porter (1977), much of our comparison of industries has focused on entry barriers, which not only play a crucial role in deterring entrants, but also enhance the sustainability of incumbents' competitive advantage (Porter 1985). While theories of entry barriers have assumed that firms must control their own resources to enter the industry, Open Innovation can increase the number of entry alternatives and thus reduce entry barriers if it allows firms to 'enter' without controlling such resources. We would predict that industries that are more 'open' in their innovation patterns would also be more open to new entrants, but this has yet to be established.

Thus, Open Innovation may allow companies to come up with new business models that do not require that the orchestrating firm physically enters the industry that will be affected. For example, Amazon and E*Trade were able to enter and innovate in retailing and financial services without investing in distribution channels, while pharmaceutical companies ship biotech products by partnering with start-up specialist biotech firms, and Intel shapes and directs competition in the computer industry even though it only makes one component of the overall systems. This suggests that the availability of new business models for commercializing innovation increases the opportunities for industry entry, but how would this be incorporated into existing industry analysis tools? Do we even know enough about the creation and success of new business models to be able to anticipate when a new business model can or cannot be created?

Finally, a key point of the existing entry barrier analysis is to assess the sustainability of profits by industry incumbents. Normally, we would expect that an industry—such as one with more Open Innovation—with lower barriers to entry would also have more competitors, lower barriers to imitation, and thus less sustainable competitive advantage. But is this pattern empirically supported? Or is it, as Grove (1996) argues, that an industry with vertical disintegration allows firms to develop horizontal specializations and economies of scale that makes it more difficult for rivals to challenge the incumbents' advantages?

14.7 National Institutions and Innovation Systems

When considering how nations (and regional economic groupings such as the European Union) differ in their institutional support for innovation, a key body of research is that on National Systems of Innovation (NSI), which emphasizes the importance of both *de jure* and *de facto* institutional structures (e.g. Lundvall 1992; Nelson 1993; Mowery and Nelson 1999). These external relationships among key actors in the system—including enterprises, universities, and government research institutes—are shaped by a set of distinct institutions which jointly and individually contribute to the development and diffusion of new technologies and which provide the framework within which governments form and implement policies to influence the innovation process (Metcalfe 1995). This literature suggests that both formal institutions and factors such as industry structure will affect the flows of innovation between firms.

This prior NSI research does not specifically address how national differences in such institutions impact Open Innovation. The original conception of Open Innovation presented by Chesbrough (2003*a*) was based on research conducted in the US context. To what extent do we see Open Innovation practices in different institutional contexts, such as Europe, Japan, Brazil, or China? Is the distribution of knowledge altered by the institutional characteristics of different countries? Do we see greater or lesser use of external technologies within firms in these countries? Are there more or fewer barriers to the external utilization of spillover technologies within firms? Are intermediate markets in particular industries more or less developed in comparison to the US, and how does this greater (lesser) development influence the adoption of more open (closed) approaches to innovation? Given that the higher education sector differs markedly across the major countries, how do these differences influence the innovation process in those countries? Do we find commonalities in these different institutional settings that spur Open Innovation practices? And, if so, can we make a policy agenda to promote Open Innovation?

As discussed in the chapters of Part II, a key institution affecting Open Innovation is a nation's IP policy. The formal appropriability provided by patent and other laws will affect the incentives provided for creating and using Open Innovations (see Chapter 6). Over time the institutions may change, as with the US decision to allow patenting of software (see Chapter 9). Nations also differ in their IP policies, so the variation of IP policies across time and national boundaries offer a potential quasi-experiment suggesting the relationship between IP policies and the practice of Open Innovation.

Other important innovation policies include government funding of innovation development, particularly the funding of public research. In most cases, such funding is direct through institutions such as the National Science Foundation or European Science Foundation, or through private institutions such as the Hughes Medical Institute or the British Heart Foundation. In the US, Fabrizio and Mowery (forthcoming) show that government funding plays a declining role in basic research, while Fabrizio (Chapter 7) links university efforts to profit from their research to increasing lags in firms' utilization of public science. This implies a declining availability of public subsidies as a source of external innovation, but more research needs to be done (both inside and outside the US) on the impacts of such policy shifts to Open Innovation.

Perhaps more significant variation can be found in the market institutions between economies, particularly in the heterogeneous roles played by firms in an Open Innovation system. Beyond the vertically integrated closed innovation exemplar, Chesbrough (2003*b*) postulates eight possible roles within an Open Innovation system. They include government agencies (acting as innovation benefactors), social movements (innovation missionaries), and capital markets (innovation investors). The other remaining types of actors play different roles in the innovation value chain (Table 14.2). While Chesbrough's inductively derived classification is supported by exemplars, further research is needed to establish whether it is mutually exclusive and exhaustive, and whether there are empirical regularities (in competencies, strategies, or outcomes) between organizations within each category. Research within other innovation systems might suggest other possible roles.

The national, sub- and supranational differences in institutions tie back to the more fundamental question of where firms should locate their innovation activities (cf. Doz et al. 2001; Iansiti and Levien 2004*b*). Open Innovation defines the process by which firms access and utilize external innovations. From other research on knowledge-based geographic clusters (e.g. Audretsch 1998; see Chapter 11) we would expect that Open Innovation processes would also benefit from geographic colocation. While we have some research supporting such locational effects in firms accessing university innovations (Chesbrough 2003*a*; see Chapter 7), and for knowledge spillovers between European firms in the chemicals industry (Verspagen and Schoenmakers 2004) there remains a broad opportunity for research on how geography affects Open Innovation activities between firms.

Table 14.2. Innovation roles for organizations

Phase	Role	Example	Using external innovations	Marketing internal innovations	Business model
Closed innovation	Fully integrated innovators	Merck	*Uses internal innovations*	*Market own innovations*	Vertically integrated
Funding	Innovation investor	Sequoia Capital	n/a	Provides capital	Financial return
	Innovation benefactor	NSF	n/a	Provides capital	*None:* goal is societal welfare
Generating	Innovation explorer	PARC Labs	n/a	Perform basic research	Licensing innovations
	Innovation merchant	Qualcomm	n/a	Perform applied research	Licensing innovations
	Innovation architect	Boeing	Source external components	Design architecture, integrate	Unique role integrating components
	Innovation missionary	Free Software Foundation	n/a	Donating innovations	*None*: goal is social cause
Commercializing	Innovation marketers	Pfizer	Incorporates them in product mix	Markets internal and external innovations	Market internal and external ideas
	Innovation one-stop centers	IBM Global Services	Incorporates them in product mix	Markets internal and external innovations	Market internal and external ideas

Source: Adapted from Chesbrough (2003*b*).

14.8 Research Designs

14.8.1 *Data Sources*

Most of the past research about Open Innovation has been based upon case studies on individual firms or projects in the firm. More extensively, Chesbrough (2002, 2003*a*) offers a comparative case study based on the history of thirty-five technology-based spin-offs from Xerox PARC. Advancing our knowledge about Open Innovation requires that researchers find new and more extensive data sources to illustrate and test different hypotheses derived from Open Innovation, and here we offer a few suggestions.

While cases have established examples of Open Innovation, additional cases could help establish the boundaries of the phenomenon. These cases could focus on particular anomalies and counterfactuals, such as why a large company is not able to generate new businesses from in-house developed technologies. Such cases could show the constraining effect that a firm's business model might have on its ability to exploit the business opportunities stemming from new technologies it developed in its R&D laboratories.

Cases also have a role to play in international theory development, as prior case studies have been biased towards large, US-based manufacturing companies. Fragmentary evidence suggests that Open Innovation is not limited to the US and Canada but is also being practiced in Europe and Asia. The international generalizability of the what and how of Open Innovation could be empirically established through rich comparative case studies from European, Asian, or Latin-American companies, considering the relationships between the differences in their NSIs and their practice (or not) of Open Innovation. Cases could also illuminate the practice of Open Innovation that crosses national borders: for example, is Open Innovation more efficient (or likely) between countries when their cultural or geographic distance is low? How would these cross-cultural differences affect the practice of Open Innovation be implemented across national boundaries but within a multinational corporation?

Surveys are one way to dramatically expand the empirical evidence on Open Innovation. To our knowledge, no large-scale survey has yet been designed to specifically analyze Open Innovation. But some existing large-scale surveys can be used to analyze the Open Innovation phenomenon if the questionnaire asks respondents about the external sources and uses of their technologies. One of the first examples of the latter was done by Laursen and Salter (2005), which analyzed responses from 2,304 manufacturing firms across the UK. The data for their analysis is drawn from the 2001 UK innovation survey, which in turn is based on the core Eurostat Community Innovation Survey (CIS) of innovation (Stockdale 2002; DTI 2003). Laursen and Salter (2005: 25) concluded that 'until more research is undertaken on the evolution of search for innovation over time, the full implications of the possible movement towards "Open Innovation" will not be fully understood.'

However, there are limits to how much can be learned from existing surveys. As with any new causal mechanism, new operationalizations are needed to measure the Open Innovation constructs. One key area is in fact defining innovation, which may include activities and outputs, a narrow definition of market introduction, or a broader definition of the entire process from R&D to product introduction (Ernst 2001; Hagedoorn and Cloodt 2003: 1367). Which of these existing (or new) attributes of innovation are relevant to measuring Open Innovation? Should measures of innovation directly incorporate the business model, or should the business model be separately measured to explain variation in the success at commercializing a potentially valuable innovation?

Cross-sectional surveys would help to establish the prevalence of Open Innovation practices within large populations of firms. More longitudinal surveys might be designed to measure the effect of external shocks (e.g. changes in regulation). Longitudinal research may also provide evidence on the causal relationship between several concepts that are used in the innovation management literature in general and in Open Innovation in particular. Take the example of the interaction between the business model and getting organized for Open Innovation: is a new business model the antecedent or consequence of developing relations with new external innovation partners?

Which unexplored sources to gather empirical evidence for Open Innovation are there? Fabrizio (Chapter 7) offers an example of the possible uses of patent data, but other possibilities remain. Do patent classes offer the possibility to make a distinction between explorative and exploitative patents so that one could analyze whether companies are involved in Open Innovations for exploring new technological areas rather than the exploitation of existing competencies? Can we use the geographical information in patents to explore the geographical networks among the innovation partners and how proximity might play a role in this? Patents also disclose information about inventors. Is it possible to use this data to link the practice of Open Innovation in different firms to the interfirm mobility of star scientists or engineers, as with Rosenkopf and Almeida (2003)?

Since patent citations offer different possibilities to quantify the knowledge flows between different firms, how can we use patent citations in the context of Open Innovation? Patent citation-based variables have proven to be valuable constructs in the analysis of corporate venturing (Schildt et al. 2005) and R&D alliances (Katila 2002). Given that Open Innovation defines innovation as both technical invention and a business model, how can this be captured using patent data?

Still other innovative approaches can be found in less obvious databases. E-mail exchange flows between researchers, engineers, and managers of different companies that are involved in Open Innovation might bring interesting insights about knowledge flows to the surface. Similarly, product catalogs offer very detailed information about the major components contained within a complex product, providing a means to assess the openness of companies in their product development—such as the sources of key components for a cross section of product models. In a similar fashion, one could study the business plans (and corresponding business models) of several ventures that are financed by (corporate) venture capital funds. Does the degree of openness embodied in each plan's business model help to explain why are some ventures financed and why others are not?

Another possibility is to combine several research methods. Data about patents can for instance be linked to a survey questionnaire to ask correspondents about their patent utilization within their company. This could provide us with some insights why in most companies patents lie fallow—neither used internally nor externally licensed—while other companies are more successful in exploiting their patents. In-depth studies—by means of interviews—can then reveal why these companies are doing much better than the industry average.

This is meant to suggest only a few of the possible ways to execute empirical research about Open Innovation. We know that researchers will come up with many more options beyond those offered above.

14.9 Mapping the Limits of Open Innovation

The Open Innovation paradigm was identified by Chesbrough (2003*a*), and extended by subsequent research both in this volume and elsewhere. The prior studies (and their choice of subjects) imply but do not establish that Open Innovation is most suitable for high-tech industries where innovation plays an important role in value creation and capture.

However, such work is far from establishing the extent of the domain of Open Innovation, that is, the boundaries between where theories of Open Innovation apply and where they do not. Identifying the limits of the paradigm is a crucial activity for both the theory of Open Innovation, and putting those theories into practice. Below are some of the areas that might be considered.

How prevalent is Open Innovation today? Where is it most often practiced? Are organizational factors that predict the use of Open Innovation related to a firm's innovation business models or rather the cognitive constraints of its scanning or integration mechanisms? Is Open Innovation really more likely in high-tech firms than low-tech ones, and in small entrants rather than established incumbents?

Where might Open Innovation be feasible in the future? Consistent with Teece (1986), discussions in Part II, Chapter 10 and earlier in this chapter have suggested Open Innovation depends on national institutions such as IP rights as well as de facto appropriability regimes. But is this the only moderator of the feasibility of Open Innovation, or are there other factors at the firm, network, industry, or national level that explain such suitability?

Is there a cycle of Open Innovation? Chesbrough (2003*e*) concludes that product and interfirm modularity increases as a technology matures, only to be reset (and supplanted by vertical integration) after a technological discontinuity. If so, then we would expect that Open Innovation would also be cyclical in a given industry.

Is Open Innovation sustainable over the long haul? As has been discussed earlier in this chapter and this volume, the ability of firms to practice Open Innovation depends on a number of factors, particularly a supply of innovations (for firms building upon innovation inflows) and markets and appropriability for innovation (for firms seeking to profit from their outflows). Competitive advantage, industry structure, national institutions, and other relevant factors change over time, so that once-successful Open Innovation strategies can eventually fail—as West (Chapter 6) argues happened to RCA's strategy of licensing its TV and radio patents. As with other forms of competitive advantage, understanding how advantages from Open Innovation are sustained may be more important than how they are created.

Under what cases does the paradigm not apply? Limits to the paradigm may be seen at either end of the spectrum. For example, Laursen and Salter (2005) found that small high-tech firms are more closed than their larger counterparts, which they explain by the firms' needs to appropriate their ideas; however, small firms may also face disadvantages of scale in searching and contracting for external innovations. At the opposite extreme, vertically integrated firms such as Samsung or Exxon continue without interruption into the twenty-first century, but we do not know how these counterexamples generalize to a critique of the limits of vertical integration—whether in Chesbrough (2003*a*), Langlois (2003*a*), or this volume. Similarly, while the value of vertical integration changed over time, based on the scarcity of technical and managerial skills (Langlois 2003*a*), conversely the value of Open Innovation might also vary over time, such that once-successful models of Open Innovation could become obsolete.

Extending Beyond Innovation. The concepts of Open Innovation are anchored explicitly in the firm's success in creating and capturing value through its business model. Does this paradigm apply to other forms of value creation or capture other than innovation? Hagel and Brown (2005) argue that all value activities (not just innovation generation) are potential candidates to get outsourced, although such normative propositions have not been empirically studied.

14.10 Conclusions

We undertook this book project because we were convinced that Open Innovation offers a new and interesting perspective to academics around the world towards understanding the processes of industrial R&D. Throughout the book, we have highlighted the many areas in which Open Innovation builds upon earlier academic work, and indicated the new contributions and emphases that Open Innovation can bring to that work. In this chapter, we have identified various areas where further research is needed, and some potential sources of data to bring insight into those areas. We acknowledge that there are limits to Open Innovation and that Open Innovation is more readily applicable in some firm or industry settings than in others, but we also recognize that what seem like limits to us may simply be research opportunities to other academic researchers.

Indeed, we do not claim to have all of the answers, or even all of the questions. What we do have is a sense that the context in which innovation occurs is evolving. Industry is changing the processes through which it innovates. Knowledge is flowing more freely and more rapidly between people and firms than ever before even though we have emphasized equally well that rent appropriation from these flows is crucial in understanding Open Innovation. The business of innovation is becoming truly global in its character, and diverse countries bring new pools of human capital and talent into play.

Accordingly, we academics must update our own understanding of the innovation process, building upon the foundations of excellent research that precedes us, and adding to that foundation when necessary. We sincerely welcome the contributions of other academics who wish to explore these areas, for we take the task of understanding innovation quite seriously. Innovation offers society the promise of increased growth and productivity. Through these, it further offers the prospect of a high and advancing standard of living. It even offers the hope of ameliorating terrible diseases and extending the number of productive years of one's life. If Open Innovation can speed up or facilitate these innovation dynamics, understanding it better will be well worth the effort.

Because a printed book necessarily becomes obsolete at some point, we have also decided to create an online website (http://www.openinnovation.net), where interested readers can find more recent updated information on research in this area (including a comprehensive bibliography of recent research). Through the site, through meetings at the Academy of Management and other research conferences, through email and phone exchanges, and through personal networks, we hope to build an academic community around Open Innovation. Please consider this an invitation to join us!

Notes

1. We appreciate the valuable feedback provided to earlier versions by Jens Frøslev Christensen, Myriam Cloodt, Kwanghui Lim, Wilfred Schoenmakers, and Simon Wakeman.
2. The Arrow Information Paradox refers to the seller's need to disclose information about the technology to the buyer, to entice the buyer into acquiring the technology. The buyer needs to know exactly what the technology is, and what it can do. However, if the seller fully discloses all this information to the buyer during the negotiation, the buyer will have effectively acquired the technology without having to pay anything for it.

References

Aaker, D. A. and George, S. D. (1986). 'The Perils of High-Growth Markets', *Strategic Management Journal*, 7(5): 409–21.

ABA (2003). 'Resources Related to Antitrust and Standard Setting'. American Bar Association Intellectual Property Committee.

Abernathy, W. J. and Utterback, J. M. (1978). 'Patterns of Industrial Innovation', *Technology Review*, 80: 40–7.

—— and Clark, K. B. (1985). 'Innovation: Mapping the Winds of Creative Destruction', *Research Policy*, 14(1): 3–22.

Abrahamson, E. (1996). 'Management Fashion, Management Fads', *Academy of Management Review*, 22(1): 254–85.

Afuah, A. (2004). *Business Models: A Strategic Management Approach*. New York: McGraw-Hill/Irwin.

Agrawal, A. and Henderson, R. (2002). 'Putting Patents in Context: Exploring Knowledge Transfer from MIT', *Management Science*, 48(1): 44–60.

Ahuja, G. (2000). 'Collaboration Networks, Structural Holes, and Innovation: A Longitudinal Study', *Administrative Science Quarterly*, 45(3): 425–55.

—— and Lampert, C. M. (2001). 'Entrepreneurship in the Large Corporation: A Longitudinal Study of How Established Firms Create Breakthrough Inventions', *Strategic Management Journal*, 22(6–7): 521–43.

Akerlof, G. A. (1970). 'The Market for Lemons: Quality Uncertainty and the Market Mechanism', *Quarterly Journal of Economics*, 84(3): 488–500.

—— Arrow, K. J., Bresnahan, T. F., Buchanan, J. M., Coase, R. H., Cohen, L. R., Friedman, M., Green, J. R., Hahn, R. W., Hazlett, T. W., Hemphill, C. S., Litan, R. E., Noll, R. G., Schmalensee, R., Shavell, S., Varian, H. R., and Zeckhauser, R. J. (2002). 'The Copyright Term Extension Act of 1998: An Economic Analysis', Brief 02-1, AEI-Brookings Joint Center for Regulatory Studies.

Almeida, P. and Kogut, B. (1999). 'Localization of Knowledge and the Mobility of Engineers in Regional Networks', *Management Science*, 45(7): 905–17.

—— Song, J., and Grant, R. (2002). 'Are Firms Superior to Alliances and Markets? An Empirical Test of Cross-border Knowledge Building', *Organization Science*, 13(2): 147–61.

—— Dokko, G., and Rosenkopf, L. (2003). 'Startup Size and the Mechanisms of External Learning: Increasing Opportunity and Decreasing Ability?', *Research Policy*, 32(2): 301–15.

Amit, R. and Zott, C. (2001). 'Value Creation in e-Business', *Strategic Management Journal*, 22(6–7): 493–520.

Anton, J. J. and Yao, D. A. (2004). 'Little Patents and Big Secrets: Managing Intellectual Property', *Rand Journal of Economics*, 35(1): 1–22.

References

Argyres, N. (1995). 'Technology Strategy, Governance Structure and Interdivisional Coordination', *Journal of Economic Behavior and Organization*, 28(3): 337–58.

—— and Silverman, B. (2004). 'R&D, Organization Structure, and the Development of Corporate Technological Knowledge', *Strategic Management Review*, 25(8–9): 929–58.

Arora, A. (2002). 'Licensing Tacit Knowledge: Intellectual Property Rights and the Market for Know-How', *Economic of Innovation and New Technology*, 4(1): 41–59.

—— Fosfuri, A., and Gambardella, A. (2001*a*). *Markets for Technology: The Economics of Innovation and Corporate Strategy*. Cambridge, MA: MIT Press.

—— —— —— (2001*b*). 'Markets for Technology and Their Implications for Corporate Strategy', *Industrial and Corporate Change*, 10(2): 419–51.

Arrow, K. J. (1962). 'Economic Welfare and the Allocation of Resources for Invention: Economics and Social Factors', in R. R. Nelson (ed.), *The Rate and Direction of Inventive Activity*. Princeton, NJ: Princeton University Press, pp. 609–25.

Arthur, W. B. (1996). 'Increasing Returns and the New World of Business', *Harvard Business Review*, 74(4): 100–9.

Ashkenas, R., Ulrich, D., Jick, T., and Kerr, S. (1995). *The Boundaryless Organization: Breaking the Chains of Prganizational Structure*. San Francisco, CA: Jossey-Bass.

Audretsch, D. B. (1998). 'Agglomeration and the Location of Innovation Activity', *Oxford Review of Economic Policy*, 14(2): 18–29.

—— Lehmann, E. E., and Warning, S. (2004). 'University Spillovers: Does the Kind of Science Matter?', *Industry and Innovation*, 11(3): 193–206.

Axelrod, R., Mitchell, W., Thomas, R., Bennett, D., and Bruderer, E. (1995). 'Coalition-Formation in Standard-Setting Alliances', *Management Science*, 41(9): 1493–508.

Baldini, N., Grimaldi, R., and Sobrero, M. (2005). 'Motivations and Incentives for Patenting within Universities: A Survey of Italian Inventors', Paper presented at Academy of Management conference, Technology and Innovation Division, August 9, Honolulu, Hawaii, US.

Bamford, J. and Ernst, D. (2002). 'Managing an Alliance Portfolio', *The McKinsey Quarterly*, 3: 29–39.

Bamford, J., Gomes-Casseres, B., and Robinson, M. (2003). *Mastering Alliance Strategy: A Comprehensive Guide to Design Management and Organization*. San Francisco, CA: Jossey-Bass John Wiley.

Baptista, R. and Swann, P. (1998). 'Do Firms in Clusters Innovate More?' *Research Policy*, 27(5): 525–40.

Barney, J. B. (1986). 'Strategic Factor Markets: Expectations, Luck and Business Strategy', *Management Science*, 32(10): 1231–41.

—— (1991). 'Firm Resources and Sustained Competitive Advantage', *Journal of Management*, 17(1): 99–120.

—— (1999). 'How Firm's Capabilities Affect Boundary Decisions', *Sloan Management Review*, 40(3): 137–45.

Baskin, E., Krechmer, K., and Sherif, M. H. (1998). 'The Six Dimensions of Standards: Contribution towards a Theory of Standardization', in L. L. R. Mason and T. Khalil (eds.), *Management of Technology, Sustainable Development and Eco-Efficiency*. Oxford: Elsevier, p. 53.

Baum, J. A. C., Calabrese, T., and Silverman, B. S. (2000). 'Don't Go It Alone: Alliance Network Composition and Startups' Performance in Canadian Biotechnology', *Strategic Management Journal*, 21(3): 267–94.

Beckman, C. M. and Haunschild, P. R. (2002). 'Network Learning: The Effects of Partners' Heterogeneity of Experience on Corporate Acquisitions', *Administrative Science Quarterly*, 47(1): 92–124.

Bekkers, R. (2001). *Mobile Telecommunications Standards: GSM, UMTS, TETRA, and ERMES*. Boston, MA: Artech House.

Bekkers, R., Duysters, G., and Verspagen, B. (2002). 'Intellectual Property Rights, Strategic Technology Agreements and Market Structure: The Case of GSM', *Research Policy*, 31(7): 1141–61.

Benjamin, B. and Podolny, J. (1999). 'Status, Quality, and Social Order in the California Wine Industry', *Administrative Science Quarterly*, 44(3): 563–89.

Bercovitz, J. and Feldman, M. (2003). 'Technology Transfer and the Academic Department: Who Participates and Why?' DRUID Summer Conference, June 2003.

Berners-Lee, T., Hendler, J., and Lassila, O. (2001). 'The Semantic Web', *The Scientific American*, 284(5): 34–43.

Besen, S. M. and Raskind, L. J. (1991). 'An Introduction to the Law and Economics of Intellectual Property', *Journal of Economic Perspectives*, 5(1): 3–27.

Bessen, J. and Hunt, R. M. (2004). 'An Empirical Look at Software Patents, Federal Reserve Bank of Philadelphia', Working Paper 03–17, http://ideas.repec.org/p/fip/fedpwp/03-17.html

Bettis, R. (1998). 'Commentary in "Redefining Industry Structure for the Information Age" by Jeffrey Sampler', *Strategic Management Journal*, 19(4): 357–61.

Betz, F. (1993). *Strategic Technology Management*. New York: McGraw-Hill.

Bjornson, B. (1998). 'Capital Market Values of Agricultural Biotechnology Firms: How High and Why?' *Agbioforum*, 1(2): 69–73.

Blumenthal, D., Gluck, M., Louis, K. S., Stoto, M. A., and Wise, D. (1986). 'University–Industry Research Relationships in Biotechnology: Implications for the University', *Science*, 232(4756): 1361–6.

Bossidy, L., Charan, R., and Burck, C. (2002). *Execution: The Discipline of Getting Things Done*. New York: Crown Books.

Boutellier, R., Gassmann, O., and von Zedtwitz, M. (2000). *Managing Global Innovation: uncovering the Secrets of Future Competitiveness*. Berlin: Springer.

Bower, J. L. (1970). *Managing the Resource Allocation Process*. Boston, MA: Harvard Business School Press.

Bozeman, B. (2000). 'Technology Transfer and Public Policy: A Review of Research and Theory', *Research Policy*, 29(4–5): 627–55.

Brandenburger, A. M. and Nalebuff, B. J. (1996). *Co-opetition*. New York: Doubleday.

Brandenburger, A. M. and Stuart, H. W. (1996). 'Value-based Business Strategy', *Journal of Economics and Management Strategy*, 5(1): 5–24.

Bresnahan, T. and Greenstein, S. (1996). 'The Competitive Crash in Large Scale Commercial Computing', in R. Landau, T. Taylor, and G. Wright (eds.), *The Mosaic of Economic Growth*. Stanford, CA: Stanford University Press.

—— —— (1999). 'Technological Competition and the Structure of the Computer Industry', *Journal of Industrial Economics*, 47(1): 1–40.

—— and Malerba, F. (1999). 'Industrial Dynamics and the Evolution of Firms' and Nations' Competitive Capabilities in the World Computer Industry', in D. C. Mowery

and R. R. Nelson (eds.), *Sources of Industrial Leadership: Studies of Seven Industries*. New York: Cambridge University Press, pp. 79–132.
Bresnahan, T. and Yin, P.-L. (2004). Standard Setting in Browsers: Technology and (Real World) Incomplete Information. Standards and Public Policy conference, Federal Reserve Bank of Chicago, May 13–14.
Brim, S. (2004). 'Guidelines for Working Groups on Intellectual Property Issues', Internet Engineering Task Force RFC 3699, http://www.ietf.org/rfc/rfc3669
Brinkley, J. (1997). *Defining Vision: The Battle for The Future of Television*. New York: Harcourt Brace.
Brockmeier, J. (2003). 'Is Open Source Apple's Salvation?', NewsFactor Network, April 21, 2003, URL: http://www.newsfactor.com/perl/story/21318.html
Brody, S. (2001). 'Interview: The Eclipse code donation', IBM developerWorks, November 1, 2001, URL: http://www.ibm.com/developerworks/linux/library/l-erick.html
Brown, J. S. and Duguid, P. (2000). 'Mysteries of the Region: Knowledge Dynamics in Silicon Valley', in C.-M. Lee, W. Miller, M. G. Hancock, and H. Rowen (eds.), *The Silicon Valley Edge*. Stanford, CA: Stanford University Press, pp. 16–39.
Brown, S. L. and Eisenhardt, K. M. (1997). 'The Art of Continuous Change: Linking Complexity Theory and Time-paced Evolution in Relentlessly Shifting Organizations', *Administrative Science Quarterly*, 42(1): 1–34.
—— —— (1998). *Competing on the Edge: Strategy as Structured Chaos*. Boston, MA: Harvard Business School Press.
Brusoni, S., Prencipe A., and Pavitt, K. (2001). 'Knowledge Specialisation, Organizational Coupling and the Boundaries of the Firm: Why Firms Know More Than They Make?' *Administrative Science Quarterly*, 46(4): 597–621.
Brynjolfsson, E. and Urban, G. (eds.) (2001). *Strategies for e-Business Success*. San Fransisco, CA: Jossey-Bass.
Bulow, J. and Klemperer, P. (1999). 'The Generalized War of Attrition.', *American Economic Review*, 89(1): 175–89.
Bunker Whittington, K., Owen-Smith, J., and Powell, W. W. (2004). 'Spillovers versus Embeddedness: The Contingent Effects of Propinquity and Social Structure', Working Paper, Stanford University.
Burgelman, R. A. (1983*a*). 'A Model of the Interaction of Strategic Behavior, Corporate Context, and the Concept of Strategy', *Academy of Management Review*, 8(1): 61–70.
—— (1983*b*). 'A Process Model of Internal Corporate Venturing in the Diversified Major Firm', *Administrative Science Quarterly*, 28(2): 223–44.
—— (1994). 'Fading Memories: A Process Theory of Strategic Business Exit in Dynamic Environments', *Administrative Science Quarterly*, 39(1): 24–56.
Burke, A. E. (1996). 'How Effective are International Copyright Conventions in the Music Industry?' *Journal of Cultural Economics*, 20(1): 51–66.
Burt, R. (1992). *Structural Holes: The Social Structure of Competition*. Cambridge, MA: Harvard University Press.
Cargill, C. F. (1989). *Information Technology Standardization: Theory, Process, and Organizations*. Bedford, MA: Digital Press.
—— (1997). *Open Systems Standardization: A Business Approach*. Upper Saddle River, NJ: Prentice-Hall PTR.

—— (2001). *Evolutionary Pressures in Standardization: Considerations on ANSI's National Standards Strategy.* Testimony Before the U.S. House of Representatives Science Committee, September 13, URL: http://www.sun.com/software/standards/Testimony-Final-reformatted.pdf

—— and Bolin, S. (2004). 'The Changing Nature of Standards Organizations: Walden Pond has been Drained'. Standards and Public Policy conference, Federal Reserve Bank of Chicago, May 13–14.

Carpenter, J. and Gianessi, L. (1999). 'Herbicide Tolerant Soybeans: Why Growers Are Adopting Roundup Ready Varieties', *Agbioforum,* 2(2): 65–72.

Castilla, E. J., Hwang, H., Granovetter, E., and Granovetter, M. (2000). 'Social Networks in Silicon Valley', in C.-M. Lee, W. Miller, M. G. Hancock, and H. Rowen (eds.), *The Silicon Valley Edge.* Stanford, CA: Stanford University Press, pp. 218–47.

Caves, R. and Porter, M. E. (1977). 'From Entry Barriers to Mobility Barriers: Conjectural Decisions and Contrived Deterrence to New Competition', *Quarterly Journal of Economics,* 91(2): 241–61.

CBO (1998). 'How Increased Competition from Generic Drugs has Affected Prices and Returns in the Pharmaceutical Industry', Washington, DC: Congressional Budget Office.

Cesaroni, F. and Piccaluga, A. (2002). 'Patenting Activity of European Universities. Relevant? Growing? Useful?' SPRU NPRnet Conference, 'Rethinking Science Policy: Analytical Framework for Evidence-Based Policy', March 21–23, 2002, SPRU, University of Sussex, Brighton.

Chandler, A. D., Jr. (1962). *Strategy and Structure: Chapters in the History of American Industrial Enterprise.* Cambridge, MA: MIT Press.

—— (1977). *The Visible Hand: The Managerial Revolution in American Business.* Cambridge, MA: Belknap Press.

—— (1990). *Scale and Scope: The Dynamics of Industrial Capitalism.* Cambridge, MA: Belknap Press.

—— (1991). 'The Functions of the HQ Unit in the Multibusiness Firm'. *Strategic Management Journal,* 12(Winter): 31–50.

—— (2001). *Inventing the Electronic Century. The Epic Story of the Consumer Electronics and Computer Industries.* New York: Free Press.

Chesbrough, H. (2002*a*). 'Making Sense of Corporate Venture Capital', *Harvard Business Review,* 80(3): March: 90–9.

—— (2002*b*). 'Graceful Exits and Missed Opportunities: Xerox's Management of its Technology Spinoff Organizations', *Business History Review,* 76(4): 803–38.

—— (2003*a*). *Open Innovation: The New Imperative for Creating and Profiting from Technology,* Boston, MA: Harvard Business School Press.

—— (2003*b*). 'The Era of Open Innovation', *Sloan Management Review,* 44(3): 35–41.

—— (2003*c*). 'Open Innovation: How Companies Actually Do It', *Harvard Business Review,* 81(7): 12–14.

—— (2003*d*). 'Open Platform Innovation: Creating Value from Internal and External Innovation', *Intel Technical Journal,* 7(3): 5–9, URL: http://www.intel.com/technology/itj/2003/volume07issue03/art01_open/p01_abstract.htm

Chesbrough, H. (2003*e*). *'Towards a Dynamics of Modularity: A Cyclical Model of Technical Advance'*, in A. Prencipe, A. Davies, and M. Hobday (eds.), *The Business of Systems Integration*. Oxford: Oxford University Press, pp. 174–98.

—— (2004). 'Managing Open Innovation: Chess and Poker', *Research & Technology Management*, 47(1): 13–16.

—— and Kusonoki, K. (2001). 'The Modularity Trap: Innovation, Technology Phase Shifts and the Resulting Limits of Virtual Organizations', in I. Nonaka and D. Teece (eds.), *Managing Industrial Knowledge*. London: Sage.

—— and Rosenbloom, R. S. (2002). 'The Role of the Business Model in Capturing Value from Innovation: Evidence from Xerox Corporation's Technology Spin-off Companies', *Industrial and Corporate Change*, 11(3): 529–55.

Chesbrough, H. W. and Teece, D. J. (1996). 'When Is Virtual Virtuous? Organizing for Innovation', *Harvard Business Review*, 74(1): 65–73.

Chiao, B., Lerner, J., and Tirole, J. (2005). 'The Rules of Standard Setting Organizations: An Empirical Analysis'. National Bureau of Economic Research, Working Paper No. 11156.

Child, S. (2003). 'Open Source and Intellectual Property', *MyITAdvisor*. National Computing Center, March 22.

Christensen, J. F. (1995). 'Asset Profiles for Technological Innovation', *Research Policy*, 24(5): 727–45.

—— (1996). 'Innovative Assets and Inter-Asset Specificity: A Resource-Based Approach to Innovation', *Economics of Innovation and New Technology*, 4(3): 193–209.

—— (2000). 'Building Innovative Assets and Dynamic Coherence in Multi-Technology Companies', in N. J. Foss and P. L. Robertson (eds.), *Resources, Technology and Strategy. Explorations in the Ressource-Based Perspective*. London and New York: Routledge.

—— . (2002). 'Incongruities as a Source of Organizational Renewal in Corporate Management of R&D'. *Reseach Policy*, 31(8–9): 1317–32.

—— and Foss, N. J. (1997). 'Dynamic Corporate Coherence and Competence-Based Competition: Theoretical Foundations and Strategic Implications', in A. Heene and R. Sanchez (eds.), *Competence-Based Strategic Management*, Oxford: Elsevier.

—— Olesen, M. H., and Kjær. J. S. (2005). 'The Industrial Dynamics of Open Innovation: Evidence from the Transformation of Consumer Electronics', *Research Policy*, 34(10): 1533–49.

Christensen, C. M. (1997). *The Innovator's Dilemma: When New Technologies Cause Great Firms to Fail*. Boston, MA: Harvard Business School Press.

—— and Bower, J. L. (1996). 'Customer Power, Strategic Investment, and the Failure of Leading Firms', *Strategic Management Journal*, 17(3): 197–218.

—— and Raynor M. E. (2003). *The Innovator's Solution: Creating and Sustaining Successful Growth*, Boston: Harvard Business School Press.

—— and Rosenbloom, R. S. (1995). 'Explaining the Attacker's Advantage: Technological Paradigms, Organizational Dynamics, and the Value Network', *Research Policy*, 24(2): 233–57.

Clark, K. and Wheelwright, S. (1992). *Revolutionizing Product Development*. New York: Free Press.

Coase, R. (1937). 'The Nature of the Firm', *Economica*, 4: 386–405.

Cockburn, I. and Henderson, R. (1998). 'Absorptive Capacity, Coauthoring Behavior, and the Organization of Research in Drug Discovery', *The Journal of Industrial Economics*, 46(2): 157–82.

—— and Henderson, R. (2000). 'Publicly Funded Science and the Productivity of the Pharmaceutical Industry', in Jaffe, Lerner, and Stern (eds.), *Innovation Policy and the Economy*, vol. 1. Cambridge, MA: MIT Press, pp. 1–34.

Cohen, S. S. and Fields, G. (2000). 'Social Capital and Capital Gains: An Examination of Social Capital in Silicon Valley', in M. Kenney (ed.), *Understanding Silicon Valley: Anatomy of an Entrepreneurial Region*. Stanford, CA: Stanford University Press, pp. 190–217.

Cohen, W. M. and Levinthal, D. A. (1989). 'Innovation and Learning: The Two Faces of R&D', *The Economic Journal*, 99(397):569–96.

—— and Levinthal, D. A. (1990). 'Absorptive Capacity: A New Perspective on Learning and Innovation', *Administrative Science Quarterly*, 35(1): 128–52.

Cohen, W., Nelson, R., and Walsh, J. P. (2000). Protecting Their Intellectual Assets: Appropriability Conditions and Why US Manufacturing Firms Patent (Or Not). National Bureau of Economic Research Working Paper No. 7552.

Cohen, W. M., Nelson, R. R., and Walsh, J. P. (2002). 'Links and Impacts: The Influence of Public Research on Industrial R&D', *Management Science*, 48(1): 1–23.

Coleman, J. S. (1988). 'Social Capital in the Creation of Human Capital', *American Journal of Sociology*, 94: S95–S120.

Collins, S. and Wakoh, H. (2000). 'Universities and Technology Transfer in Japan: Recent Reforms in Historical Perspective', *Journal of Technology Transfer*, 25: 213–22.

Colyvas, J., Crow, M., Gelijns, A., Richard, R. M., Nelson, R., Rosenberg, N., and Sampat, B. N. (2002). 'How Do University Inventions Get Into Practice?', *Management Science*, 48(1): 61–72.

Conant, J. (2002). *Tuxedo Park*. New York: Simon & Schuster.

Cooke, P. (1992). Regional Innovation Systems: Competitive Regulation in the New Europe, *Geoforum*, 23: 365–82.

—— (1998). *Introduction: Origins of the Concept*, in H. Braczyk, P. Cooke, and M. Heidenreich (eds.), *Regional Innovation Systems*. London: UCL Press.

—— (2004*a*). 'Introduction: Origins of the Concept', in H. Braczyk, P. Cooke, and M. Heidenreich (eds.), *Regional Innovation Systems*, 2nd edn. London: UCL Press, pp. 2–25.

—— (2004*b*). 'Knowledge Capabilities, Open Innovation and Cluster Competitiveness', Paper presented at The Competitiveness Institute Conference 'Building Innovative Clusters for Competitive Advantage', Ottawa, September 27–October 1, 2004.

—— (2005). 'Regional Knowledge Capabilities and Open Innovation: Regional Innovation Systems and Clusters in the Asymmetric Knowledge Economy', in S. Breschi and F. Malerba (eds.), *Clusters, Networks & Innovation*. Oxford: Oxford University Press.

Coombs, R. and Richards, A. (1993). 'Strategic Control of Technology in Diversified Companies with Decentralized R&D', *Technology Analysis and Strategic Management*, 5: 385–96.

Cooper, G., Edgett, S. J., and Kleinschemidt, E. J. (2002). Optimizing the Stage-Gate Process: What Best Practice Companies Do. *Research and Technology Management,* 45(6): 21–7.

'Corporate Overview', Open Source Development Labs, April 1, 2004, URL: http://www.osdl.org/about_osdl/OSDL_overview_website.pdf

Cowan, R., David, P. A, and Foray, D. (2000). 'The Explicit Economics of Knowledge Codification and Tacitness', *Industrial and Corporate Change,* 9(2): 211–53.

Cusumano, M. A. (2004). *The Business of Software.* New York: Free Press.

—— and Yoffie, D. B. (1998). *Competing on Internet Time: Lessons from Netscape and Its Battle with Microsoft.* New York: Free Press.

Cusumano, M. A., Mylonadis, Y., and Rosenbloom, R. S. (1992). 'Strategic Maneuvering and Mass-market Dynamics: The Triumph of VHS over Beta', *Business History Review,* 66(1): 51–94.

Cyert, R. M. and March, J. G. (1963). *A Behavioral Theory of the Firm.* Englewood Cliffs, NJ: Prentice-Hall.

Dasgupta, P. and David, P. (1994). 'Toward a New Economics of Science', *Research Policy,* 23(5): 487–521.

David, P. A. (1998). 'Common Agency Contracting and the Emergence of "Open Science" Institutions', *The American Economic Review,* 88(2): 15–21.

—— (2002). 'The Economic Logic of "Open Science" and the Balance between Private Property Rights and the Public Domain in Scientific Data and Information: A Primer', Stanford Institution for Economic Policy Research, Discussion Paper No. 02–30, Stanford University, URL: http://siepr.stanford.edu/papers/pdf/02-30.pdf

—— (2005). 'Can "Open Science" be Protected from the Evolving Regime of IPR Protections?', Economics Working Paper Archive, Working Paper 0502010, Washington University in St. Louis, URL: http://ideas.repec.org/p/wpa/wuwpio/0502010.html

—— and Greenstein, S. (1990). 'The Economics of Compatibility Standards: An Introduction to Recent Research', *Economics of Innovation and New Technology,* 1: 3–42.

—— Mowery, D., and Steinmueller, E. E. (1992). 'Analysing the Economic Payoffs from Basic Research', *Economics of Innovation and New Technology,* 2: 73–90.

—— Hall, B. H., and Toole, A. A. (2000). 'Is Public R&D a Complement or Substitute for Private R&D? A Review of the Econometric Evidence', *Research Policy,* 29(4–5): 497–529.

Davis, J. and Harrison, S. (2001). *Edison in the Boardroom.* New York: John Wiley.

Day, D. (1994). 'Raising Radicals: Different Processes for Championing Innovative Corporate Ventures', *Organization Science,* 5(2): 149–72.

De Laat, P. B. (1999). 'Systemic Innovation and the Virtues of Going Virtual: The Case of the Digital Video Disc', *Technology Analysis & Strategic Management,* 11(2): 159–80.

Debackere, K. and Veugelers, R. (2005). 'The Role of Academic Technology Transfer Organizations in Improving Industry Science Links', *Research Policy,* 34(3): 321–42.

Dedrick, J. and Kraemer, K. L. (1998). *Asia's Computer Challenge: Threat or Opportunity for the United States and the World*? New York: Oxford University Press.

—— and West, J. (2004). 'An Exploratory Study into Open Source Platform Adoption', Proceedings of the 37th Hawaii International Conference on Systems Sciences, p. 80265b.

Deng, Z., Baruch L., and Narin, F. (1999). 'Science and Technology as Predictors of Stock Performance', *Financial Analysts Journal*, 55(3): 20–32.

Dertouzos, M. L., Lester, R. K., and Solow, R. M., (1989). *Made in America: Regaining the Productivity Edge*. Cambridge, MA: MIT Press.

Dhebar, A. (1995). 'Complementarity, Compatibility, and Product Change: Breaking with the Past?' *Journal of Product Innovation Management*, 12(2): 136–52.

DiBona, C., Sam, O., and Mark, S. (eds.) (1999). *Open Sources: Voices from the Open Source Revolution*. Sebastopol, CA: O'Reilly.

Dierickx, I. and Cool, K. (1989). 'Asset Stock Accumulation and Sustainability of Competitive Advantage', *Management Science*, 35(12): 1504–11.

DiMasi J. A. (2001). 'Risks in New Drug Development: Approval Success Rates for Investigational Drugs', *Clinical Pharmacology & Therapeutics*, 69(5): 297–307.

Dokko, G. and Rosenkopf, L. (2003). 'Job Mobility of Technical Professionals and Firm Centrality in Wireless Standards Committees', Proceedings of the Academy of Management Conference, Seattle, pp. A1–A6.

Dougherty, D. (1992). 'Interpretive Barriers to Successful Product Innovation in Large Firms', *Organization Science*, 3(2): 179–202.

—— (1995). 'Managing Your Core Incompetencies for Corporate Venturing', *Entrepreneurship: Theory and Practice*, 19(3): 113–35.

—— and Heller, T. (1994). 'The Illegitimacy of Successful Product Innovations in Established Firms', *Organization Science*, 5(2): 200–18.

Doz, Y. and Hamel, G. (1998). *Alliance Advantage*, Cambridge, MA: Harvard Business School Press.

—— Santos, J., and Williamson, P. (2001). *From Global To Meta-National: How Companies Win in the Knowledge Economy*. Boston, MA: Harvard Business School Press.

DTI (2003). *3rd Community Innovation Survey*, London: Department of Trade and Industry.

Dushnitsky, G. (2004). 'Limitations to Inter-organizational Knowledge Acquisition: The Paradox of Corporate Venture Capital', *Proceedings of the 2004 Academy of Management Conference*, New Orleans, LA, pp. C1–C6.

—— and Lenox, M. (2005). 'When Do Incumbents Learn from Entrepreneurial Ventures? Corporate Venture Capital and Investing firm Innovation Rates', *Research Policy*, 34(5): 615–39.

Duysters, G. and Vanhaverbeke, W. (1996). 'Strategic Interactions in DRAM and RISC Technology: A Network Approach', *Scandinavian Journal of Management*, 12(4): 437–61.

Dyer, J. (1996). 'Does Governance Matter? Keiretsu Alliances and Asset Specificity as Sources of Japanese Competitive Advantage', *Organization Science*, 7(6): 649–66.

—— (1997). 'Effective Interfirm Collaboration. How Firms Minimize Transaction Cost and Maximize Transaction Value', *Strategic Management Journal*, 18(7): 553–56.

—— and Nobeoka, K. (2000). 'Creating and Managing a High-performance Knowledge-sharing Network: The Toyota Case', *Strategic Management Journal*, 21(3): 345–67.

—— and Singh, H. (1998). 'The Relational View: Cooperative Strategy and Sources of Interorganizational Competitive Advantage', *Academy of Management Review*, 23(4): 660–79.

—— Kale, P. and Singh, H. (2004). 'When to Ally & When to Acquire', *Harvard Business Review*, 82(7–8): 109–15.

'Eclipse Forms Independent Organization', press release, February 2, 2004, URL: http://www.eclipse.org/org/index.html

Eisenhardt, K. M. (1989). 'Building Theories from Case Study Research', *Academy of Management Review,* 14(4): 532–50.

—— and Martin, J. (2000). 'Dynamic Capabilities: What Are They?' *Strategic Management Journal,* 21(10–11): 1105–21.

—— and Schoonhoven, C. (1996). 'Resource-based View of Strategic Alliance Formation: Strategic and Social Effects in Entrepreneurial Firms', *Organisation Science,* 7(2): 136–50.

Enriquez, J. and Goldberg, R. (2000). 'Transforming Life, Transforming Business: the Life Science Revolution', *Harvard Business Review,* 78(2): 67–78.

Erickson, I. (2005). 'Network Dynamics and the Problem of Trust: A Structurationist Approach to Change in Interorganizational Collaboration', Stanford University, Working Paper.

Ernst, D. (2003). 'Digital Information Systems and Global Flagship Networks: How Mobile Is Knowledge in the Global Network Economy?', in J. F. Christensen and P. Maskell (eds.), *The Industrial Dynamics of the New Digital Economy.* Cheltenham: Edward Elgar.

Ernst, H. (2001). 'Patent Applications and Subsequent Changes of Performance: Evidence from Time-series Cross-section Analyses on the Firm Level', *Research Policy,* 30(1): 143–57.

European Commission (1997). 'Promoting Innovation Through Patents: Green Paper on the Community Patent and the Patent System in Europe', June, URL: http://europa.eu.int/comm/internal_market/en/indprop/patent/paten.pdf

Fabrizio, K. R. (2005*a*). 'Opening the Dam or Building Channels: University Patenting and the Use of Public Science in Industrial Innovation', Working Paper.

—— (2005*b*). 'Absorptive Capacity and Innovation: Evidence from Pharmaceutical and Biotechnology Firms', Working Paper, Goizueta Business School, Emory University.

—— and Mowery, D. C. (forthcoming) 'Defense-Related R&D and the Growth of the Postwar Information Technology Industrial Complex in the United States', *Revue d'économie industrielle.*

—— and Mowery, D. C. (2005). 'The Federal Role in Financing Major Innovations: Information Technology During the Postwar Period', in N. Lamoreaux and K. Sokoloff (eds.), *Financing Innovation.*

Farrell, J. (1996). 'Choosing the Rules for Formal Standardization', Working Paper, U.C. Berkeley Department of Economics.

—— and Saloner, G. (1988). 'Coordination through Committees and Markets', *Rand Journal of Economics,* 19(2): 235–52.

—— and Shapiro, C. (1992). 'Standard Setting in High-Definition Television', *Brookings Papers on Economic Activity,* 1–93.

—— and Shapiro, C. (2004). 'Intellectual Property, Competition, and Information Technology', Working Paper CPC04-45, Competition Policy Center, University of California, Berkeley. URL: http://repositories.cdlib.org/iber/cpc/CPC04-045

—— Choi, J. P., Edlin, A. S., Greenstein, S., Hall, B. H. and Saloner, G. (2004). Brief Amicus Curiae of Economics Professors and Scholars in the Matter of Rambus, Inc.: U.S. Federal Trade Commission Docket No. 9302.

Fast, N. D. (1978). 'New Venture Departments: Organizing for Innovation', *Industrial Marketing Management*, 7: 77–88.

Feldman, M. (2003). 'The Locational Dynamics of the US Biotech Industry: Knowledge Externalities and the Anchor Hypothesis', *Industry and Innovation*, 10(3): 311–29.

Feyerabend, P. K. (1981). *Problems of Empiricism*. New York: Cambridge University Press.

Fine, C. (1998). *Clockspeed: Winning Industry Control in the Age of Temporary Advantage*. London: Little Brown and Co.

Fleming, L. and Waguespack, D. (2005). 'Penguins, Camels, and Other Birds of a Feather: Brokerage, Boundary Spanning, and Leadership in Open Innovation Communities'. Social Science Research Network, http://ssrn.com/abstract=710641

Foray, D. (1994). 'Users, Standards and the Economics of Coalition and Committees', *Information Economics and Policy*, 6(3–4): 269–93.

Foss, N. and Pedersen, T. (2002). 'Transferring Knowledge in MNCs: The Roles of Sources of Subsidiary Knowledge and Organizational Context', *Journal of International Management*, 8: 1–19.

Franke, N. and von Hippel, E. (2003). 'Satisfying Heterogeneous User Needs via Innovation Toolkits: The Case of Apache Security Software', *Research Policy*, 32(7): 1199–215.

Free Software Foundation (1991). 'GNU General Public License Version 2', http://www.fsf.org/licensing/licenses/gpl.txt

Friedman, T. (2005). *The World is Flat: A Brief History of the 21st Century*. New York: Farrar, Straus and Giroux.

Fruin, W. M. (2006). 'Business Groups and Interfirm Networks', in G. Jones and J. Zeitland (eds.), *Oxford Handbook of Business History*. Oxford: Oxford University Press.

FTC (2002). 'Welcome and Overview of Hearings', Federal Trade Commission, http://www.ftc.gov/opp/intellect/detailsandparticipants.htm#February%2028

FTC (2003). *In the Matter of Rambus Incorporated*. Federal Trade Commission, Docket No. 9302.

Funk, J. L. (2003). 'Standards, Dominant Designs and Preferential Acquisition of Complementary Assets Through Slight Information Advantages', *Research Policy*, 32(8): 1324–41.

Galbraith, J. R. (1982). 'Designing the Innovative Organization', *Organizational Dynamics*, 11(1): 5–25.

Gallagher, S. and Park, S. H. (2002). 'Innovation and Competition in Standard-based Industries: A Historical Analysis of the U.S. Home Video Game Market.' *IEEE Transactions on Engineering Management*, 49(1): 67–82.

Gallini, N. T. (2002). 'The Economics of Patents: Lessons from Recent U.S. Patent Reform'. *Journal of Economic Perspectives*, 16(2): 131–54.

Galunic, P. C. and Eisenhardt, K. M. (2001). 'Architectural Innovation and Modular Corporate Forms'. *Academy of Management Journal*, 44(6): 1229–49.

Gambardella, A. (1992). 'Competitive Advantages from In-house Scientific Research: The U.S. Pharmaceutical industry in the 1980s', *Research Policy*, 21(5): 391–407.

—— and Torrisi, S. (1998). 'Does Technological Convergence Imply Convergence in Markets? Evidence from the Electronics Industry', *Research Policy*, 27(5): 445–63.

Gandal, N., Salant, D., and Waverman, L. (2003). 'Standards in Wireless Telephone Networks', *Telecommunications Policy*, 27(5–6): 325–32.

Gans, J. S. and Stern, S. (2003). 'The Product Market and the Market for "Ideas": Commercialization Strategies for Technology Entrepreneurs', *Research Policy,* 32(2): 333–50.

Gans, J., Hsu, D., and Stern, S. (2001). 'When Does Startup Innovation Spur the Gale of Creative Destruction', DRUID conference presentation, June 2001, NBER Working Paper No. 7851.

Garcia, R. and Calantone, R. (2002). 'A Critical Look at Technological Innovation Typology and Innovativeness Terminology: a Literature Review', *Journal of Product Innovation Management,* 19(2): 110–32.

Gargiulo, M. and Benassi, M. (2000). 'Trapped in Your Own Net? Network Cohesion, Structural Holes, and the Adaptation of Social Capital', *Organization Science,* 11(2): 183–96.

Garud, R., Jain, S. and Kumaraswamy, A. (2002). 'Institutional Entrepreneurship in the Sponsorship of Common Technological Standards: The Case of Sun Microsystems and Java', *Academy of Management Journal,* 45(1): 196–214.

—— —— and Langlois, R. N. (eds.) (2003). *Managing in the Modular Age: Architectures, Networks, and Organizations.* Oxford: Blackwell.

Gassmann, O. and von Zedtwitz, M. (2002*a*). 'Market versus Technology Drive in R&D Internationalization: Four Different Patterns of Managing Research and Development', *Research Policy,* 31(4): 569–88.

—— and von Zedtwitz, M. (2002*b*). 'Managing Customer Oriented Research', *International Journal of Technology Management,* 24(2–3): 165–93.

Gawer, A. and Cusumano, M. A. (2002). *Platform Leadership: How Intel, Microsoft, and Cisco Drive Industry Innovation.* Boston, MA: Harvard Business School Press.

Gelijns, A. C. and Their, S. O. (2002). 'Medical Innovation and Institutional Independence: Rethinking University-Industry Connections', *JAMA,* 287(1): 72–7.

Gemunden, H. G., Ritter, T. and Heydebreck, P. (1996). 'Network Configuration and Innovation Success: An Empirical Analysis in German High-tech Industries', *International Journal of Research in Marketing,* 13(5): 449–62.

Gerlach, M. (1992). *Alliance Capitalism.* Berkeley, CA: University of California Press.

Gerybadze, A. and Reger, G. (1999). 'Globalization of R&D: Recent Changes in the Management of Innovation in Transitional Corporations', *Research Policy,* 28(2–3): 251–74.

Geuna, A. and Nesta, L. (2003). 'University Patenting and its Effects on Academic Research: The Emerging European Evidence', SPRU Electronic Working Paper No. 99.

Ghemawat, P., Collis, D. J. Pisano, G. P. and Rivkin, J. W. (1999). *Strategy and the Business Landscape: Text and Cases.* Reading, MA: Addison-Wesley.

Giarratana, M. S. (2004). 'The Birth of a New Industry: Entry by Start-ups and the Drivers of Firm Growth: The Case of Encryption Software', *Research Policy,* 33(5): 707–806.

Gilbert, C. G. (2002*a*). 'Beyond Resource Allocation: Towards a Process Model of Response to Disruptive Change', Harvard Business School Working Paper No. 03–018. Boston, MA: Harvard Business School.

—— (2002*b*). 'Can Competing Frames Co-exist? The Paradox of Threatened Response', Working Paper Paper No. 02–056. Boston, MA: Harvard Business School.

Gilbert, R. J., Newbery, D. M., and Reinganum, J. F. (1984). 'Uncertain Innovation and the Persistence of Monopoly', *American Economic Review,* 74(1): 238–46.

Gilson, R. J. (1999). 'The Legal Infrastructure of High Technology Industrial Districts: Silicon Valley, Route 128, and Covenants Not To Compete', *New York University Law Review,* 74(3): 575–629.

Glaser, B. and Strauss, A. (1967). *The Discovery of Grounded Theory: Strategies of Qualitative Research.* London: Wiedenfeld & Nicholson.

Goerzen, A. and Beamish, P. W. (2005). 'The Alliance Portfolio Effect on MNE Performance', *Strategic Management Journal,* 26(4): 333–54.

Goldberg, R. and Gourville, J. (2002). *Calgene, Inc.* Harvard Business School Case No. 9-502-041. Boston, MA: Harvard Business School Press.

Golder, P. N. and Tellis, G. J. (1993). 'Pioneer Advantage: Marketing Logic or Marketing Legend?' *Journal of Marketing Research,* 30(2): 158–70.

Gomes, L. (2004). 'Avalanche Project is Clearing the Path for Tech Cooperation', *Wall Street Journal,* April 12, p. B1.

Gomes-Casseres, B. (1994). 'Group versus Group: How Alliance Networks Compete', *Harvard Business Review,* 72(4): 62–74.

—— (1996). *The Alliance Revolution: The New Shape of Business Rivalry.* Boston, MA: Harvard University Press.

—— (2003). 'Competitive Advantage in Alliance Constellations', *Strategic Organization,* 1(3): 327–35.

Gompers, P. A. and Lerner, J. (1999). *The Venture Capital Cycle.* Cambridge, MA: MIT Press.

Gonsalves, A. and Coffee, P. (1998). 'Jikes! More Open Source Code', *PC Week,* December 7, p. 6.

Goolsbee, A. (1998). 'Does Government R&D Policy Mainly Benefit Scientists and Engineers?', *American Economic Review,* 88(2): 298–302.

Gottfredson, M., Puryear, R., and Phillips, S. (2005). 'Strategic Sourcing: from Periphery to the Core', *Harvard Business Review,* 83(2): 132–39.

Graff, G., Heiman, A., and Zilberman, D. (2002). 'University Research and Offices of Technology Transfer', *California Management Review,* 45(1): 88–111.

Graham, S. J. H. (2004). 'Continuation, Complementarity, and Capturing Value: Three Studies Exploring Firms' Complementary Uses of Appropriability Mechanisms in Technological Innovation', unpublished Ph.D. dissertation, Haas School of Business, U.C. Berkeley.

—— and Mowery, D. C. (2003). 'Intellectual Property Protection in the Software Industry', in W. M. Cohen and S. A. Merrill (eds.), *Patents in the Knowledge-based Economy.* Washington, DC: National Academies Press, pp. 219–58.

—— —— (2004). 'Submarines in Software? Continuations in U.S. Software Patenting in the 1980s and 1990s', *Economics of Innovation and New Technology,* 13(5): 443–56.

—— —— (2005). 'Software Patents: Good News or Bad News?', in Robert W. Hahn (ed.), *Intellectual Property Rights in Frontier Industries: Software and Biotechnology.* Washington, DC: AEI-Brookings Joint Center, pp. 45–80.

——, Hall, B. H., Harhoff, D., and Mowery, D. C. (2003). 'Post-issue Patent "Quality Control": A Comparative Study of U.S. Patent Reexaminations and European Patent Oppositions', in W. M. Cohen and S. A. Merrill (eds.), *Patents in the Knowledge-Based Economy.* Washington, DC: National Academies Press.

Grand, S., von Krogh, G., Leonard, D., and Swap, W. (2004). 'Resource Allocation Beyond Firm Boundaries: A Multi-level Model for Open Source Innovation', *Long Range Planning,* 37(6): 591–610.

Granovetter, M. (1985). 'Economic Action and Social Structure: The Problem of Embeddedness', *American Journal of Sociology* 91(3): 481–510.

Granstrand, O. (1982). *Technology Management and Markets*. London: Pinter.

—— and Sjölander, S. (1990). 'Managing Innovation in Multi-technology Corporations', *Research Policy,* 19(1): 35–60.

—— Patel, P. and Pavitt, K. (1997). 'Multi-Technology Corporations: Why They Have "Distributed" Rather than "Distinctive" Core Competence', *California Management Review,* 39(4): 8–25.

Grant, R. M. (1991). 'The Resource-Based Theory of Competitive Advantage: Implications for Strategy Formulation', *California Management Review,* 33(3): 114–35.

Grant, R. and Baden-Fuller, C (2002). 'The Knowledge-based View of Strategic Alliance Formation: Knowledge Accessing Versus Organizational Learning', in F. Contractor and P. Lorange (eds.), *Cooperative Strategies and Alliances*. Oxford: Elsevier Science.

—— —— (2004). 'A Knowledge Accessing Theory of Strategic Alliances', *Journal of Management Studies,* 41(1): 61–84.

Griliches, Z. (1957). 'Hybrid Corn: An Exploration in the Economics of Technological Change', *Econometrica,* 25(4): 501–22.

—— (1986). 'Productivity, R&D, and Basic Research at the Firm Level in the 1970s', *American Economic Review,* 76(1): 141–54.

Grindley, P. C. and Teece, D. J. (1997). 'Managing Intellectual Capital: Licensing and Cross-licensing in Semiconductors and Electronics', *California Management Review,* 39(2): 8–41.

Grove, A. S. (1996). *Only the Paranoid Survive: How to Exploit the Crisis Points that Challenge Every Company and Career.* New York: Doubleday.

Gulati, R. (1995*a*). 'Does Familiarity Breed Trust? The Implications of Repeated Ties for Contractual Choice in Alliances', *Academy of Management Journal,* 38(1): 85–112.

—— (1995*b*). 'Social Structure and Alliance Formation Patterns: A Longitudinal Analysis', *Administrative Science Quarterly,* 40: 619–52.

—— (1998). 'Alliances and Networks', *Strategic Management Journal,* 19(4): 293–317.

—— (1999). 'Network Location and Learning: The Influence of Network Resources and Firm Capabilities on Alliance Formation', *Strategic Management Journal,* 20(5): 397–420.

—— and Westphal, J. D. (1999). 'Cooperative or Controlling? The Effects of CEO-Board Relations and the Content of Interlocks on the Formation of Joint Ventures', *Administrative Science Quarterly,* 44(3): 473–506.

—— , Nohria, N. and Zaheer, A. (2000). Strategic Networks, *Strategic Management Journal,* 21(3): 203–15.

Guth, R. A. and Clark, D. (2004). Behind Secret Settlement Talks: New Power of Tech Customers, *Wall Street Journal,* April 5, p. 1.

Hage, J. (1980). *Theories of Organizations*. New York: John Wiley.

Hagedoorn, J. and Cloodt, M. (2003). 'Measuring Innovative Performance: Is There an Advantage in Using Multiple Indicators?', *Research Policy,* 32(8): 1365–79.

—— and Duysters, G. (2002). 'External Appropriation of Innovative Capabilities: The Choice Between Strategic Partnering and Mergers and Acquisitions', *Journal of Management Studies*, 39(2): 167–88.

Hagel, J., III and Brown, J. (2005). *The Only Sustainable Edge; Why Business Strategy Depends on Productive Friction and Dynamic Specialization*. Boston, MA: Harvard Business School Press.

Hall, B. H. and Ziedonis, R. H. (2001). 'The Patent Paradox Revisited: An Empirical Study of Patenting in the US Semiconductor Industry, 1979–1995', *Rand Journal of Economics*, 32(1): 101–28.

—— , Jaffe, A., and Trajtenberg, M. (2000). 'Market Value and Patent Citations: A First Look', National Bureau of Economic Research, Working Paper No. 7741.

—— , Link, A. N., and Scott, J. T. (2001). 'Barriers Inhibiting Industry from Partnering with Universities: Evidence from the Advanced Technology Program', *Journal of Technology Transfer*, 26(1): 87–98.

Hambrick, D. C., Cho, T. S., and Chen, M.-J. (1996). 'The Influence of Top Management Team Heterogeneity on Firms' Competitive Moves', *Administrative Science Quarterly*, 41(4): 659–84.

Hamel, G. (1991). 'Competition for Competence and Inter-Partner Learning Within International Strategic Alliances', *Strategic Management Journal*, 12 (Summer): 83–103.

—— (2000). *Leading the Revolution*. Boston, MA: Harvard Business School Press.

Hansen, E. (2003). 'AOL lays off Netscape developers', CNET News.com, July 15, URL: http://news.com.com/2100-1032_3-1026078.html.

Hansen, M. T. (1999). 'The Search-Transfer Problem: The Role of Weak Ties in Integrating Knowledge Across Subunits', *Administrative Science Quarterly*, 44(1): 82–111.

—— (2002). 'Knowledge Networks: Explaining Effective Knowledge Sharing in Multiunit Companies', *Organization Science*, 13(3): 232–48.

—— and Nohria, N. (2004). 'How to Build Collaborative Advantage', *Sloan Management Review*, 46(1): 22–30.

Hargadon, A. B. and Douglas, Y. (2001). 'When Innovations Meet Institutions: Edison and the Design of the Electric Light', *Administrative Science Quarterly*, 46(3): 476–501.

—— and Sutton, R. (1997). 'Technology Brokering and Innovation in a Product Development Firm', *Administrative Science Quarterly*, 42(4): 716–49.

Harris, S. G. and Sutton, R. I. (1986). 'Functions of Parting Ceremonies in Dying Organizations', *Academy of Management Journal*, 29(1): 5–30.

Hars, A. and Ou, S. (2002). 'Working for Free? Motivations for Participating in Open-source Projects', *International Journal of Electronic Commerce*, 6(3): 25–39.

Hayek, F. (1945). 'The Use of Knowledge in Society', *American Economic Review*, 35(4): 19–30.

Healy, K. and Schussman, A. (2003). 'The Ecology of Open-Source Software Development', Working Paper, Department of Sociology, University of Arizona, January 29, URL: http://opensource.mit.edu /papers/healyschussman.pdf

Hedlund, G. (1994). 'A Model of Knowledge Management and the N Form Corporation', *Strategic Management Journal*, 15(Summer): 73–90.

Heimeriks, K. (2005). 'Managing Portfolios of Strategic Alliances: The Effect of Alliance Mechanisms on Portfolio Performance', Working Paper, Copenhagen Business School.

Heller, M. A. and Eisenberg, R. S. (1998). 'Can Patents Deter Innovation? The Anticommons in Biomedical Research', *Science*, 280: 698–701.

Henderson, R. and Clark, K. B. (1990). 'Architectural Innovation: The Reconfiguration of Existing Product Technologies and The Failure of Established Firms', *Administrative Science Quarterly*, 35(1): 9–30.

—— and Cockburn, I. (1994). 'Measuring Competence? Exploring Firm Effects in Pharmaceutical Research', *Strategic Management Journal*, 15(Winter): 63–84.

——, Jaffe, A. B., and Trajtenberg, M. (1998*a*). 'Universities As A Source Of Commercial Technology: A Detailed Analysis Of University Patenting, 1965–1988', *The Review of Economics and Statistics*, 80(1): 119–27.

—— —— ——. (1998*b*). 'University Patenting Amid Changed Incentives for Commercialization', in Navaretti, Dasgupta, Maler, and Siniscalco (eds.), *Creation and Transfer of Knowledge: Institutions and Incentives*. Berlin: Springer, pp. 88–114.

Hertel, G., Niedner, S., and Herrmann, S. (2003). 'Motivation of Software Developers in Open Source Projects: An Internet-based Survey of Contributors to the Linux Kernel', *Research Policy*, 32(7): 1159–77.

Hill, C. W. L. and Rothaermel, F. T. (2003). 'The Performance of Incumbent Firms in the Face of Radical Technological Innovation', *Academy of Management Review*, 28(2): 257–74.

Hitt, M., Ireland, D., Camp, M. and Sexton, D. (2002). *Strategic Entrepreneurship*. Oxford: Blackwell.

Hobday, M. (1998). 'Product Complexity, Innovation and Industrial Organization', *Research Policy*, 26(6): 689–710.

—— (2000). 'The Project-based Organisation: An Ideal Form for Managing Complex Products and Systems', *Research Policy*, 27(7–8): 871–93.

Hoffmann, W. (2005). 'How to Manage a Portfolio of Alliances', *Long Range Planning*, 38(2): 121–43.

—— and Schaper-Rinkel, W. (2001). 'Acquire or Ally?—A Strategic Framework for Deciding between Acquisitions and Cooperation', *Management International Review*, 41(2): 131–59.

Hwang, P. and Burgers, W. P. (1997). 'The Many Faces of Multi-firm Alliances: Lessons for Managers', *California Management Review*, 39(2): 101–17.

Iansiti, M. (1998). *Technology Integration: Making Critical Choices in a Dynamic World*. Boston, MA: Harvard Business School Press.

—— and Levien, R. (2004*a*). *The Keystone Advantage: What the New Dynamics of Business Ecosystems Mean for Strategy, Innovation and Sustainability*. Boston, MA: Harvard Business School Press.

—— —— (2004*b*). 'Strategy as Ecology', *Harvard Business Review*, 82(3): 68–78.

Iversen, E. J. (1999). 'Standardisation and Intellectual Property Rights: ETSI's Controversial Search for New IPR Procedures', Proceedings of the 1st IEEE Conference on Standardisation and Innovation in Information Technology, Aachen, Germany, September 15–17: 55–63.

Jacobides, M. G. and Winter, S. G. (2005). 'The Co-evolution of Capabilities and Transaction Costs: Explaining the Institutional Structure of Production', *Strategic Management Journal*, 26(5): 395–413.

Jacobs, I. (2005). 'My Life on the Wireless Frontier', Videotaped Public Lecture, Computer History Museum, Mountain View, Calif., May 25, URL: http://www.computerhistory.org/events/index.php?id=1099686446

Jaffe, A. B. (1996). 'Trends and Patterns in Research and Development Expenditures in the United States', *Proceedings of the National Academy of Sciences,* 93(23): 12658–63.

—— (2000). 'The U.S. Patent System in Transition: Policy Innovation and the Innovation Process', *Research Policy,* 29(4–5): 531–57.

—— and Lerner, J. (2004). *Innovation and Its Discontents: How Our Broken Patent System is Endangering Innovation and Progress, and What to Do About It.* Princeton, NJ: Princeton University Press.

—— and Trajtenberg, M. (1996). 'Flows of Knowledge from Universities and Federal Labs: Modeling the Flow of Patent Citations Over Time and Across Institutions and Geographic Boundaries', *Proceeding of the National Academy of Science,* 93(23): 12671–7.

——, Trajtenberg, M. and Henderson, R. (1993). 'Geographic Localization of Knowledge Spillovers as Evidenced by Patent Citations', *Quarterly Journal of Economics,* 108(3): 577–98.

—— —— and Fogarty, M. S. (2000). 'Knowledge Spillovers and Patent Citations: Evidence from a Survey of Inventors', *American Economic Review,* 90(2): 215–18.

Jelinek, M. and Schoonhoven, C. B. (1990). *The Innovation Marathon.* Oxford: Blackwell.

Johnson, B., Lorenz, E., and Lundvall, B.-Å. (2002). 'Why all this Fuss about Codified and Tacit Knowledge?', *Industrial and Corporate Change,* 11(2): 245–62.

Judge, G., Hill, R. C., Griffiths, W. E., Lutkepohl, H., and Lee, T. C. (1985). *The Theory and Practice of Econometrics,* 2nd edn. New York: John Wiley.

Kano, S. (2000). 'Technical Innovations, Standardization and Regional Comparison: A Case Study in Mobile Communications', *Telecommunications Policy,* 24(4): 305–21.

Kanter, R. (1994). 'Collaborative Advantage: The Art of Alliances', *Harvard Business Review,* 72(4): 96–108.

Kaplan, J. (1996). *Startup: A Silicon Valley Adventure,* Paperback edn. New York: Penguin.

Katila, R. (2002). 'New Product Search Over Time: Past Ideas in their Prime?', *Academy of Management Journal,* 45(5): 995–1010.

Katz, M. L. (1986). 'An Analysis of Cooperative Research-and-Development'. *Rand Journal of Economics,* 17(4): 527–43.

—— and Shapiro, C. (1985). 'Network Externalities, Competition and Compatibility', *American Economic Review,* 75(3): 424–40.

Katz R. and Allen, T. (1985). 'Organizational Issues in the Introduction of New Technologies', in P. Kleindorfer (ed.), *The Management of Productivity and Technology in Manufacturing.* New York: Plenum Press.

Keil, T. (2002*a*). *External Corporate Venturing: Strategic Renewal in Rapidly Changing Industries*. WestPort, CT: Quorum Books.

—— (2002*b*). 'De-facto Standardization Through Alliances—Lessons from Bluetooth', *Telecommunications Policy,* 26(3–4): 205–13.

Kelley, D., Peters, L., and O'Connor, G. C. (2005). 'Leveraging the Organizational Network for Radical Innovation: Three Broker Models', Working Paper, Lally School of Management, Rensselaer Polytechnic Institute.

Kenney, M. (ed.) (2000*a*). *Understanding Silicon Valley: Anatomy of an Entrepreneurial Region.* Stanford, CA: Stanford University Press.

Kenney, M. (2000*b*). 'Introduction', in M. Kenney (ed.), *Understanding Silicon Valley*. Stanford, CA: Stanford University Press, pp. 1–12.

—— and Florida, R. (2000). 'Venture Capital in Silicon Valley: Fueling new Firm Formation', in Martin Kenney (ed.), *Understanding Silicon Valley*. Stanford, CA: Stanford University Press, pp. 98–123.

—— and Goe, W. R. (2003). 'The Role of Social Embeddedness in Professional Entrepreneurship: A Comparison of Electrical Engineering and Computer Science at UC Berkeley and Stanford', *Research Policy*, 33(3): 691–707.

—— and Patton, D. (2005). 'Entrepreneurial Geographies: Support Networks in Three High-Tech Industries', *Economic Geography*, 81(2): 201–28.

Kerner, S. M. (2005). 'IBM: Open Source is more than just Linux', *Internet News*, April 20.

Kessler, E. H., Bierly, P. E., and Gopalakrishnan, S. S. (2000). 'Internal vs. External Learning in new Product Development: Effects of Speed, Costs and Competitive Advantage', *R&D Management*, 30(3): 213–24.

Khanna, T., Gulati, R., and Nohria, N. (1998). 'The Dynamics of Learning Alliances: Competition, Cooperation, and Relative Scope', *Strategic Management Journal*, 19(3): 193–210.

Kim, K., Park, J.-H., and Prescot, J. (2003). 'The Global Integration of Business Functions: A Study of Multinational Businesses in Integrated Global Industries', *Journal of International Business Studies*, 34(4): 327–44.

Kirschbaum, R. (2005). 'Open Innovation in Practice', *Research-Technology Management*, 48(4): 24–28.

Kodama, F. (1992). 'Technology Fusion and the New R&D', *Harvard Business Review*. 70(4): 70–8.

Kogut, B. and Metiu, A. (2001). 'Open-source Software Development and Distributed Innovation', *Oxford Review of Economic Policy*, 17(2): 248–64.

—— and Zander, U. (1992). 'Knowledge of the Firm, Combinative Capabilities, and the Replication of Technology', *Organization Science*, 3(3): 383–97.

—— (1993). 'Knowledge of the Firm and the Evolutionary Theory of the Multinational Corporation', *Journal of International Business Studies*, 24(4): 625–46.

—— (1996). 'What Do Firms Do? Coordination, Identity and Learning', *Organization Science*, 7(5): 502–18.

Koka, B. R. and Prescott, J. E. (2002). 'Strategic Alliances as Social Capital: A Multidimensional View', *Strategic Management Journal*, 23(9): 795–816.

Kortum, S. and Lerner, J. (1999). 'What is Behind the Recent Surge in Patenting?', *Research Policy*, 28(1): 1–22.

—— and Lerner, J. (2000). 'Assessing the Contribution of Venture Capital to Innovation', *Rand Journal of Economics* 31(4): 674–92.

Koza, M. P. and Lewin, A. Y. (1998). 'The Coevolution of Strategic Alliances', *Organization Science*, 9(3): 255–64.

Kozmetsky, G. (1993). 'The Growth and Internationalization of Creative and Innovative Management', in R. L. Kuhn (ed.), *Generating Creativity and Innovation in Large Bureaucracies*, Westport, CT: Quorum Books.

Kraemer, K. L. and Dedrick, J. (1998). 'Globalization and Increasing Returns: Implications for the U.S. Computer Industry', *Information Systems Research*, 9(4): 303–22.

Krechmer, K. (2005). 'Communications Standards and Patent Rights: Conflict or Coordination?', *The Economics of the Software and Internet Industries Conference*. Tououse, France, January.

—— (2006). 'The Meaning of Open Standards', *International Journal of IT Standards & Standardisation Research*, 4/1.

Krueger, C. W. (1992). 'Software Reuse', *ACM Computing Surveys*, 24(2): 132–83.

Krugman, P. (1991). *Geography and Trade*. Cambridge, MA: MIT Press.

Kuan, J. W. (2001). 'Open Source Software as Consumer Integration into Production', SSRN Working Paper No. 259648.

Kuemmerle, W. (1998). 'Optimal Scale for Research and Development in Foreign Environments: An Investigation into Size and Performance of Research and Development Laboratories Abroad', *Research Policy*, 27(2): 111–26.

Kuhn, T. (1962). *The Structure of Scientific Revolutions*. Chicago: University of Chicago Press.

Kunda, G. (1992). *Engineering Culture: Control and Commitment in a High-Tech Corporation*. Philadelphia: Temple University Press.

Lagerstrom, K. and Andersson, M. (2003). 'Creating and Sharing Knowledge within a Transnational Team—The Development of a Global Business System', *Journal of World Business*, 38(2): 84–95.

Lakhani, K. R. and von Hippel, E. (2003). 'How Open Source Software Works: 'Free' User-to-User Assistance', *Research Policy*, 32(6): 923–43.

Lammers, D. and Ohr, S. (2003). 'Startup Bets on Drop-in Digital Amps', *Electronic Engineering Times*. (October 31), URL: http://www.eetimes.com/showArticle.jhtml?articleID=18309565

Lamoreaux, N. and Sokoloff, K. (1999). 'Inventive Activity and the Market for Technology in the United States, 1840–1920', National Bureau of Economic Research, Working Paper No. 7107.

—— Raff, D., and Temin, P. (1999). *Learning by Doing in Markets, Firms, and Countries*. Chicago: University of Chicago Press.

Lane, N. (2003). 'Risk of Patent Poker too High a Gamble for Nokia', *3G Mobile*, 5(2): 1–2.

Lane, P. J. and Lubatkin, M. (1998). 'Relative Absorptive Capacity and Interorganizational Learning', *Strategic Management Journal*, 19(8): 461–77.

Langlois, R. N. (2003*a*). 'The Vanishing Hand: The Changing Dynamics of Industrial Capitalism', *Industrial and Corporate Change*, 12(2): 351–85.

—— (2003*b*). 'Modularity in Technology and Organization', *Journal of Economic Behavior and Organization*, 49(1): 19–37.

—— and Robertson, P. L. (1995). *Firms, Markets, and Economic Change: A Dynamic Theory of Business Institutions*. London: Routledge.

Laursen, K. and Salter, A. J. (2004). 'Searching High and Low: What Types of Firms Use. Universities as a Source of Innovation?' *Research Policy*, 33(8): 1201–15.

—— —— (2006). 'Open for innovation: the role of openness in explaining innovation performance among UK manufacturing firms', *Strategic Management Journal*, 27(2): 131–50.

—— (2005). 'The Paradox of Openness of Knowledge for Innovation', Paper presented for the All-academy Symposium Open Innovation: Locating and Incorporating

External Innovations, August 9, 2005, Academy of Management Conference 2005, Honolulu, Hawaii.

Lawler, E. E. (1971). *Pay and Organizational Effectiveness: A Psychological View.* New York: McGraw-Hill.

Layne-Farrar, A. (2005). 'Defining Software Patents: A Research Field Guide', AEI-Brookings Joint Center Working Paper No. 05–14, URL: http://aei-brookings.org/admin/authorpdfs/page.php?id=1177

Lazerson, M. H. (1995). 'A New Phoenix? Modern Putting-out in the Modena Knitwear Industry', *Administrative Science Quarterly,* 40(1): 34–59.

Leamon, A. and Hardymon, G. F. (2000). Intel® 64 Fund. Harvard Business School Case 9-800-351. Boston, MA: Harvard Business School Press.

Lecuyer, C. (2000). 'Fairchild Semiconductor and Its Influence', in C.-M. Lee, W. Miller, M. G. Hancock, and H. Rowen (eds.), *The Silicon Valley Edge.* Stanford, CA: Stanford University Press, pp. 158–83.

Lee, G. K. and Cole, R. E. (2003). 'From a Firm-Based to a Community-Based Model of Knowledge Creation: The Case of the Linux Kernel Development', *Organization Science,* 14(6): 633–49.

Leifer, R., McDermott, C. M., O'Connor, G. C., Peters, L. S., Rice, M. P., Veryzer, R. W., and Rice, M. (2000). *Radical Innovation: How Mature Companies Can Outsmart Upstarts.* Boston, MA: Harvard Business School Press.

Lemley, M. (2001). 'Rational Ignorance at the Patent Office', UC Berkeley School of Law Public Law and Legal Theory Working Paper No. 46 (February).

—— (2002). 'Intellectual Property Rights and Standard Setting Organizations', *California Law Review,* 90: 1889–1981.

Leonard-Barton, D. (1992). 'Core Capabilities and Core Rigidities: A Paradox on Managing New Product Development', *Strategic Management Journal,* 13 (Summer): 111–25.

Lerner, J. (2000). '150 Years of Patent Office Practice', National Bureau of Economic Research, Working Paper No. 7477.

—— and Tirole, J. (2002). 'Some Simple Economics of Open Source', *Journal of Industrial Economics,* 50(2): 197–234.

—— —— (2004*a*). 'A Model of Forum Shopping with Special Reference to Standard Setting Organizations', National Bureau of Economic Research, Working Paper No. 10664.

—— —— (2004*b*). 'Efficient Patent Pools', *American Economic Review,* 94(3): 691–711.

—— —— (2005). 'The Scope of Open Source Licensing', *Journal of Law Economics & Organization,* 21(1): 20–56.

Leslie, S. (2000). 'The Biggest "Angel" of Them All: The Military and the Making of Silicon Valley', in M. Kenney (ed.), *Understanding Silicon Valley.* Stanford, CA: Stanford University Press, pp. 48–67.

Lessig, L. (2004). *Free Culture: How Big Media Uses Technology and the Law to Lock Down Culture and Control Creativity.* New York: Penguin Press.

Levin, R., Klevorick, A., Nelson, R., and Winter, S. (1988). 'Appropriating the Returns from Industrial R&D', Cowles Foundation Discussion Papers 862, Yale University.

Lieberman, M. B. and Montgomery, D. B. (1988). 'First-Mover Advantages', *Strategic Management Journal,* 9 (Summer): 41–58.

—— —— (1998). 'First Mover (Dis)advantages: Retrospective and Link with the Resource-Based View', *Strategic Management Journal*, 19(12): 1111–25.

Liebeskind, J. P., Oliver, A. L., Zucker, L., and Brewer, M. (1996). 'Social Networks, Learning, and Flexibility: Sourcing Scientific Knowledge in New Biotechnology Firms', *Organizational Science*, 7(4): 428–43.

Lilien, G. L., Morrison, P. D., Searls, K., Sonnack, M., and von Hippel, E. (2002). 'Performance Assessment of the Lead User Idea-Generation Process for New Product Development', *Management Science*, 48(8): 1042–59.

LizardTech (2001). 'LizardTech's Statement on DjVu Open Source Licensing', http://djvulibre.djvuzone.org/lti-licensing.html

Lo, A., M. (2002). 'A Need for Intervention: Keeping Competition Alive in the Networking Industry in the Face of Increasing Patent Assertions Against Standards', FTC/DOJ Hearings on Competition and Intellectual Property Law and Policy in the Knowledge-Based Economy, URL: http://www.ftc.gov/opp/intellect/0204181o.pdf

Lohr, S. (2005). 'I.B.M. to Give Free Access to 500 Patents', *New York Times*, January 11.

Loomis, T. (2005). 'Cell Break', *IP Law & Business*, July, URL: http://www.ipww.com/texts/0705/europhony0705.html

Lorenzoni, G. and Baden-Fuller, C. (1995). 'Creating a Strategic Center to Manage a Web of Alliances', *Calfornia Management Review*, 37(2): 146–63.

Louis, K. S., Jones, L. M., Anderson, M. S., Blumenthal, D., Campbell, E. G. (2001). 'Entrepreneurship, Secrecy, and Productivity: A Comparison of Clinical and Non-clinical Life Sciences Faculty', *Journal of Technology Transfer*, 26(3): 233–45.

Lundvall, B. Å. (ed.) (1992). *National Systems of Innovation. Towards a Theory of Innovation and Interactive Learning*. London: Pinter.

Lynch, R. P. (1993). *Business Alliance Guide*. New York: John Wiley.

Lynn, G., Morone, J. G., and Paulson, A. S. (1996). 'Marketing and Discontinuous Innovation: The Probe and Learn Process', *California Management Review*, 38(3): 8–37.

Lyons, D. (2005). 'Open Source Smackdown', *Forbes.com*. (June 15), URL: http://www.forbes.com/2005/06/15/jboss-ibm-linux_cz_dl_0615jboss.html

Machlup, F. (1983). 'Semantic Quirks in Studies of Information', in F. Machlup and U. Mansfield (eds.), *The Study of Information*. New York: John Wiley, pp. 641–71.

Madhok, A. (1997). 'Cost, value and foreign market entry mode: The transaction and the firm', *Strategic Management Journal*, 18(1): 39–61.

—— Tallman, S. (1998). 'Resources, Transactions and Rents: Managing Value through Interfirm Collaborative Relationships', *Organization Science*, 9(3): 326–39.

Maidique, M. A. and Zirger, B. J. (1985). 'The New Product Learning Cycle', *Research Policy*, 14(6): 299–313.

Malone, T. W. and Laubacher, R. J. (1998). 'The Dawn of the e-lance Economy', *Harvard Business Review*, 76(5): 145–52.

Maltsbarger, R. and Kalaitzandonakes, N. (2000). 'Direct and Hidden Costs in Identity Preserved Supply Chains', *Agbioforum*, 3(4): 236–42.

Mansfield, E. (1968). *Industrial Research and Technical Innovation*. New York: Norton.

—— (1991). 'Academic Research and Industrial Innovation', *Research Policy*, 20(1): 1–12.

—— (1998). 'Academic Research and Industrial Innovation: An Update of Empirical Findings', *Research Policy*, 26(7–8): 773–6.

March, J. G. (1991). 'Exploration and Exploitation in Organizational Learning', *Organization Science*, 2(1): 71–87.

Markides, C. C. and Williamson, P. J. (1994). 'Related Diversification, Core Competences and Corporate Performance', *Strategic Management Journal*, 15 (Summer): 149–65.

Marshall, A. (1920). *Industry and Trade*. London: Macmillan.

Maskell, P. (2001). 'Toward a Knowledge-Based Theory of the Geographical Cluster', *Industrial and Corporate Change*, 10(4): 921–43.

Mason, I. 'IT Giants Accused of Exploiting Open Source', CNET News.com, May 31, URL: http://news.com.com/2100-7344_3-5726714.html

Maula, M. and Murray, G. (2001). 'Corporate Venture Capital and the Creation of US Public Companies: The Impact of Sources of Venture Capital on the Performance of Portfolio Companies', in M. A. Hitt, R. Amit, C. Lucier, and R. D. Nixon (eds.), *Creating Value: Winners in the New Business Environment*. Oxford: Blackwell, pp. 164–87.

—— Salmenkaita, J. P., Keil, T., and Väisänen, K. (2003). 'Dynamic Complementarities of Startups and Incumbents in the Creation of a Systemic Innovation', Paper presented at the 23rd Annual International Conference of the Strategic Management Society, Baltimore, Maryland.

Mayer, D. and Kenney, M. (2004). 'Economic Action Does Not Take Place in a Vacuum: Understanding Cisco's Acquisition and Development Strategy', *Industry and Innovation*, 11(4): 299–325.

McDermott, C. M. and O'Connor, G. C. (2002). 'Managing Radical Innovation: An Overview of Emergent Strategy Issues', *Journal of Product Innovation Management*, 19(6): 424–38.

McEvily, B. and Zaheer, A. (1999). 'Bridging Ties: A Source of Firm Heterogeneity in Competitive Capabilities', *Strategic Management Journal*, 20(12): 133–56.

McKelvey, M. (1991). 'How do National Systems of Innovation Differ: A Critical Analysis of Porter, Freeman, Lundvall and Nelson', in G. M. Hodgson and E. Screpanti (eds.), *Rethinking Economics: Markets, Technology, and Economic Evolution*. Aldershot, Hants: Edward Elgar, pp. 117–37.

McKenna, R. (2000). 'Free Advice: Consulting the Silicon Valley Way', in C.-M. Lee, W. Miller, M. G. Hancock, and H. Rowen (eds.), *The Silicon Valley Edge*. Stanford, CA: Stanford University Press, pp. 370–9.

McMillan, S. G., Narin, F., and Deeds, D. L. (2000). 'An Analysis of the Critical Role of Public Science in Innovation: The Case of Biotechnology', *Research Policy*, 29(1): 1–8.

Meller, P. (2004). 'Europeans Rule Against Microsoft: Appeal Is Promised', *New York Times*, March 25.

Merges, R. P. (1996). 'A Comparative Look at Intellectual Property Rights and the Software Industry', in D. C. Mowery (ed.), *The International Computer Software Industry: A Comparative Study of Industry Evolution and Structure*. New York: Oxford University Press.

Metcalfe, J. S. (1995). 'The Economic Foundations of Technology Policy: Equilibrium and Evolutionary Perspectives', in P. Stoneman (ed.), *Handbook of the Economics of Innovation and Technological Change*. Oxford: Blackwell.

—— and Gibbons, M. (1989). 'Technology, Variety and Organisation: A Systematic Perspective on the Competitive Process', in R. Rosenbloom (ed.), *Research on Technological Innovation, Management and Policy*, Vol. 4. Greenwich: JAI Press, pp. 153–93.

Meyers, P. W. and Tucker, F. G. (1989). 'Defining Roles for Logistics During Routine and Radical Technological Innovation', *Journal of the Academy of Marketing Science*, 17(1): 73–82.

Miller, R., Hobday, M., Leroux-Demers, T., and Olleros, X. (1995). 'Innovation in Complex Systems Industries: The Case of Flight Simulation', *Industrial and Corporate Change*, 4(2): 363–400.

Miner, A. S., DeVaughn, M., Eesley, D., and Rua, T. (2001). 'The Magic Beanstalk Vision of University Venture Formation', in K. Schoonhoven and E. Romanell (eds.), *The Entrepreneurship Dynamic*. Stanford, CA: Stanford University Press, pp. 109–46.

Miotti, L. and Sachwald, F. (2003). 'Co-operative R&D: Why and with Whom?: An Integrated Framework of Analysis', *Research Policy*, 32(8): 1481–99.

Mock, D. (2005). *The Qualcomm Equation*. New York: AMACOM.

Mockus, A., Fielding, R. T., and James D. H. (2002). 'Two Case Studies of Open Source Software Development: Apache and Mozilla', *ACM Transactions on Software Engineering and Methodology*, 11(3): 309–46.

Montobbio, F. (1999). 'National Innovation Systems: A Critical Survey', ESSY Working Paper, Bocconi University, Milan.

Moore, G. A. (1991). *Crossing The Chasm: Marketing and Selling Technology Products To Mainstream Customers*. New York: HarperBusiness.

Moore, G. and Davis, K. (2004). 'Learning the Silicon Valley Way', in T. Bresnahan and A. Gambardella (eds.), *Building High-Tech Clusters: Silicon Valley and Beyond*. Cambridge, MA: Cambridge University Press.

Morone, J. G. (1993). *Winning in High-Tech Markets*. Boston: Harvard Business School Press.

Morris, C. R. and Ferguson, C. H. (1993). 'How Architecture Wins Technology Wars', *Harvard Business Review*, 71(2): 86–96.

Moschella, D. C. (1997). *Waves of Power: Dynamics of Global Technology Leadership, 1964–2010*. New York: AMACOM.

Mowery, D. C. (1983). 'The Relationship Between Contractual and Intrafirm Forms of Industrial Research in American Manufacturing, 1900–1940', *Explorations in Economic History*, 20(4): 351–74.

—— (ed.) (1996). *The International Computer Software Industry: A Comparative Study of Industry Evolution and Structure*. New York: Oxford University Press.

—— and Nelson, R. R. (eds.) (1999). *Sources of Industrial Leadership: Studies of Seven Industries*. New York: Cambridge University Press.

—— and Sampat, B. N. (2001). 'University Patents and Patent Policy Debates in the US, 1925–1980', *Industrial and Corporate Change*, 10(3): 781–814.

—— —— (2004). 'Growth of U.S. University Patenting since the 1970s', Paper presented at the Economic History Association, San Jose, California, September.

—— and Simcoe, T. (2002*a*). 'Is the Internet a US Invention?—An Economic and Technological History of Computer Networking', *Research Policy*, 31(8–9): 1369–87.

—— (2002*b*). 'The Origins and Evolution of the Internet', in B. Steil, R. Nelson, and D. Victor (eds.), *Technological Innovation and National Economic Performance*. Princeton, NJ: Princeton University Press.

—— and Ziedonis, A. A. (2001). 'Numbers, Quality, and Entry: How Has the Bayh-Dole Act Affected U.S. University Patenting and Licensing?', in Jaffe, Lerner, and Stern (eds.), *Innovation Policy and the Economy*. Cambridge, MA: MIT Press, 187–220.

—— Oxley, J. E., and Silverman, B. S. (1996). 'Strategic Alliances and Interfirm Knowledge Transfer', *Strategic Management Journal*, 17 (Winter): 77–91.

Mowery, D. C. Oxley, J. E., and Silverman, B. S.(1998). 'Technological Overlap and Interfirm Cooperation: Implications for the Resource-Based View of the Firm', *Research Policy*, 27(5): 507–23.

—— Nelson, R. R., Sampat, B. N., and Ziedonis, A. A. (2001). 'The Growth of Patenting and Licensing by U.S. Universities: As Assessment of the Effect of the Bayh-Dole Act of 1980', *Research Policy*, 20(1): 99–119.

—— Sampat, B. N., and Ziedonis, A. A. (2002). 'Learning to Patent: Institutional Experience, Learning, and the Characteristics of U.S. University Patents after the Bayh-Dole Act, 1981–1992', *Management Science*, 48(1): 73–89.

Mozilla.org. 1999. 'Mozilla Public License Version 1.1', URL: http://www.mozilla.org/MPL/MPL-1.1.html

Murray, F. (2002). 'Innovation as Co-evolution of Scientific and Technological Networks: Exploring Tissue Engineering', *Research Policy*, 31(8–9): 1389–403.

—— and Stern, S. (2005). 'Do Formal Intellectual Property Rights Hinder the Free Flow of Scientific Knowledge?', National Bureau of Economic Research, working paper 11465.

Nahapiet, J. and Ghoshal, S. (1998). 'Social Capital, Intellectual Capital, and the Organizational Advantage', *Academy of Management Review*, 23(2): 242–66.

Narin, F. (1994). 'Patent Bibliometrics', *Scientometrics*, 30(1): 147–55.

—— and Olivastro, D. (1991). 'Status Report: Linkage between Technology and Science', *Research Policy*, 21(3): 237–49.

—— Rosen, M., and Olivastro, D. (1989). 'Patent Citation Analysis: New Validation Studies and Linkage Statistics', in A. F. J. Van Raan (ed.), *Science and Technology Indicators: Their Use in Science Policy and Their Role in Science Studies*. Leiden, The Netherlands: DSWO Press.

—— Hamilton, K. S., and Olivastro, D. (1997). 'The Increasing Linkage between U.S. Technology and Public Science', *Research Policy*, 26(3): 317–30.

Nelson, R. R. (1959). 'The Simple Economics of Basic Scientific Research', *Journal of Political Economy*, 67(3): 297–306.

—— (1982). 'The Role of Knowledge in R&D Efficiency', *The Quarterly Journal of Economics*, 97(3): 453–70.

—– (ed.) (1993). *National Innovation Systems: A Comparative Analysis*. New York: Oxford University Press.

—— (1998). 'The Co-evolution of Technology Industrial Structure and Supporting Institutions', in G. Dosi, D. J. Teece and J. Chytry (eds.), *Technology Organization and Competitiveness: Perspectives on Industrial and Corporate Change*. Oxford: Oxford University Press.

—— (2001). 'Observations on the Post-Bayh-Dole Rise of Patenting at American Universities', *Journal of Technology Transfer*, 26(1–2): 13–19.

—— and Winter, S. G. (1982). *An Evolutionary Model of Economic Change*. Cambridge, MA: Harvard University Press.

Nesta, L. and Dibiaggio, L. D. (2003). 'Technology Strategy and Knowledge Dynamics: The Case of Biotech', *Industry and Innovation*, 10(3): 329–47.

'Netscape Navigator', Wikipedia, April 16, 2004, URL: http://en.wikipedia.org/wiki/Netscape_Navigator

Nichols, K. (1998). *Inventing Software: The Rise of 'Computer-Related' Patents*. Westport, CT: Quorum Books.

Noam, E. (1992). *Telecommunications in Europe*. New York: Oxford University Press.
Noda, T. and Bower, J. L. (1996). 'Strategy Making as Iterated Processes of Resource Allocation', *Strategic Management Journal*, 17(Summer): 159–92.
Nohria, N. and Garcia-Pont, C. (1991). 'Global Strategic Linkages and Industry Structure', *Strategic Management Journal*, 12(Summer): 105–24.
Nonaka, I. (1994). 'A Dynamic Theory of Organizational Knowledge Creation', *Organization Science*, 5(1): 14–37.
—— and Takeuchi, H. (1995). *The Knowledge Creating Company*. Oxford: Oxford University Press.
Nooteboom, B. (1999). *Inter-Firm Alliances: Analysis and Design*. London: Routledge.
Nooteboom, B., Berger, H. and Noorderhaven, N. G. (1997). 'Effects of Trust and Governance on Relational Risk', *Academy of Management Journal*, 40(2): 308–38.
Normann, R. (2001). *Reframing Business: When the Map Changes the Landscape*. Chichester, UK: John Wiley.
—— and Ramírez, R. (1993). 'From Value Chain to Value Constellation: Designing Interactive Strategy', *Harvard Business Review*, 71(4): 65–77.
North, D. C. (1990). *Institutions, Institutional Change, and Economic Performance*. New York: Cambridge University Press.
O'Connor, G. C. (1998). 'Market Learning and Radical Innovation: A Cross Case Comparison of Eight Radical Innovation Projects', *Journal of Product Innovation Management*, 15(2): 151–66.
—— (2005). 'A Systems Approach to Building a Radical Innovation Dynamic Capability', Working Paper, Lally School of Management, Rensselaer Polytechnic Institute.
—— and Ayers, A. D. (2005). 'Building a Radical Innovation Competency: A Mid-Study Review', *Research-Technology Management*, 48(1): 23–31.
—— and DeMartino, R. (2005). 'Organizing for Radical Innovation: An Exploratory Study of the Structural Aspects of RI Management Systems in Large Established Firms', Working Paper, Lally School of Management, Rensselaer Polytechnic Institute.
—— and Rice, M. P. (2005). 'Business Models and Market Development: Key Oversights in the Radical Innovation Process', Working Paper, Lally School of Management, Rensselaer Polytechnic Institute.
—— —— Peters, L., and Veryzer, R. W. (2003). 'Managing Interdisciplinary, Longitudinal Research on Radical Innovation: Methods for the Study of Complex Phenomena', *Organization Science*, 14(4): 1–21.
O'Mahony, S. (2003). 'Guarding the Commons: How Community Managed Software Projects Protect Their Work', *Research Policy*, 32(7): 1179–98.
—— and West, J. (2005). 'The Participation Architecture of Online Production Communities', Working Paper, Harvard Business School.
Ouchi, W. G. and Bolton, M. K. (1988). 'The Logic of Joint Research and Development', *California Management Review*, 30(3): 9–34.
Owen-Smith, J. and Powell, W. W. (2004). 'Knowledge Networks as Channels and Conduits: The Effects of Spillovers in the Boston Biotechnology Community', *Organization Science*, 15: 5–21.
—— and Powell, W. W. (forthcoming). 'Accounting for emergence and novelty in Boston and Bay Area biotechnology', in P. Braunerhjelm and M. P. Feldman (eds.), *Cluster Genesis: The Emergence of Technology Clusters and Their Implications for Government Policies*. Oxford: Oxford University Press.

Ozcan, P. and Eisenhardt, K. (2005). 'Start-ups in an Emerging Market: Building a Strong Alliance Portfolio from a Low-power Position', Paper presented at the 10th DRUID Conference, Copenhagen.

Parise, S. and Casher, A. (2003). 'Alliance Portfolios: Designing and Managing Your Network of Business-Partner Relationships', *Academy of Management Executive,* 17(4): 25–39.

Parolini, C. (1999). *The Value Net: A Tool for Competitive Strategy.* Chichester, UK: John Wiley.

Patel, P. and Pavitt, K. (1997). 'The Technological Competencies of the World's Largest Firms: Complex and Path-Dependent, but not much Variety', *Research Policy,* 26(2): 141–56.

Pavitt, K. (1984). 'Sectoral Patterns of Technical Change: Towards a Taxonomy and a Theory', *Research Policy,* 13(6): 343–73.

—— (1998). 'Technologies, Products and Organization in the Innovating Firm: What Adam Smith Tells Us and Joseph Schumpeter Doesn't', *Industrial and Corporate Change,* 7(3): 433–51.

—— (2000). 'Why European Union Funding of Academic Research Should be Increased: A Radical Proposal', *Science and Public Policy,* 27(6): 455–60.

—— (2003). 'What Are Advances in Knowledge Doing to the Large Industrial Firm in the "New Economy"?' in J. F. Christensen and P. Maskell (eds.), *The Industrial Dynamics of the New Digital Economy.* Cheltenham: Edward Elgar, pp. 103–20.

Pavitt, K., Robson, M., and Towsend, J. (1989). 'Technological Accumulation, Diversification and Organisation in UK Companies, 1945–1983', *Management Science* 35(1): 81–99.

Pelled, L. H., Eisenhardt, K., and Xin, K. (1999). 'Exploring the Black Box: An Analysis of Work Group Diversity, Conflict, and Performance', *Administrative Science Quarterly,* 44: 1–28.

Penrose, E. (1959/1995). *The Theory of the Growth of the Firm.* Oxford: Oxford University Press.

Peteraf, M. A. (1993). 'The Cornerstone of Competitive Advantage: A Resource-based View', *Strategic Management Journal,* 14(3): 179–91.

Pfeffer, J. and Salncik, G. R. (1978). *The External Control of Organization: A Resource-Dependence Perspective.* New York: Harper & Row.

Piore, M. J. and Sabel, C. F. (1984). *The Second Industrial Divide.* New York: Basic Books.

Pisano, G. P. (1990). 'The R&D Boundaries of the Firm: An Empirical Analysis', *Administrative Science Quarterly,* 35(1): 153–76.

—— (1991). 'The Governance of Innovation: Vertical Integration and Collaborative Arrangements in the Biotechnology Industry', *Research Policy,* 20(3): 237–49.

Podolny, J. and Paige, K. (1998). 'Network Forms of Organizations', *Annual Review of Sociology,* 24: 57–76.

Polanyi, M. (1967). *The Tacit Dimension.* London: Routledge.

Porter, K. (2004). 'You Can't Leave Your Past Behind: The Influence of Founders' Career Histories on Their Firms', Unpublished doctoral dissertation, Stanford University, Department of Management Science and Engineering.

—— Whittington, K. B., and Powell, W. W. (2006). 'The Institutional Embeddedness of High-tech Regions: Relational Foundations of the Boston Biotechnology Community', in S. Breschi and F. Malerba (eds.), *Clusters, Networks, and Innovation*. Oxford: Oxford University Press.

Porter, M. E. (1980). *Competitive Strategy: Techniques for Analyzing Industries and Competitors*. New York: Free Press.

—— (1985). *Competitive Advantage: Creating and Sustaining Superior Performance*. New York: Free Press.

—— (1987). 'From Competitive Advantage to Corporate Strategy', *Harvard Business Review*, 65(3): 43–59.

—— (1990). *The Competitive Advantage of Nations*. New York: Free Press.

—— (1991). 'Towards a Dynamic Theory of Strategy', *Strategic Management Journal*, 12 (Winter): 95–117.

—— (1996). 'What is Strategy?' *Harvard Business Review*, 74(6): 61–78.

—— (1998). 'Clusters and the New Economics of Competition', *Harvard Business Review*, 76(6): 77–90.

—— and Solvell, O. (2002). 'Finland and Nokia', Harvard Business School case study 9-702-427.23 pp.

Powell, W. W. (1990). 'Neither Market Nor Hierarchy: Network Forms of Organization', in B. M. Staw and L. L Cummings (eds.), *Research in Organizational Behavior*, Vol. 12. Greenwich, CT: JAI Press.

—— (2001). 'Patterns in Western Enterprise', in (P. DiMaggio (ed.), *The Twenty-First Century Firm*. Princeton, NJ: Princeton University Press.

—— and Smith-Doerr, L. (2005). 'Networks and Economic Life', in N. Smelser and R. Swedberg (eds.), *Handbook of Economic Sociology*. Princeton, NJ: Princeton University Press, pp. 368–402.

—— Koput, K. W., and Smith-Doerr, L. (1996). 'Interorganizational Collaboration and the Locus of Control of Innovation: Networks of Learning in Biotechnology', *Administrative Science Quarterly*, 41(1): 116–45.

—— —— —— and Owen-Smith, J. (1999). Network Position and Firm Performance: Organizational Returns to Collaboration in the Biotechnology Industry', *Research in the Sociology of Organizations*, 16:129–59.

Prahalad, C. K. and Bettis, R. A. (1986). 'The Dominant Logic: A New Linkage Between Diversity and Performance', *Strategic Management Journal*, 7(6): 485–511.

—— and Hamel, G. (1990). 'The Core Competence of the Corporation', *Harvard Business Review*, 68(3): 79–91.

Prencipe, A. (1997). 'Technological Competencies and Product's Evolutionary Dynamics: A Case Study from the Aero-engine Industry', *Research Policy*, 25: 1261–76.

—— (2000). 'Breadth and Depth of Technological Capabilities in Complex Product Systems: The Case of the Aircraft Engine Control System', *Research Policy*, 29(7–8): 895–911.

—— Davies, A., and Hobday, M. (2003). *The Business of Systems Integration*. Oxford: Oxford University Press.

Pugh, E. (1995). *Building IBM*. Cambridge, MA: MIT Press.

Putnam, R. (1993). *Making Democracy Work: Civic Traditions in Modern Italy*. Princeton, NJ: Princeton University Press.

Quinn, J. B. (1985). 'Managing Innovation: Controlled Chaos', *Harvard Business Review*, 63(3): 73–84.

Ramirez, R. (1999). 'Value Co-production: Intellectual Origins and Implications for Practice and Research', *Strategic Management Journal*, 20: 49–65.

—— and Wallin, J. (2000). *Prime Movers*. Chichester, UK: John Wiley.

Rappert, B., Webster, A., and Charles, D. (1999). 'Making Sense of Diversity and Reluctance: Academic–Industrial Relations and Intellectual Property', *Research Policy*, 28(8): 873–90.

Reuer, J. and Ragozzino, R. (forthcoming). 'Agency Hazards and Alliance Portfolios', *Strategic Management Journal*.

Rice, M. P., Leifer, R., and O'Connor, G. C. (2002). 'Commercializing Discontinuous Innovations: Bridging the Gap from Discontinuous Innovation Project to Operations', *IEEE Transactions on Engineering Management*, 49(4): 330–40.

Rindova, V. and Kotha, S. (2001). 'Continuous Morphing: Competing through Dynamic Capabilities, Form, and Function', *Academy of Management Journal*, 44(6): 1263–80.

Ring, P. and van de Ven, A. (1992). 'Structuring Cooperative Relationships between Organizations', *Strategic Management Journal*, 13(7): 483–98.

Rivette, K. and Kline, D. (2000). *Rembrandts in the Attic: Unlocking the Hidden Value of Patents*. Cambridge, MA: Harvard Business School Press.

Roberts, E. B. (1977). 'Generating Effective Corporate Innovation', *Technology Review*, 80(1): 27–33.

Roberts, E. and Liu, W. K. (2001). 'Ally or Acquire? How Technology Leaders Decide', *Sloan Management Review*, 43(1): 26–34.

Rodman & Renshaw (2003). 'Tripath Technology, Media Technology, Market Outperform/Speculative Risk', Analyst report, Rodman & Renshaw LLC, December 30.

Rodrik, D. (2000). 'How Far Will International Economic Integration Go?', *Journal of Economic Perspectives*, 14(1): 177–86.

Romer, P. M. (1987). 'Growth Based on Increasing Returns Due to Specialization', *American Economic Review*, 77(2): 56–62.

Rosen, L. (2004). *Open Source Licensing: Software Freedom and Intellectual Property Law*. Upper Saddle River, NJ: Prentice-Hall PTR.

Rosenberg, N. (1982). *Inside the Black Box: Technology and Economics*. Cambridge: Cambridge University Press.

—— (1990). 'Why Do Firms Do Basic Research (With Their Own Money)?', *Research Policy*, 19(2): 165–74.

—— (1994). *Exploring the Black Box: Technology, Economics, and History*. Cambridge: Cambridge University Press.

—— (2000). 'America's University–Industry Interfaces, 1945–2000', Working Paper, Department of Economics, Stanford University.

—— and Steinmueller, W. E. (1988). 'Why Are Americans such Poor Imitators?', *American Economic Review*, 78(2): 229–35.

Rosenbloom, R. and Cusumano, M. A. (1987). 'Technological Pioneering and Competitive Advantage: The Birth of the VCR Industry', *California Management Review*, 29(4): 51–76.

—— and Spencer, W. (1996). *Engines of Innovation: Industrial Research at the End of an Era*. Boston, MA: Harvard Business School Press.

Rosenkopf, L. and Almeida, P. (2003). 'Overcoming Local Search Through Alliances and Mobility', *Management Science*, 49(6): 751–66.

Rotemberg, J. J. and Saloner, G. (1990). 'Competition and Human Capital Accumulation: A Theory of Interregional Specialization and Trade', National Bureau of Eeconomic Research working paper W3228.

Rothaermel, F. T. and Deeds, D. L. (2004). 'Exploration and Exploitation Alliances in Biotechnology: A System of new Product Development', *Strategic Management Journal* 25(3): 201–22.

Rothwell, R. (1994). 'Towards the Fifth-Generation Innovation Process', *International Marketing Review,* 11(1): 7–31.

—— Freeman, C., Horlsey, A., Jervis, V. T. P., Robertson, A. B., and Townsend, J. (1974). 'SAPPHO Updated: Project SAPPHO Phase II', *Research Policy,* 3(3): 258–91.

Rowley, T., Behrens, D., and Krackhardt, D. (2000). 'Redundant Governance Structures: An Analysis of Structural and Relational Embeddedness in the Steel and Semiconductor Industries', *Strategic Management Journal,* 21(3): 369–86.

Rysman, M. and Simcoe, T. (2005). 'Patents and the Performance of Voluntary Standard Setting Organizations', Rotman Working Papers, University of Toronto.

Sakakibara, M. (1997). 'Heterogeneity of Firm Capabilities and Cooperative Research and Development: An Empirical Examination of Motives', *Strategic Management Journal,* 18 (Summer): 143–64.

—— (2001). 'Cooperative Research and Development: Who Participates and in Which Industries do Projects Take Place?', *Research Policy,* 30(7): 993–1018.

Sakkab, N. (2002). 'Connect & Develop Complements Research & Develop at P&G', *Research-Technology Management,* 45(2): 38–45.

Salz, P. A. (2004). 'Pay by the Rules', *Mobile Communications International.* April, p. 109.

Sampat, B. N., Mowery, D. C., and Arvids, A. Z. (2003). 'Changes in University Patent Quality after the Bayh-Dole Act: A Re-examination', *International Journal of Industrial Organization,* 21(9): 1371–90.

Samuelson, P. (1990). 'Benson Revisited: The Case Against Patent Protection for Algorithms and Other Computer Program-Related Inventions', *Emory Law Journal,* 39: 1025–122.

Sanchez, R. (2004). 'Creating Modular Platforms for Strategic Flexibility', *Design Management Review,* 15(1): 58–67.

—— and Mahoney, J. T. (1996). 'Modularity, Flexibility and Knowledge Management in Organizational Design', *Strategic Management Journal,* 17 (Winter): 63–76.

Sarkar, M., Preet, A., and Madhok, A. (2004). 'A Process View of Alliance Capability: Generating Value in Alliance Portfolios', Social Science Research Network, URL: http://ssrn.com/abstract=611241

Saxenian, A. (1994). *Regional Advantage: Culture and Competition in Silicon Valley and Route 128.* Cambridge, MA: Harvard University Press.

—— (1996). 'Inside-out: Regional Networks and Industrial Adaptation in Silicon Valley and Route 128', *Cityscape: A Journal of Policy Development and Research,* 2(2): 41–60.

Schildt, H. A., Maula, M. V. J. and Keil, T. (2005). 'Explorative and Exploitative Learning from External Corporate Ventures', *Entrepreneurship: Theory & Practice,* 29(4): 493–515.

Schnaars, S. P. (1994). *Managing Imitation Strategies: How Later Entrants Seize Markets from Pioneers.* New York: Free Press.

Schon, D. A. (1967). *Technology and Change.* New York: Delacorte.

Schumpeter, J. A. (1934). *The Theory of Economic Development.* Cambridge, MA: Harvard University Press.

—— (1942). *Capitalism, Socialism, and Democracy.* New York: Harper & Row.

—— (1950). *Capitalism, Socialism, and Democracy,* 3rd edn. New York: Harper & Row.

Searls, D. (2003). 'Surprise: Apple's New Browser Is a Sister to Konqueror', LinuxJournal.com, January 11, URL: http://www.linuxjournal.com/article.php?sid=6565

Shah, S. (2004). 'Understanding the Nature of Participation and Coordination in Open and Gated Source Software Development Communities', *Proceedings of the Academy of Management Conference,* New Orleans, August 9–11, B1–6.

Shan, W., Walker, G., and Kogut, B. (1994). 'Interfirm Cooperation and Startup Innovation in the Biotechnology Industry', *Strategic Management Journal,* 15(5): 387–94.

Shane, S. (2004). 'Encouraging University Entrepreneurship? The Effect of the Bayh-Dole Act on University Patenting in the United States', *Journal of Business Venturing,* 19(1): 127–51.

Shapiro, C. and Varian, H. R. (1999). *Information Rules: A Strategic Guide to the Network Economy.* Boston, MA: Harvard Business School Press.

Sheremata, W. A. (2004). 'Competing Through Innovation in Network Markets: Strategies for Challengers', *Academy of Management Review,* 29(3): 359–77.

Shimoda, S. (1998). 'Agricultural Biotechnology—Master of the Universe?', *Agbioforum,* 1(2): 62–68.

Siegel, D. S., Waldman, D., and Link, A. (2003). 'Assessing the Impact of Organizational Practices on the Productivity of University Technology Transfer Offices: An Exploratory Study', *Research Policy,* 32(1): 27–48.

—— —— Atwater, L. E., and Link, A. N. (2003). 'Commercial Knowledge Transfers from Universities to Firms: Improving the Effectiveness of University-Industry Collaboration', *Journal of High Technology Management Research,* 14(1): 111–33.

Simard, C. (2004). 'From Weapons to Cell-Phones: Knowledge Networks in the Creation of San Diego's Wireless Valley'. Unpublished dissertation, Stanford University, Department of Communication.

Simcoe, T. (2005). 'Committees and the Creation of Technical Standards'. Unpublished working paper, University of Toronto.

—— (2006). 'Delays and *de jure* Standards: What Caused the Slowdown in Internet Standards Development?', in S. Greenstein and V. Stango (eds.), *Standards and Public Policy,* Cambridge: Cambridge University Press.

Slaton, J. (1999). 'A Mickey Mouse Copyright Law?' Wired.com, Jan. 13, URL: http://www.wired.com/news/politics/0,1283,17327,00.html

Slywotsky, A. J., Morrison, D. J., and Andelman, B. (2001). *The Profit Zone: How Strategic Business Design Will Lead You to Tomorrow's Profits.* New York: Times Books/Random House.

Smith, D. K. and Alexander, R. C. (1988). *Fumbling the Future.* New York: William Morrow.

Smith-Doerr, L. and Powell, W. W. (2003). 'Networks and Economic Life', in N. J. Smelsen and R. Swedberg (eds.), *The Handbook of Economic Sociology,* 2nd edn. Princeton, NJ: Princeton University Press, pp. 379–402.

Softletter (2001). 'The 2001 Softletter 100', *Softletter Trends & Strategies in Software Publishing,* 3: 1–8.

Software & Information Industry Alliance (2005). 'Packaged Software Industry Revenue and Growth', URL: www.siia.net/software/pubs/growth_software05.pdf

Sorenson, O. and Audia, P. G. (2000). 'The Social Structure of Entrepreneurial Activity: Geographic Concentration of Footwear Production in the United States, 1940–1989', *American Journal of Sociology*, 106(2): 324–62.

Southwick, K. (2004). 'Big Blue's Mr. Web Services', March 17, CNET News.com, URL: http://news.com.com/2008-7345_3-5173667.html

Spence, A. M. (1973). 'Job Market Signaling', *Quarterly Journal of Economics,* 87(3): 355–74.

Spender, J. C. (1996). 'Making Knowledge the Basis of a Dynamic Theory of the Firm', *Strategic Management Journal,* 17 (Winter): 45–62.

Stabell, C. and Fjeldstad, Ø. (1998). 'Confiduring Value for Competitive Advantage: On Chains, Shops, and Networks', *Strategic Management Journal*, 19(5): 413–37.

Stallman, R. (1999). 'The GNU Operating System and the Free Software Movement', in C. DiBona, S. Ockman, and M. Stone (eds.), *Open Sourcess: Voices from the Open Source Revolution.* Sebastopol, CA: O'Reilly, pp. 53–70.

Stark, D. (2001). 'Ambiguous Assets for Uncertain Environments: Heterarchy in Post-socialist Firms', in P. DiMaggio (ed.), *The 21st-Century Firm.* Princeton, NJ: Princeton University Press.

Staudenmayer, N., Tripsas, M., and Tucci, C. L. (2000). 'Development Webs: A New Paradigm for Product Development', in R. K. F. Bresser, M. A. Hitt, R. D. Nixon, and D. Heuskel (eds.), *Winning Strategies in a Deconstructing World.* New York: John Wiley.

Steers, R. M., Shin, Y. K. and Ungson, G. R. (1989). *Chaebol: Korea's New Industrial Might.* New York: HarperCollins.

Steinmueller, W. E. (1996). 'The U.S. Software Industry: An Analysis and Interpretive History', in D. C. Mowery (ed.), *The International Computer Software Industry: A Comparative Study of Industry Evolution and Structure.* New York: Oxford University Press, pp. 15–52.

—— (2003). 'The Role of Technical Standards in Coordinating the Division of Labour in Complex System Industries', in A. Prencipe, A. Davies, and M. Hobday (eds.), *The Business of Systems Integration.* Oxford: Oxford University Press.

Stern, S. (2004). 'Do Scientists Pay to Be Scientists?', *Management Science,* 50(6): 835–53.

Stockdale, B. (2002). *UK Innovation Survey.* London: Department of Trade and Industry.

Stokes, D. (1997). *Pasteur's Quadrant: Basic Science and Technological Innovation.* Washington, DC: Brookings Institution.

Stuart, T. (2000). 'Interorganizational Alliances and the Performance of Firms: A Study of Growth Rates in a High-technology Industry', *Strategic Management Journal,* 21(8): 791–811.

—— and Podolny, J. M. (1996). Local Search and the Evolution of Technological Capabilities', *Strategic Management Journal,* Special Issue, 17 (Summer): 21–38.

—— and Sorenson, O. (2003). 'The Geography of Opportunity: Spatial Heterogeneity in Founding Rates and the Performance of Biotechnology Firms', *Research Policy,* 32(2): 229–53.

Sturgeon, T. J. (2002). 'Modular Production Networks: A New American Model of Industrial Organization', *Industrial and Corporate Change,* 11(3): 451–96.

Sweeney, D. (2004). 'Emerging Markets for Class D Power Amplification', Forward Concepts Report No: 4030, URL: http://www.fwdconcepts.com/classdamp.htm

Szulanski, G. (1996). 'Exploring Internal Stickiness: Impediments to the Transfer of Best Practice within the Firm', *Strategic Management Journal*, 17 (Winter): 27–43.

Taft, D. K. (2003). 'Sun Mulls Joining Java Eclipse Effort', *eWeek*, September 1, URL: http://www.eweek.com/article2/0,1759,1236123,00.asp

Takahashi, D. (2002). *Opening the XBox: Inside Microsoft's Plan to Unleash an Entertainment Revolution*. Roseville, CA: Prima.

Tansey, R., Mark, N., and Carroll, R. (2005). ' "Get Rich, or Die Trying": Lessons from Rambus' High-Risk Predatory Litigation in the Semiconductor Industry', *Industry and Innovation*, 12(1): 93–115.

Taylor, D. and Terhune, A. (2001). *Doing E-business: Strategies for Thriving in an Electronic Market Place*. New York: John Wiley.

Teece, D. J. (1986). 'Profiting from Technological Innovation—Implications for Integration, Collaboration, Licensing and Public-Policy', *Research Policy*, 15(6): 285–305.

—— (1989). 'Interorganizational Requirements of the Innovation Process', *Managerial and Decision Economics*, 10(1): 35–42.

—— (1996). 'Firm Organization, Industrial Structure, and Technological Innovation', *Journal of Economic Behavior & Organization*, 31(2): 193–224.

—— (2000). *Managing Intellectual Capital*. New York: Oxford University Press.

—— and Chesbrough, H. (1996). 'When is Virtual Virtuous?', *Harvard Business Review*, 74(1): 65–73.

—— and Sherry, E. F. (2003). 'Standards Setting and Antitrust', *Minnesota Law Review*, 87(6): 1913–94.

—— Rumelt, R., Dosi, G. and Winter, S. G. (1994). 'Understanding Corporate Coherence: Theory and Evidence', *Journal of Economic Behavior & Organization*, 23(1): 1–30.

—— Pisano, G. and Shuen, A. (1997). 'Dynamic Capabilities and Strategic Management', *Strategic Management Journal*, 18(7): 509–33.

Tennenhouse, D. (2003). 'Innovation Breeds Success at Intel', *IEE Engineering Management*, 13(6): 44–7.

Testa, B. M. (2005). 'Zander's Balancing Act', *Electronic Business* (August): 31–8.

Thurbsy, J. G., Jensen, R., and Thursby, M. C. (2001). 'Objectives, Characteristics, and Outcomes of University Licensing: A Survey of Major U.S. Universities', *Journal of Technology Transfer*, 26(1): 59–72.

Tidd, J. and Trewhella, M. J. (1997). 'Organizational and Technological Antecedents for Knowledge Acquisition and Learning', *R&D Management*, 27(4): 359–75.

——, Bessant, J. and Pavitt, K. (2005). *Managing Innovation. Integrating Technological, Market and Organizational Change*, 3rd edn. New York: John Wiley.

Todd, B. (2004). 'The Whys of Modding', *Computer Games*, 163(June): 38–40.

Toivanen, O. (2004). 'Choosing Standards', Helsinki Center of Economic Research, Discussion Paper 28.

Trajtenberg, M. (1990). 'A Penny for Your Quotes: Patent Citations and the Value of Innovations', *RAND Journal of Economics*, 21(1): 172–87.

—— Henderson, R., and Jaffe, A. (1997). 'University Versus Corporate Patents: A Window on the Basicness of Invention', *Economic Innovation and New Technology*, 5(1): 19–50.

Traxler, G. and Falck-Zepeda, J. (1999). 'The Distribution of Benefits from the Introduction of Transgenic Cotton Varieties', *Agbioforum*, 2(2): 94–8.

Tsai, W. and Ghoshal, S. (1998). 'Social Capital and Value Creation: The Role of Intrafirm Networks', *Academy of Management Journal*, 41(4): 464–76.

Tulloch, J. (2002). 'Clear as Mud', *Mobile Communications International*, December.

Tushman, M. L. and Anderson, P. (1986). 'Technological Discontinuities and Organizational Environments', *Administrative Science Quarterly*, 31(3): 439–65.

—— and O'Reilly III, C. A. (1996). 'The Ambidextrous Organization: Managing Evolutionary and Revolutionary Change', *California Management Review*, 38(4): 1–23.

—— —— (1997). *Winning through Innovation*. Boston, MA: Harvard Business School Press.

U.C. Regents. (1999). 'The BSD License', URL: http://www.opensource.org/licenses/bsd-license.php

U.S. Patent and Trademark Office, (2002). 'Patenting by Organizations, 2001', Report, URL: http://www.uspto.gov/go/taf/topo_01.pdf

Utterback, J. M. (1994). *Mastering the Dynamics of Innovation: How Companies can Seize Opportunities in the Face of Technological Change*. Boston, MA: Harvard Business School Press.

Uzzi, B. (1994). 'The Dynamics of Organizational Networks: Structural Embeddedness and Economic Behavior'. Unpublished Ph.D. dissertation, Department of Sociology, The State University of New York at Stony Brook.

—— (1996). 'The Sources and Consequences of Embededdness for the Economic Performance of Organizations: The Network Effect', *American Sociological Review*, 61(4): 674–98.

—— (1997). 'Social Structure and Competition in Interfirm Networks: The Paradox of Embeddedness', *Administrative Science Quarterly*, 42(1): 35–67.

—— and Gillespie, J. J. (1999*a*). 'Corporate Social Capital and the Cost of Financial Capital: An Embeddedness Approach', in R. Leenders and S. Gabbay (eds.), *Corporate Social Capital and Liability*. Boston, MA: Kluwer Academic Publishers, pp. 446–59.

—— and Gillespie, J. J. (1999*b*). 'Inter-firm Ties and the Organization of the Firm's Capital Structure in the Middle Financial Market', in D. Knoke and S. Andrews (eds.), *Research in the Sociology of Organizations*, Vol. 16. Stamford, CT: JAI Press, pp. 107–26.

Väisänen, K., Maula, M. V. J., and Salmenkaita, J. P. (2003). *'Systemic Innovation Process: Organizational Boundaries and Process Elements'*. Paper presented at the 23rd Annual International Conference of the Strategic Management Society, Baltimore, MD.

Välimäki, M. (2003). 'Dual Licensing in Open Source Software Industry', *Systemes d'Information et Management*, 8/1: 63–75, URL: http://opensource.mit.edu/papers/valimaki.pdf

Vanhaverbeke, W. and Peeters, N. (2005). 'Embracing Innovation as Strategy: Corporate Venturing, Competence Building and Corporate Strategy Making', *Creativity and Innovation Management*, 14(3): 262–73.

Verspagen, B. and Schoenmakers, W. (2004). 'The Spatial Dimension of Patenting by Multinational Firms in Europe', *Journal of Economic Geography*, 4(1): 23–42.

von Burg, U. (2001). *The Triumph of Ethernet: Technological Communities and the Battle for the LAN Standard*. Stanford, CA: Stanford University Press.

von Hippel, E. (1988). *The Sources of Innovation*. New York: Oxford University Press.

—— (2005). *Democratizing Innovation*. Cambridge, MA: MIT Press.

—— and von Krogh, G. (2003). 'Open Source Software and the "Private-Collective" Innovation Model: Issues for Organization Science', *Organization Science,* 14(2): 209–23.

Walsh, J. P., Arora, A., and Cohen, W. M. (2003). 'Effects of Research Tool Patents and Licensing on Biomedical Innovation', in W. M. Cohen and S. A. Merrill (eds.), *Patents in the Knowledge-based Economy.* Washington, DC: The National Academies Press, pp. 285–340.

Wasserman, S. and Faust, K. (1994). *Social Network Analysis.* Cambridge: Cambridge University Press.

Weber, A. (1928). *Theory of The Location of Industries.* Translated by C. J. Friedrich: University of Chicago Press.

Weick, K. E. (1995). *Sensemaking in Organizations.* Thousand Oaks, CA: Sage.

Weiss, M. and Sirbu, M. (1990). 'Technological Choice in Voluntary Standards Committees: An Empirical Analysis', *Economics of Innovation and New Technology,* 1(1): 111–34.

Wernerfelt, B. (1984). 'A Resource-based View of the Firm', *Strategic Management Journal,* 5(2): 171–80.

West, J. (1995). 'Software Rights and Japan's Shift to an Information Society', *Asian Survey,* 35(12): 1118–39.

—— (2000). 'Institutional Constraints in the Initial Deployment of Cellular Telephone Service on Three Continents', in K. Jakobs (ed.), *Information Technology Standards and Standardization: A Global Perspective.* Philadelphia: Idea Group, pp. 198–221.

—— (2002). 'Qualcomm 2000: CDMA Technologies', European Case Clearinghouse, Case 302-069-1.

—— (2003). 'How Open is Open Enough? Melding Proprietary and Open Source Platform Strategies', *Research Policy,* 32(7): 1259–85.

—— (2005). 'The Fall of a Silicon Valley Icon: Was Apple Really Betamax Redux?', in R. A. Bettis (ed.), *Strategy in Transition.* Oxford: Blackwell, pp. 274–301.

—— (2006). 'The Economic Realities of Open Standards: Black, White and Many Shades of Gray', in S. Greenstein and V. Stango (eds.), *Standards and Public Policy.* Cambridge: Cambridge University Press.

—— and Dedrick, J. (2001). 'Open Source Standardization: The Rise of Linux in the Network Era', *Knowledge, Technology and Policy,* 14(2): 88–112.

—— —— (2005). 'The Effect of Computerization Movements Upon Organizational Adoption of Open Source', Social Informatics Workshop: Extending the Contributions of Professor Rob Kling to the Analysis of Computerization Movements, Irvine, California, March 11. URL: http://www.crito.uci.edu/2/si/resources/westDedrick.pdf

—— and Fomin, V. (2001). 'National Innovation Systems in the Mobile Telephone Industry, 1946–2000', Academy of Management Conference, Washington, DC.

—— and Graham, J. L. (2004). 'A Linguistic-Based Measure of Cultural Distance and its Relationship to Managerial Values', *Management International Review,* 44(3): 239–60.

—— and O'Mahony, S. (2005). 'Contrasting Community Building in Sponsored and Community Founded Open Source Projects', *Proceedings of the 38th Annual Hawai'i International Conference on System Sciences,* Waikoloa, Hawaii.

—— and Tan, J. (2002). 'Qualcomm in China (B)', *Asian Case Research Journal,* 6(2): 101–28.

Williams, M. (2005). 'Sony Loses Playstation Patent Case, Must Pay $91 Million', *PC World,* March.

Williamson, O. E. (1975). *Markets and Hierarchies, Analysis and Antitrust Implications: A Study in the Economics of Internal Organization.* New York: Free Press.

—— (1983). 'Credible Commitments: Using Hostages to Support Exchange', *American Economic Review,* 73(4): 519–40.

—— (1985). *The Economic Institutions of Capitalism: Firms, Markets, Relational Contracting.* New York: Free Press.

—— (1999). 'Strategy Research: Governance and Competence Perspectives', *Strategic Management Journal,* 20(12): 1087–108.

Yin, R. (1988). *Case Study Research: Design and Methods.* Newbury Park, CA: Sage.

—— (1994). *Case Study Research.* Thousand Oaks, CA: Sage.

Yoffie, D. B. and Cusumano, M. A. (1999). 'Judo Strategy: The Competitive Dynamic of Internet Time', *Harvard Business Review,* 77(1): 71–81.

Young, R. and Rohm, W. G. (1999). *Under the Radar: How Red Hat Changed the Software Business—and Took Microsoft by Surprise.* Scottsdale, AZ: The Coriolis Group.

Zaheer, A., McEvily, B., and Perrone, V. (1998). 'Does Trust Matter? Exploring the Effects of Interorganizational and Interpersonal Trust on Performance', *Organization Science,* 9(2): 141–59.

Zajac, E. and Olsen, C. (1993). 'From Transaction Cost to Transactional Value Analysis: Implications for the Study of Interorganizational Strategies', *Journal of Management Studies,* 30(1): 131–45.

Zucker, L. G., Darby, M. R. (1997). 'Present at the Biotechnological Revolution: Transformation of Technological Identity for a Large Incumbent Pharmaceutical Firm', *Research Policy,* 26(4): 429–46.

—— —— and Brewer, M. B. (1998). 'Intellectual Human Capital and the Birth of U.S. Biotechnology Enterprises', *The American Economic Review,* 88(1): 290–306.

—— —— and Armstrong, J. S. (2002). 'Commercializing Knowledge: University Science, Knowledge Capture, and Firm Performance in Biotechnology', *Management Science,* 48(1): 138–52.

Index